Stroke Genomics

METHODS IN MOLECULAR MEDICINE™

John M. Walker, SERIES EDITOR

Stroke Genomics

Methods and Reviews

Edited by

Simon J. Read

AstraZeneca Pharmaceuticals
Macclesfield, Cheshire, UK

David Virley

Neurology Centre
for Excellence in Drug Discovery
GlaxoSmithKline Pharmaceuticals,
Harlow, Essex, UK

HUMANA PRESS ✳ TOTOWA, NEW JERSEY

© 2005 Humana Press Inc.
999 Riverview Drive, Suite 208
Totowa, New Jersey 07512

www.humanapress.com

This publication is printed on acid-free paper. ∞

ANSI Z39.48-1984 (American Standards Institute) Permanence of Paper for Printed Library Materials.

Cover design by Patricia F. Cleary

Cover illustrations: Imaging evidence of small vessel disease, axial T2-weighted MRI scan (*foreground*; Fig. 2A, Chapter 9; *see* full caption on p. 235 and discussion on p. 229). Double immunofluorescence for Hsp27 and glial fibrillary acidic protein (*background*; Fig. 7, Chapter 6; *see* discussion and full caption on pp. 143, 144).

For additional copies, pricing for bulk purchases, and/or information about other Humana titles, contact Humana at the above address or at any of the following numbers: Tel.: 973-256-1699; Fax: 973-256-8341; E-mail: humana@humanapr.com; or visit our Website: www.humanapress.com

Photocopy Authorization Policy:
Authorization to photocopy items for internal or personal use, or the internal or personal use of specific clients, is granted by Humana Press Inc., provided that the base fee of US $25.00 per copy is paid directly to the Copyright Clearance Center at 222 Rosewood Drive, Danvers, MA 01923. For those organizations that have been granted a photocopy license from the CCC, a separate system of payment has been arranged and is acceptable to Humana Press Inc. The fee code for users of the Transactional Reporting Service is: [1-58829-333-5/05 $25.00].

Printed in the United States of America. 10 9 8 7 6 5 4 3 2 1

eISBN: 1-59259-836-6
ISSN: 1543-1894

Library of Congress Cataloging-in-Publication Data

Stroke genomics : methods and reviews / edited by Simon J. Read, David Virley.
        p. ; cm. — (Methods in molecular medicine ; 104)
    Includes bibliographical references and index.
    ISBN 1-58829-333-5 (hardcover : alk. paper)
    1. Cerebrovascular disease—Genetic aspects. 2. Cerebrovascular disease—Molecular aspects. 3. Genomics. [DNLM: 1. Cerebrovascular Accident—genetics. WL 355 S921307 2005] I. Read, Simon J. II. Virley, David. III. Series.
    RC388.5.S85254 2005
    616.8'1042—dc22
                                                                                                2004012019

Preface

With sequencing of the human genome now complete, deciphering the role of gene function in human neurological pathophysiology is a promise that has yet to be realized. More than most diseases, stroke has been keenly studied from a genomic perspective. Studies are numerous and incorporate data on stroke inheritance, chromosomal loci of risk, preclinical models of stroke, and differential gene expression of brain injury, repair, and recovery. The problem is no longer a lack of information but one of interpretation and prioritization of what we do know.

The aims of *Stroke Genomics: Methods and Reviews* are twofold. First, it aims to provide the reader with cutting-edge reviews of clinical and preclinical genomics, written by leading experts in the field. In particular, the authors of certain chapters relate gene expression changes to physiological end points, such as functional imaging paradigms. Thus, a more holistic approach to gene expression is described, one in which molecular biology goes hand in hand with stroke pathophysiology.

Second, detailed methods for study of the molecular biology of stroke are also included. Following the format of the *Methods in Molecular Medicine*™ series, these chapters will enable the reader to employ each technique without recourse to other methods texts. In its entirety, this book should provide the reader with the knowledge needed to design, execute, and interpret preclinical and clinical studies of stroke genomics.

Simon J. Read
David Virley

Contents

Contributors

STUART MCRAE ALLAN • *School of Biological Sciences, University of Manchester, Manchester, United Kingdom*

FRANK C. BARONE • *High-Throughput Biology, Discovery Research, GlaxoSmithKline Pharmaceuticals, King of Prussia, PA*

BRIAN C. BOND • *Statistical Sciences, GlaxoSmithKline Pharmaceuticals, Harlow, Essex, United Kingdom*

HILARY V. O. CARSWELL • *Division of Clinical Neuroscience, Wellcome Surgic al Institute, University of Glasgow, Glasgow, Scotland, United Kingdom*

RICHARD J. DAVIS • *Pharmagene plc, Royston, Hertfordshire, United Kingdom*

ANNA F. DOMINICZAK • *BHF Glasgow Cardiovascular Research Centre, University of Glasgow, Glasgow, Scotland, United Kingdom*

RICHARD FAGAN • *Target Discovery, Inpharmatica Ltd., London, United Kingdom*

ANDREW B. GOULTER • *Exploratory Target Profiling, Pharmagene plc, Royston, Hertfordshire, United Kingdom*

DELYTH GRAHAM • *BHF Glasgow Cardiovascular Research Centre, University of Glasgow, Glasgow, Scotland, United Kingdom*

ALEX J. HARPER • *Neurology and Gastroenterology Centre of Excellence for Drug Discovery, GlaxoSmithKline Pharmaceuticals, Harlow, Essex, United Kingdom*

DAVID C. HARRISON • *Neurology and Gastroenterology Centre of Excellence for Drug Discovery, GlaxoSmithKline Pharmaceuticals, Harlow, Essex, United Kingdom*

AHAMAD HASSAN • *Clinical Neurosciences, St. George's Hospital Medical School, London, United Kingdom*

HELEN HODGES • *Department of Psychiatry, Institute of Psychiatry at Maudsley; Functional Assessment, ReNeuron Ltd., Surrey Research Park, Guildford, Surrey, United Kingdom*

I. MHAIRI MACRAE • *Division of Clinical Neuroscience, Wellcome Surgical Institute, University of Glasgow, Glasgow, Scotland, United Kingdom*

HUGH S. MARKUS • *Clinical Neurosciences, St. George's Hospital Medical School, London, United Kingdom*

MARTIN W. MCBRIDE • *BHF Glasgow Cardiovascular Research Centre, University of Glasgow, Glasgow, Glasgow, Scotland, United Kingdom*

DAVID MICHALOVICH • *Target Discovery, Inpharmatica Ltd., London, United Kingdom*

PAUL R. NELSON • *Prism Training and Consultancy Ltd., Cambridge, United Kingdom*

SIMON J. READ • *AstraZeneca Pharmaceuticals, Macclesfield, Cheshire, United Kingdom*

TOBY JOHN ROBERTS • *Neuroimaging Research Group, Institute of Psychiatry, London, United Kingdom*

ROBERT M. SAPOLSKY • *Department of Biological Science, Stanford University, Stanford, CA*

PAUL STROEMER • *Functional Assessment, ReNeuron Ltd., Surrey Research Park, Guildford, Surrey, United Kingdom*

ZOLTÁN SZOLNOKI • *Department of Neurology and Neurophysiology, Pándy Kálmán County Hospital, Gyula, Hungary*

DAVID VIRLEY • *Neurology Centre of Excellence for Drug Discovery, GlaxoSmithKline Pharmaceuticals, Harlow, Essex, United Kingdom*

XINKANG WANG • *Department of Thrombosis Research, Bristol Myers Squibb Company, Princeton, NJ*

MIDORI A. YENARI • *Departments of Neurosurgery and Neurology, Stanford University School of Medicine, Stanford, CA*

I

INTRODUCTION

1

Introduction to Stroke Genomics

Simon J. Read and Frank C. Barone

Summary

Translation of the explosion in knowledge of acute ischemic stroke into satisfactory treatment regimens has yet to happen. At present tPA, intra-arterial prourokinase and low-molecular-weight heparin form the vanguard for therapeutic intervention, yet these treatments have a limited therapeutic window.

Part of this expansion in understanding has been driven by the contribution of stroke genetics and genomics. However, despite the enormous preclinical and clinical information of receptors, enzymes, second messenger systems, and so forth, that are implicated in stroke pathophysiology, delivery of novel drug treatment has been slow.

This introductory chapter discusses the multiple sources of clinical and preclinical genetic information. It will describe the importance of integrating expression information into multiple preclinical models with temporal and spatial roles in lesion pathology and, furthermore developing an understanding of function in the clinic before claiming a role in ischemic stroke.

The goal of such a holistic integration of information is to increase the yield from current datasets of gene expression and ultimately to help expand the choice of treatment available to the physician and patient.

KEY WORDS

Twin studies; Mendelian populations; candidate gene studies; preclinical models; differential gene expression; variation and confidence; protein confirmation.

1. Introduction

In the United States, stroke is the third largest cause of death, ranking only behind heart disease and all forms of cancer. It is the leading cause of disability and has the highest disease cost. About 775,000 strokes are estimated to occur per year, and there are about 4 million surviving patients who are at increased risk of a secondary cardiovascular event. Stroke is expensive in the United States, with annual health care costs of $30–50 billion. Estimates indicate that

From: *Methods in Molecular Medicine, Vol. 104: Stroke Genomics: Methods and Reviews*
Edited by: S. J. Read and D. Virley © Humana Press Inc., Totowa, NJ

stroke is responsible for half of all patients hospitalized for acute neurological disease *(1,2)*.

The recent completion of the human genome project has refocused attention on the contribution to ischemic stroke of abnormalities in gene function. The completion of this project has raised more questions than answers, and we are just beginning to understand the magnitude of the challenge. The completed map reveals that humans have about 26,500–39,000 genes, not the 100,000 estimated previously. Protein diversity is achieved by alternative splicing of genes, with the consequent production of an average of 2.6–3.2 transcripts per gene. Further complexity is added by posttranslational modification of proteins, which change the function, activity, location, or even the combination of parts of separate genes and combine to form novel, unique proteins *(3)*. It appears that the question of genetic contribution to ischemic stroke may have to be addressed at the protein rather than the gene level.

Historically, the early epidemiological studies of the 1970s provided the initial evidence for a genetic influence in stroke. The Framingham study was one of the first to suggest that a positive parental history of stroke contributed a significant risk to the offspring *(4)*. Thirty years later, stroke remains an area of substantially unmet medical need. The complexity of stroke undoubtedly reflects the heterogeneity of the human stroke population, the contribution of monogenic and polygenic disorders to this disease process, and the interactions of these with a multitude of environmental factors.

2. The Contribution of Genetic Risk

2.1. Twin Studies

The strongest evidence for a genetic risk in stroke is derived from twin studies. Proband concordance rates have long been used to identify the hereditability of a trait or disorder. Initial studies by Brass et al. *(5)* identified an elevated probandwise concordance rate for stroke risk in monozygotic twins over dizygotic twins (17.7% vs 3.6%), confirming a genetic predisposition to stroke. Subsequent reassessment of this cohort of patients 6 yr later reported a risk attributable to genetic influence but with an increase in the role of environmental factors as patients age *(6)*. However, defining the genetic contribution to stroke depends on accurate classification of phenotype *(7)*. A meta-analysis of the standardization of human stroke studies by Meschia *(7)* emphasized that approximately one-third of all studies make no attempt to characterize stroke variability in phenotype. Moreover, nonstandard systems of subtyping stroke phenotype are used as often as standard systems. When more robust attempts are made to define phenotype accurately, for example, by using magnetic resonance imaging criteria of white matter hyperintensity volumes, genetic factors can be found to account for up to 71% of the variability in this endpoint.

2.2. Mendelian Populations

Identification of possible genetic determinants of stroke risk has been hampered by the lack of homologous patient populations. One plausible approach for dealing with this heterogeneity has been the use of mendelian disorders with specific stroke-like phenotypes to explore genetic models of the more general population. These disorders are numerous and include cerebral autosomal dominant arteriopathy with subcortical infarcts and leukoencephalopathy (CADASIL) *(8)*, mitochondrial encephalopathy, lactic acidosis, and stroke-like syndrome (MELAS) *(9)*, familial hemiplegic migraine *(10)*, and hereditary coagulopathies (for reviews, *see* **ref. *11***). Although these subgroups contribute little to the overall prevalence of stroke, it is hoped that genes identified from them will identify potential commonalities in the wider patient population.

The best described of these conditions is CADASIL *(12)*, an autosomally dominant dementia with multiple infarcts that has been identified sporadically in Europe *(13,14)* and North America *(15,16)*. In these patients, small periventricular white matter lesions are identified, often involving the internal capsule *(17)*. The so-called CADASIL gene has been identified as *Notch 3*: approx 90% of patients have missense mutations in extracellular domains of the protein product, whereas in about 70% of patients the mutation is located within exons 3 and 4 *(18)*. All known mutations associated with CADASIL result in removal or addition of cysteine residues, and it is proposed that expression of these mutated *Notch 3* proteins results in cerebral vascular smooth muscle dysfunction *(19)*.

Whether abnormalities in *Notch* signaling impact on the broader stroke population is at present unknown, although the pathogenesis of CADASIL, which is characterized by progressive disruption of vascular endothelium, secondary fibrosis, and thrombosis, is typical of some stroke subpopulations *(20)*. Interestingly, anticoagulant therapy has been tried in CADASIL without positive results *(21)*. This may exemplify how narrowing heterogeneity and studying single gene/mendelian disorders may limit the application to the broader patient population.

2.3. Candidate Gene Studies

Typically, "casting the net wider" by investigating broader patient populations has been accomplished by candidate gene studies in heterogeneous stroke populations. This approach utilizes the *a priori* choice of a functionally relevant gene and explores its relationship with a particular phenotype. This "association" is a statistical measure of the dependence of a particular phenotype on the presence of a particular candidate gene/allele. Candidate gene choice is frequently driven by accepted stroke risk factors (e.g., hypertension, hemostasis, and abnormalities in lipid metabolism), and indeed significant positive associations of numerous markers with ischemic stroke have been identified, including specific ApoE alleles, D/D genotype of angiotensin-converting enzyme 1, and polymorphisms of fibrinogen, prothrombin, and atrial natriuretic peptide (ANP; *22–28*). However, there are equally numerous examples of negative association. Further levels of complexity are currently being explored with interactions between candidate genes that lead to exacerbated risk (*see* Chap. 10).

3. Preclinical Models of Spontaneous Stroke

Controlling environmental and genetic variability requires the use of preclinical animal models. Bioinformatic approaches using synteny can facilitate the matching of "stroke loci" found in stroke-prone rats to candidate genes on the human chromosome. Homogeneous populations of stroke-prone rats have been isolated from the incompletely inbred, spontaneously hypertensive rat (SHR), and then inbred further to achieve cohorts of stroke-prone rats. Spontaneous stroke in these rats is relatively similar to that in humans (*29*). In addition, recent data indicate that stroke-prone rats are not only more sensitive to ischemia but also exhibit greater neurological deficits with less recovery of function following stroke and reduced endogenous brain protection or tolerance following ischemic preconditioning (*30,31*).

Two separate groups have utilized these inbred populations for identification of genes associated with manifestation of specific stroke phenotypes. Rubattu et al. (*32*) performed a genome-wide screen in an F_2 cross obtained by mating stroke-prone and SHR rats, in which latency to stroke was used as a phenotype. Three major quantitative trait loci (QTLs) were isolated in which possible candidate genes encoded ANP and brain natriuretic peptide (BNP). Furthermore, interactions between alleles from within STR1 and STR2 suggested that this phenotype was a reasonable model of the polygenicity of stroke in humans. Follow-up sequencing to characterize ANP and BNP as candidates for stroke revealed point mutations in ANP and no differences in BNP. In vitro functional studies indicated lower ANP promoter activation in endothelial cells from stroke-prone rats vs SHR, with significantly lower ANP expression in the brain and no difference in BNP expression (*26–28*). To determine the in vivo significance of the STR2-lowered ANP promoter activation in stroke-prone animals, in comparison with stroke-resistant animals, Rubattu et al. (*26–28*) performed a cosegregation analysis of stroke occurrence in SHR stroke-prone rats/SHR stroke-resistant F_2 descendents and ANP expression. It was found that reduced expression of ANP did cosegregate with the appearance of "early" strokes in F_2 animals, and therefore although lowered ANP expression may be part of the phenotype of the "protective" STR2 QTL, it is unlikely that this is the primary protective mechanism in these animals. Parallel human studies of the role of ANP in cerebrovascular disease have confirmed that variation in the ANP gene may represent an independent risk factor for "cerebrovascular accidents" in humans *(32)*, emphasizing the utility of this cohort of animals as a model of ANP dysfunction in multiple subtypes of stroke.

Two other groups have utilized a modified model of the stroke-prone animal, employing F_2 hybrids derived from crossing the stroke prone SHR with Wistar-Kyoto (WKY) rats *(33,34)*. Ikeda et al. *(33)* used brain weight post stroke as the phenotype for linkage analysis, after the discovery that F_2 animals had higher levels of brain edema formation post stroke. Evidence of the linkage of phenotype to a gene on chromosome 4 was found that contributed to the severity of edema and was independent of blood pressure and the STR3 identified by Rubattu et al. *(32)*. Jeffs et al. *(34)* designed studies to identify the genetic component responsible for large infarct volumes in the stroke-prone rat in response to a focal ischemic insult. To do this, they performed a genome scan in an F_2 cross, derived from the stroke-prone rat and the normotensive WKY rat. In contrast to Rubattu and coworkers *(32)*, they were only able to identify one major QTL responsible for large infarct volumes. This QTL was located on rat chromosome 5; like STR2, it colocalized with ANP and BNP and was blood pressure-independent. Unlike STR2, this locus showed a much higher significance (lod 16.6) and accounted for 67% of phenotypic variance. Subsequent studies that infarct volumes in the F_1 rats were approximately identical to those of the stroke-prone animals, suggesting a dominant mode of inheritance *(35)*.

Authors have argued over the significance of the overlap of STR2 identified by Rubattu et al. *(32)* with the QTL identified by Jeffs et al. *(34)* on chromosome 5. It is unclear how the two phenotypes studied (latency to stroke, i.e., relative risk) *[32]*, and size of infarct after occlusion, i.e., sensitivity to focal ischemia *[34]*) would relate to each other physiologically. This may only become apparent when individual genes can be cosegregated with each phenotype. At present, altered ANP expression seems to play a role in the phenotype described by Rubattu et al. *(32)* but has been excluded from a role in the colony used by Jeffs et al. *(34)* (also *see* **ref. 36** and Chap. 3).

What can be concluded from each of these stroke-prone rat models? Certainly each represents a unique and valid "model" of stroke for the study of inheritance and for the role of candidate genes in particular stroke phenotypes. Neither colony represents a definitive model of human stroke, although progress has been made with titrating identified candidate genes in these stroke prone colonies to the human population *(26–28)*. One such research strategy that has been utilized is the analysis of genomic synteny between the rat and human genome. This bioinformatic approach seeks to align regions of homology using evolutionarily conserved markers and has been applied with some success in relating animal models to human genetics of other disease paradigms, e.g., non-insulin-dependent diabetes *(37)*. Relating identified loci from stroke-prone animals to the human genome offers a strategy for potential identification of candidate genes. For example, the STR2 region of rat chromosome 5 shows a conserved gene order and synteny with the human chromosome region 1p35-36. The high level of synteny between these regions makes this region ideal for rat–human comparative analysis. Sequence-tagged sites localized to this region have been identified and mapped to human transcript clusters.

Interestingly, only a few candidate genes identified at 1p35-36 have been examined in association studies. ANP has recently been assessed for association with multiple subtypes of stroke *(26–28)*. The polymorphism G664A, responsible for a valine–methionine substitution in proANP peptide, was found to be positively associated with the occurrence of stroke. In contrast, methylenetetrahydrofolate, another marker located at 1p35-36, was negatively associated with occurrence of stroke. Further studies may elucidate the predictability of markers of 1p35-36 and association with stroke.

In contrast, rat–human synteny in regions of the rat STR1 and STR3 loci are not well conserved, as several disruptions of synteny appear to have been introduced during evolution. It may be difficult to determine exact regions of synteny between these rat loci and human chromosomal loci, and thus it may be difficult to extrapolate the candidate genes from rat to human. Human chromosomal regions syntenic with STR1 span regions of two human chromosomes, around 16p11 and 19q13. Human synteny with the STR3 region also appears to be disrupted, with regions of synteny mapping telomerically to opposite arms of chromosome 7 (7p21 and 7q35). Of course, this is a problem of animal modeling of human diseases in general and is not restricted only to ischemic stroke.

4. Gene Expression in the Evolution of Post-Stroke Brain Injury

Cerebral ischemia is a powerful stimulus for the *de novo* expression and upregulation of numerous gene systems *(38,39)*. In terms of isolation of gene candidates for a neuroprotection strategy, interpretation of expression changes has proved difficult. The multitude of animal models of ischemia with varying genetic heterogeneities and infarct pathophysiologies, is also complicated by spatial and temporal variations that have largely confounded interpretation (for reviews, *see* **refs.** *40* and *41*). Furthermore, assays of differential expression have varying sensitivities to the relative fold increase or decrease in mRNA expression. As a result, "fishing" exercises will often result in "catches" of differential gene expression that vary depending on the assay employed.

With this bewildering array of complexities, the appropriate assessment of which animal model(s) that might be utilized, the appropriate assays available for differential gene expression analysis, the confirmation methodology and ultimately, the functional assessment of genes in the disease process are all important (*see* Chaps. 2 and 11).

5. Methodologies for Differential Gene Expression in Focal Stroke

Genes that are differentially expressed owing to stroke can be identified by using the simpler (i.e., well-established and straightforward) techniques such as Northern blotting, reverse transcription polymerase chain reaction (RT-PCR), *in situ* hybridization, and others. These techniques involve the selection and study of a specific gene of interest (i.e., based on previous data that provide a biological rationale for study in stroke or another specific disease). However, several more complex screening techniques are now available that can identify groups of differentially expressed genes, both known and unknown. These complex screening techniques include subtractive libraries/subtractive hybridization, differential hybridization, serial analysis of gene expression, representational differential analysis (RDA), and differential display; many of these are discussed in this book (*see* Chap. 11). Choice of assay is defined by variations in assay and threshold of detection and will often result in the isolation of different gene sets. Therefore, to ensure maximum confidence in the detection of adaptive upregulation of gene expression, a pragmatic approach should be adopted. For example, commonalities in identified genes and pathways should be identified across several differential expression assays and/or independent cross-validation of a gene's upregulation should be emphasized, rather than reliance on the results of a single differential expression technique.

5.1. Between Assay Variation and Increasing Confidence in Identified Gene Targets

With respect to generating and comparing gene expression data, several issues warrant discussion, including the significant variability between techniques and the identification of false-positive and false-negative results. Assays of differential expression have an inherent variability dependent on assay methodology, sensitivity, and reaction efficiency. When one is exploring disease paradigms that are powerful stimulators of gene expression, such as cerebral ischemia, the tendency is to highlight vast gene sets that are upregulated and differentially expressed. Given the large numbers of genes identified, it is difficult to confirm all expressed genes, and false-positive and false-negative differential gene expression becomes an issue. To manage these issues, multiple assays of differential expression on the same RNA pool should be employed.

Confidence in particular products can then be increased by identifying commonalities in expression across assays. Using this approach, follow-up analysis by RT-PCR often confirms that robust "hits," i.e., differential gene expression identified across all assays, have high fold increases in expression vs naive animals.

5.1.1. Confirmation

Quantitative RT-PCR techniques, TaqMan probes, or SYBR green *(42)* can be employed to monitor an accumulating PCR product in real time. TaqMan RT-PCR analysis has been extensively applied to the temporal profiling of caspase expression following middle cerebral artery occlusion in rats *(43,44)*, and SYBR green has been used to confirm differentially expressed genes identified by RDA *(45)*.

More sensitive than other confirmatory technologies such as Northern blotting, TaqMan RT-PCR uses approx 50 ng of total RNA per gene and quantitates gene expression over five to six orders of magnitude without multiple dilution series, which are needed in other assays *(46)*. The sensitivity of PCR-based methodologies means that sufficient RNA can be isolated from a single animal to allow the simultaneous assessment of several hundred genes without variation between studies. However, changes in gene expression have to be understood in the context of the evolving cell types present at the time. Ultimately, this has to involve techniques such as *in situ* hybridisation and immunohistochemistry, which allow the localization of expression to be viewed in relation to the structure of the evolving lesion and the identification of the types of cells in which expression is occurring. Application and statistical interpretation of cutting edge TaqMan RT-PCR techniques are discussed in further detail in Chapter 12.

5.2. Confirmation and Localization of Protein Expression: Transcription-to-Translation Verification

Confirmation of protein expression and the time-course of translation is of primary importance. This is especially pertinent given the severe energy compromise in the ischemic brain. During this state, transcription and translation can become uncoupled owing to the energetic demand of assembling functional protein. This will be temporally and spatially dependent and has been extensively reviewed by Koistinaho and Hokfelt *(39)* and Sharp et al. *(40)*. We have previously reported that this phenomenon is aptly demonstrated in the transcription and translation of cytokines during experimental stroke. Protein determination by enzyme-linked immunosorbent assay (ELISA) at 6 h after permanent middle cerebral artery occlusion fails to confirm a concomitant increase in interleukin-1β (IL-1β) protein, although IL-6 protein levels are significantly elevated in ischemic animals at this time point, as expected *(47)*. Therefore, from a family of related proteins sampled from an identical cortical region, with an apparently similar mRNA expression profile over the first 6 h after permanent middle cerebral artery occlusion, completely opposing protein profiles can be obtained. When further levels of complexity are superimposed on these observations, such as protein diversity achieved by alternative splicing of genes, post-translational modification of proteins changing their function, activity, location, and so on, or even the combination of parts of separate genes combining to form novel, unique proteins *(3)*, one cannot overemphasize the importance of protein confirmation.

6. Models for Probing Differential Gene Expression

Numerous in vitro and in vivo models of ischemia, cell stress, excitotoxicity, free radical damage, and preconditioning have been employed both to explore novel gene function and to probe differential expression during stroke. Barone and Feuerstein *(38)* have found that focal ischemia elicits temporal episodes or "waves" of expression of different groups of genes. Initial phases of focal ischemic gene expression are largely composed of increased expression of inflammatory cytokines driving leukocyte infiltration and associated secondary brain injury. Many other classes of gene expression typically follow in a time and spatially dependent manner *(40)* including apoptotic genes and growth factors (for review, *see* **ref.** *47*). Elucidating the function of these genes during the pathophysiological process of stroke has necessitated the use of adjuvant models such as transgenic animals (Chap. 7), models of preconditioning (Chap. 6), free radical stress (3-NPA) (Chap. 8), and numerous rodent models of evoked focal ischemia (Chap. 2). Complementary cutting edge technologies such as whole animal cellular and molecular imaging will also help to elucidate gene function.

7. The Future

One of the chief goals of this book is to emphasize that exploration of stroke genomics cannot be meaningfully approached without taking a holistic approach to interpreting differential expression. This requires the evaluation and comparison of different experimental models of stroke-induced brain injury and of endogenous brain protection, with the eventual comparison with human data. Studies of expression need to be approached with multiple technologies (differential display, subtractive hybridization, representational difference analysis) confirmed at the protein level and assessed in disease- and mechanism-relevant models for function. Furthermore, this expression must be put into context with lesion pathology by the use of clinically relevant endpoints such as magnetic resonance imaging. We are now in the post-human genome era, with an index of genes and little functional knowledge. The way forward will be by integrating the expression information and relating it to the human disease, rather than by making lists.

References

1. Stephenson, J. (1998) Rising stroke rates spur efforts to identify risks, prevent disease. *JAMA* **279,** 1239–1240.
2. Fisher, M. and Bogousslavsky, J. (1998) Further evolution toward effective therapy for acute ischemic stroke. *JAMA* **279,** 1298–1303.
3. Alberts, M. J. (2001) Genetics update : impact of the human genome projects and identification of a stroke gene. *Stroke* **32,** 1239–1241.
4. Kannel, W. B., Wolf, P. A., Verter, J., McNamara, and P. M. (1970) Epidemiologic assessment of the role of blood pressure in stroke. The Framingham Study. *JAMA* **214,** 301–310.
5. Brass, L. M., Isaacsohn, J. L., Merikangas, K. R., and Robinette, C. D. (1992) A study of twins and stroke. *Stroke* **23,** 221–223.
6. Brass, L. M. (2000) The impact of cerebrovascular disease. *Diabetes Obes. Metab.* **2(suppl. 2),** S6–S10.
7. Meschia, J. F. (2002) Addressing the heterogeneity of the ischemic stroke phenotype in human genetics research. *Stroke* **33,** 2770–2774.
8. Tournier-Lasserve, E., Joutel, A., Melki, J., et al. (1993) Cerebral autosomal dominant arteriopathy with subcortical infarcts and leukoencephalopathy maps to chromosome 19q12. *Nat. Genet.* **3,** 256–259.
9. Hirano, M. and Pavlakis, S. G. (1994) Mitochondrial myopathy, encephalopathy, lactic acidosis, and stroke-like episodes (MELAS): current concepts. *J. Child Neurol.* **9,** 4–13.
10. Joutel, B., Bousser, M.G., Biousse, V., et al. (1993) A gene for hemiplegic migraine maps to chromosome 19. *Nat. Genet.* **5,** 40–45.
11. Hassan, A. and Markus, H. S. (2000) Genetics and ischemic stroke. *Brain* **123,** 1784–1812.
12. Sourander, P. and Walinder, J. (1977) Hereditary multi-infarct dementia. *Lancet* **1,** 1015.
13. Chabriat, H., Tournier-Lasserve, E., Vahedi, K., et al. (1995) Autosomal dominant migraine with MRI white-matter abnormalities mapping to the CADASIL locus. *Neurology* **45,** 1086–1091.
14. Dichgans, M., Mayer, M., Uttner, I., et al. (1998) The phenotypic spectrum of CADASIL: clinical findings in 102 cases. *Ann. Neurol.* **44,** 731–739.
15. Hedera, P. and Friedland, R. P. (1997) Cerebral autosomal dominant arteriopathy with subcortical infarcts and leukoencephalopathy: study of two American families with predominant dementia. *J. Neurol. Sci.* **146,** 27–33.

16. Desmond, D. W., Moroney, J. T., Lynch, T., et al. (1998) CADASIL in a North American family: clinical, pathologic, and radiologic findings. *Neurology* **51,** 844–849.
17. Chabriat, H., Levy, C., Taillia, H., et al. (1998) Patterns of MRI lesions in CADASIL. *Neurology* **51,** 452–457.
18. Joutel, A., Vahedi, K., Corpechot, C., et al. (1997) Strong clustering and stereotyped nature of *Notch3* mutations in CADASIL patients. *Lancet* **350,** 1511–1515.
19. Joutel, A., Tournier-Lasserve, E. (1998) *Notch* signalling pathway and human diseases. *Stem Cell Develop. Biol.* **9,** 619–625.
20. Ruchoux, M. M. and Maurage, C. A. (1998) Endothelial changes in muscle and skin biopsies in patients with CADASIL. *Neuropathol. Appl. Neurobiol.* **24,** 60–65.
21. Viitanen, M. and Kalimo, H. (2000) CADASIL: hereditary arteriopathy leading to multiple brain infarcts and dementia. *Ann. NY Acad. Sci.* **903,** 273–284.
22. Couderc, R., Mahieux, F., Bailleul, S., Fenelon, G., Mary, R., and Fermanian, J. (1993) Prevalence of apolipoprotein E phenotypes in ischemic cerebrovascular disease. A case-control study. *Stroke* **24,** 661–664.
23. Nakata, Y., Katsuya, T., Rakugi, H., et al. (1997) Polymorphism of angiotensin converting enzyme, angiotensinogen, and apolipoprotein E genes in a Japanese population with cerebrovascular disease. *Am. J. Hypertens.* **10,** 1391–1395.
24. Kessler, C., Spitzer, C., Stauske, D., et al. (1997) The apolipoprotein E and β-fibrinogen G/A-455 gene polymorphisms are associated with ischemic stroke involving large-vessel disease. *Arterioscl. Thromb. Vasc. Biol.* **17,** 2880–2884.
25. DeStefano, V., Chiusolo, P., Paciaroni, K., et al. (1998) Prothrombin G20210A mutant genotype is a risk factor for cerebrovascular ischemic disease in young patients. *Blood* **91,** 3562–3565.
26. Rubattu, S., Ridker, P., Stampfer, M.J., Volpe, M., Hennekens, C.H., and Lindpaintner, K. (1999) The gene encoding atrial natriuretic peptide and the risk of human stroke. *Circulation* **100,** 1722–1726.
27. Rubattu, S., Giliberti, R., Ganten, U., and Volpe, M. (1999) Differential brain atrial natriuretic peptide expression co-segregates with occurrence of early stroke in the stroke-prone phenotype of the spontaneously hypertensive rat. *J. Hyperten.* **17,** 1849–1852.
28. Rubattu, S., Lee-Kirsch, M.A., DePaolis, P., et al. (1999) Altered structure, regulation, and function of the gene encoding the atrial natriuretic peptide in the stroke-prone spontaneously hypertensive rat. *Circ. Res.* **85,** 900–905.
29. Yamori, Y., Horie, R., Handa, H., Sato, M., and Fukase, M. (1976) Pathogenetic similarity of strokes in stroke-prone spontaneously hypertensive rats and humans. *Stroke* **7,** 46–53.
30. Barone, F. C., Maguire, S., Strittmatter, R., et al. (2001) Longitudinal MRI measures brain injury and its resolution: reduced neurological recovery post-stroke and decreased brain tolerance following ischemic preconditioning in stroke-prone rats. *J. Cereb. Blood Flow Metab.* **21(suppl. 1),** S230.

31. Purcell, J. E., Lenhard, S. C., White, R. F., Schaeffer, T., Barone, F. C., and Chandra, S. (2003) Strain-dependent response to cerebral ischemic preconditioning: Differences between spontaneously hypertensive and stroke-prone spontaneously hypertensive rats. *Neurosci. Lett.* **339,** 151–155.
32. Rubattu, S., Volpe, M., Kreutz, R., Ganten, U., Ganten, D., Lindpaintner, K. (1996) Chromosomal mapping of quantitative trait loci contributing to stroke in a rat model of complex human disease. *Nat. Genet.* **13,** 429–434.
33. Ikeda, K., Nara, Y., Matumoto, C., et al. (1996) The region responsible for stroke on chromosome 4 in the stroke-prone spontaneously hypertensive rat. *Biochem. Biophys. Res. Comm.* **229,** 658–662.
34. Jeffs, B., Clark, J. S., Anderson, N. H., et al. (1997) Sensitivity to cerebral ischemic insult in a rat model of stroke is determined by a single genetic locus. *Nat. Genet.* **16,** 364–367.
35. Gratton, J.A., Sauter, A., Rudin, M., et al. (1998) Susceptibility to cerebral infarction in the stroke-prone spontaneously hypertensive rat is inherited as a dominant trait. *Stroke* **29,** 690–694.
36. Brosnan, M. J., Clark, J. S., Jeffs, B., et al. (1999) Genes encoding atrial and brain natriuretic peptides as candidates for sensitivity to brain ischemia in stroke-prone hypertensive rats. *Hypertension* **33,** 290–297.
37. Ktorza, A., Bernard, C., Parent, V., et al. (1997) Are animal models of diabetes relevant to the study of the genetics of non-insulin-dependent diabetes in humans? *Diabetes Metab.* **23(suppl. 2),** 38–46.
38. Barone, F. C. and Feuerstein, G. Z. (1999) Inflammatory mediators and stroke: new opportunities for novel therapeutics. *J. Cereb. Blood Flow Metab.* **19,** 819–834.
39. Koistinaho, J. and Hokfelt, T. (1997) Altered gene expression in brain ischemia. *Neuroreport* **8,** i–viii.
40. Sharp, F. R., Lu, A., Tang, Y., and Millhorn, D. E. (2000) Multiple molecular penumbras after focal cerebral ischemia. *J. Cereb. Blood Flow Metab.* **20,** 1011–1032.
41. Iadecola, C. and Ross, M. E. (1997) Molecular pathology of cerebral ischemia: delayed gene expression and strategies for neuroprotection. *Ann. New York Acad. Sci.* **835,** 203–217.
42. Gibson, U. E., Heid, C. A., and Williams, P. M. (1996) A novel method for real time quantitative RT-PCR. *Genome Res.* **6,** 995–1001.
43. Harrison, D. C., Medhurst, A. D., Bond, B. C., Campbell, C. A., Davis, R. P., Philpott, K. L. (2000) The use of quantitative RT-PCR to measure mRNA expression in a rat model of focal ischemia—caspase-3 as a case study. *Mol. Brain Res.* **75,** 143–149.
44. Harrison, D. C., Davis, R. P., Bond, B. C., et al. (2001) Caspase mRNA expression in a rat model of focal cerebral ischemia. *Mol. Brain Res.* **89,** 133–146.
45. Bates, S., Read, S. J., Harrison, D. C., et al. (2001) Characterisation of gene expression changes following permanent MCAO in the rat using subtractive hybridisation. *Brain Res. Mol. Brain Res.* **93,** 70–80.
46. Medhurst, A. D., Harrison, D. C., Read, S. J., Campbell, C. A., Robbins, M. J., and Pangalos, M. N. (2000) The use of TaqMan RT-PCR assays for

semiquantitative analysis of gene expression in CNS tissues and disease models. *J. Neurosci. Meth.* **98,** 9–20.

47. Read, S. J., Parsons, A. A., Harrison, D. C., et al. (2001) Stroke genomics: approaches to identify, validate, and understand ischemic stroke gene expression. *J. Cere. Blood Flow Metab.* **21,** 755–778.

II

PRECLINICAL MODELS AND GENE MANIPULATION

2

Choice, Methodology, and Characterization of Focal Ischemic Stroke Models

The Search for Clinical Relevance

David Virley

Summary

To develop novel neuroprotective or neurorestorative agents for clinical application, the appropriate selection and characterization of preclinical focal stroke models is required to provide confidence in predicting therapeutic efficacy. Compelling evidence for novel therapies derived from the pathological and functional consequences of models of cerebral ischemia in the rat (and higher species) is an essential prerequisite before large expensive clinical trials are begun. This chapter provides an overview of focal ischemic models, with an emphasis on objective functional assessment of pathological mechanisms and efficacy of novel therapeutic strategies. The ability to predict functional consequences from structural abnormalities is a critical theme that can be extrapolated from the preclinical to the clinical setting, in that certain brain regions are inextricably linked to specific behavioral functions. This underlying approach is highly relevant, as monitoring the dynamic pathological and functional changes attributed to focal stroke will reveal new insights into novel mechanisms and targets that play a role in the evolution of cell death and impaired function. The utility of novel genomic technologies that are aligned with methods to determine structure–function relationships in preclinical models will facilitate a greater understanding of the pathophysiological process and potentially generate new targets that may ultimately be used to predict or offer clinical benefit.

Key Words

Focal cerebral ischemia; middle cerebral artery occlusion; ischemic stroke; behavior; functional impairments; functional recovery; neuroprotection; neurorestoration; neuroregeneration; animal models.

From: *Methods in Molecular Medicine, Vol. 104: Stroke Genomics: Methods and Reviews*
Edited by: S. J. Read and D. Virley © Humana Press Inc., Totowa, NJ

1. Introduction and Review

Stroke or focal cerebral ischemia is a leading cause of death and permanent disability for which there is currently no effective treatment; hence a large unmet medical need exists. In most western populations, 0.2% of the population (2000 per million) suffer a stroke each year *(1)*, of whom one-third die over the next year, one-third remain permanently disabled, and one-third make a reasonable recovery *(2)*. The purpose of this chapter is to highlight the utility of animal models of focal stroke and to identify strategies with the objective of clinical relevance/application in mind. The chapter attempts to outline:

1. Appropriate choice of animal models to identify novel pathophysiological mechanisms and therapeutic efficacy.
2. Appropriate methods of characterization that will elucidate functional and pathological changes predictive of human stroke, and how one could attempt to extrapolate to the clinical situation using this information.

Methods for inducing middle cerebral artery occlusion (MCAO) in rats via the intraluminal suture technique are explained in detail in the Materials and Methods sections, as well as valid and robust objective behavioral tests to quantify the functional impairments.

In terms of putting the above in the context of novel target identification for stroke, stroke genomic methodologies will be alluded to (which are discussed further in accompanying chapters), in order to glean comprehensive multifactorial in vivo information to help translate preclinical data to the clinical setting.

1.1. Anatomical Considerations

The middle cerebral artery (MCA) in humans is the largest branch of the internal carotid artery (ICA) and is considered the artery most often occluded following a thromboembolic obstruction. The etiology of vessel occlusion in humans relates to a thrombotic cause as a result of a local arteriopathy *(3,4)*, or an embolic cause through circulating emboli passing from the ICA to the MCA to occlude the vessel because of its proximity and size and the proportion of blood to the area. The MCA can be anatomically divided into four main segments: M1, M2, M3, and M4. The M1 segment comprises the main MCA trunk from which approx 12 deep penetrating vessels, the lenticulostriate arteries, arise in two groups: the larger lateral and smaller medial. From the M2 segment, situated in the sylvian fissure, arise the two main divisions of the MCA and all cortical branches that subsequently comprise the M3, or opercular, segment. Once over the cortical surface, the M4 segment is formed. The cerebral cortex, basal ganglia, and internal capsule are supplied by the MCA and its small penetrating branches. These regions are especially prone to infarction.

Focal infarcts of the MCA territory represent at least 25% of first-time ischemic strokes, although some studies have associated up to 80% of cerebral infarcts with ischemic damage to the MCA territory *(5,6)*. This area is susceptible because the small penetrating arteries to the brain are not supported by a good collateral circulation, and therefore occlusion of one of these arteries is likely to cause uncompromised infarction *(7,8)*. The stroke "lesion" can therefore be considered to consist of a central core of densely ischemic tissue (the focus) and of "perifocal" or "penumbral" areas with less dense ischemia *(9–12)*. The focus, usually encompassing the lateral part of the caudate putamen or striatum (part of the basal ganglia) and the adjacent neocortex, represents tissues that depend heavily on the perfusion from the occluded MCA by end-arterial branches. The periphery of the stroke lesion and the perifocal areas are perfused by the anterior cerebral (ACA) and posterior cerebral arteries (PCA), most of which are leptomeningeal. Clearly, in all but the most central parts of the lesion, perfusion depends on the adequacy of the collateral circulation.

However, human ischemic stroke is seldom permanent; it usually involves some degree of vessel recanalization and spontaneous reperfusion. Spontaneous, or drug-induced (fibrinolytic) reperfusion has the potential to be detrimental (beyond the therapeutic window), as well as beneficial. To elucidate the benefits associated with reperfusion-related tissue salvage or the pathological processes associated with reperfusion injury, it is important to characterize not only the severity and distribution of ischemia but also the effects of the extent and duration of reperfusion on outcome.

The vascular territories of the major cerebral arteries supplying the cerebral cortex, subcortical structures, cerebellum, and brainstem in humans are relatively well mapped. It has been stated that, depending on the location of the infarct, clinical syndromes vary in stroke patients, implying that functional impairment is a feature contingent on the degree of compromised flow within a particular portion of the vascular territory *(13)*. Hemiparesis, or muscle weakness to one side of the body, is the most common deficit after stroke, affecting more than 80% of patients acutely and more than 40% chronically *(14)*. The clinical manifestations of MCA infarction depend on the location of the occlusion. Therefore, distinct clinical syndromes are associated with the involvement of the main proximal arterial trunk, the lenticulostriate arteries, and the superior and inferior divisions of the MCA.

The limitations of large-scale human studies become more apparent when one tries to consider events immediately following the insult and before irreversible brain injury. Elucidation of the cause–effect relationships in the progression of the infarct pathology requires immediate access to localized events in the affected area to acquire insight into the ischemic process. Frequently, the stroke patient is admitted many hours after the symptoms have developed and generally only postmortem tissue or cerebrospinal fluid is available for biochemical analysis of the events. Because very limited human studies are feasible, investigations of the mechanisms and endogenous agents involved in the cascade of events that lead to brain injury following an ischemic insult have been primarily (or at least initially) pursued in experimental models of cerebral ischemia. To yield novel information on pathophysiological mechanisms that may contribute to the cell-death process(es) or impairments in regeneration, it is necessary to select and characterize animal models of focal stroke using appropriate methodologies, to facilitate the development of new and improved therapeutic strategies for human stroke.

1.2. Appropriate Choice of Animal Models of Focal Ischemia

Recently a report was published that outlined recommendations for standards regarding preclinical neuroprotective and restorative drug development, involving a roundtable of key worldwide academic, clinical, and industrial experts from the stroke research area (the Stroke Therapy Academic Industry Roundtable [STAIR]) *(15)*. It was highlighted that robust rodent models would be "early, go no-go" predictors of therapeutic efficacy for agents and that functional assessment over a longer period should be implemented to investigate the long-term benefit of an approach in order to improve extrapolation to the clinical setting. Furthermore, STAIR *(15)* has encouraged the development of nonhuman primate models as a necessary intermediate step for predicting the efficacy of treatment strategies before they enter the clinic. Primate models are important not only for evaluating clinically relevant functional outcome but also for scaling up dosing regimens from rodents in order to improve the definition of dose and duration of drug administration for sustained efficacy. Significant progress has been made in developing clinically relevant primate models of focal stroke *(16–18)*, which provide a greater level of confidence in progressing up the phylogenetic tree from rodents to humans when novel targets are developed further for large-scale clinical trials *(19–21)*. Furthermore, developments in nonhuman primate transgenesis *(22)* and comparative genomics via microarray analysis *(23)* in disease models will provide an even more compelling link between basic research and clinical application for neurological diseases, such as stroke. This chapter focuses on rodent models of focal stroke, which are routinely used world-wide and more accessible to identify novel mechanisms and targets for drug development.

In theory, there are two types of models that address the mechanistic and therapeutic aspects of focal cerebral ischemia. If the objective of the study is to understand precisely the event following an ischemic insult and to identify endogenous mediators that may participate, then there is a need to use a model in which variables such as duration and extent of ischemia can be controlled and a reproducible outcome guaranteed. Moreover, if the aim of the investigation is to determine the efficacy of a particular therapeutic intervention then importance should be placed on incorporating into the model design conditions that may be observed in humans, e.g., reperfusion. In practice, most models in current use have been developed to try and satisfy both ideals or at least provide an acceptable compromise. It is well recognized that "rodent" models provide a useful controlled environment to study the mechanisms involved in ischemic pathology as well as an environment in which to assess potential therapeutic interventions. However, many differences exist between the models and the clinical setting, and extrapolation of the results from the models to humans should be carried out with utmost caution. The most widely used models of focal cerebral ischemia can be broadly categorized into two types, namely, permanent MCA occlusion (pMCAO) and transient MCA occlusion (tMCAO) which are described next.

1.3. Models of Focal Cerebral Ischaemia

1.3.1. Unilateral Permanent Electrocoagulation of the MCA

Ligation of the *distal* portion of the MCA in the Sprague-Dawley rat using a frontoparietal approach to produce a focal ischemic lesion was developed nearly 30 yr ago *(24)*. Although this method produced infarcts in several layers of the cortex (2–5 mm in diameter), infarcts of a reproducible size were not always demonstrated, nor was damage involving the striatum. Refinement of this method was undertaken by Tamura et al. *(10,11)*. Using a subtemporal craniectomy, these investigators were able to gain access to the more *proximal* portions of the MCA. The coronoid process of the mandible and zygoma were removed, and a burr hole was opened lateral to the foramen ovale. Once the dura had been opened, the MCA could be visualized through the burr hole. Electrocoagulation of the main trunk of the MCA was conducted just medial to the olfactory tract, at a point proximal to the origin of the lateral lenticulostriate branches that supply the lateral portion of the anterior basal ganglia. This refinement produced an area of damage in the MCA territory that involved the cortex and striatum and also had a low mortality. With further slight variation in the surgical approach, involving preservation of the zygomatic arch, the MCA was still accessed and occluded, without compromising recovery from surgery. Lesion size examined many hours after the initial insult from this modified approach was equivalent to that obtained from the original Tamura approach *(25–27)*. In addition, bisecting or severing the coagulated MCA has been demonstrated to be necessary to ensure blood flow cessation completely, as coagulation alone has been shown to make the thrombus unstable and hence reduce the homogeneity in size of the infarction *(28)*. Step-by-step methods for conducting the permanent distal electrocoagulation of the MCA are comprehensively explained in the Chapter 3.

An excellent study *(29)* confirmed that electrocoagulation of the MCA proximal to the lenticulostriate branch was associated with lesions in the striatum. Furthermore, it was necessary to extend the length of the occlusion proximal to the inferior cerebral vein to elicit consistent, reproducible infarcts involving the cortex and striatum. The significant contribution of the lenticulostriate branches and other branches of the MCA to perfusion of the cortex and striatum in rats as well as humans has been demonstrated *(30–34)*. As there may be variation in the length of the vessel occluded, there has been considerable disparity (47–330 mm^3) in the volume of the hemispheric damage induced by proximal lesions *(26,29,35)*. In more recent years, a hemispheric infarct volume of approx 100–150 mm^3 is consistently reported following proximal electrocoagulation of the MCA in normotensive rats *(36)*. The extent and location of the lesion reflect the severity of the neurological deficit in both animal models and the stroke patient. The infarct volume of 100–150 mm^3 in a rodent would represent a large hemispheric infarct observed in humans.

Although this permanent model of MCAO has provided a well-controlled and reproducible environment to study the pathology and novel treatments applicable to ischemic stroke, like any model there are some drawbacks. The most obvious relate to the need for surgical craniectomy under anesthesia and the permanent nature of the occlusion. Indeed, exposure of the brain to air during craniectomy may alter intracranial pressure and blood–brain barrier permeability *(32,37)*. Exposing the brain to occlude the MCA can lead to thermal damage from drilling and dessication of the tissue around the crainectomy site. Frequent irrigation with saline during drilling and expedient surgery can reduce this damage. Furthermore, cauterizing the proximal MCA may cause damage to autonomic nerves around the MCA, and autoregulation of the cerebral blood flow (CBF) may be lost *(38)*. In addition to being a clinical strategy, reperfusion of a previously ischemic area is observed in humans spontaneously, through resolution of the emboli; however, it cannot be incorporated into the study design of permanent MCAO. Reperfusion presents its own clinical and therapeutic challenges, and other models have been developed to investigate the consequences of reperfusion of ischemically compromised tissue.

1.3.2. Unilateral Intraluminal Thread Occlusion of the MCA

The introduction of a model that allows both permanent (p) and transient (t) MCAO in the rat without craniectomy has heralded a new era in the experimental study of focal cerebral ischemia and reperfusion injury.

An intraluminal occlusion method with subsequent reperfusion was pioneered by Koizumi et al. *(39,40)*, to study the progression of edema with reperfusion. These authors described the introduction of a suture into the ICA at the bifurcation of the common carotid artery (CCA) and external carotid artery (ECA) following ligation of the latter vessels. This suture was advanced intraluminally beyond the origin of the PCA and past the origin of the MCA. At this level, the intraluminal device prevents blood flow to the MCA from the ICA and PCA, with the help of restricting any anterograde blood flow from ligation of the CCA. The greatest advantage of this model is the ease with which recirculation can be instigated. Recirculation can be initiated by simply withdrawing the thread and re-exposing the origin of the MCA. In Koizumi's technique, the ipsilateral CCA is ligated, as is the ECA, so recirculation is achieved by the complete circle of Willis via a retrograde flow to the MCA. This technique has led several investigators to conclude that this is a feasible method of inducing reversible ischemia. This method also obviates what are perceived as the surgically more demanding and, some would construe, damaging aspects of the subtemporal exposure of the MCA developed by Tamura et al. *(10,11)*. The utility of this model to examine reperfusion injury has been fully exploited, and its suitability for inducing pMCAO has been used by a number of groups to reconcile the temporal profile of events between this method and the original Tamura surgical technique.

Several types of coated or uncoated sutures *(39,41–46)* and rat strains *(47,48)* have been used in studies utilizing the intraluminal thread approach, and it has become increasingly clear that these factors significantly influence outcome. In the original Koizumi method, the 4/0 suture was coated with silicone at the distal 5 mm (0.25–0.30 mm in diameter, for animals weighing 280–350 g) to provide a soft and malleable coating, which gently dilates the vessels through which it passes. This leads to a greatly reduced risk of vessel perforation and a much tighter fit (reducing the incidence of subarachnoid hemorrhage), preventing any residual blood flow around the thread. However, more recently the reproducibility and reliability of Koizumi's silicone-coated suture method has been tested further, illustrated by the fact that coating the thread makes the diameter of the occluder more consistent and offsets minor variations of commercially prepared suture diameters, which may ultimately affect infarct volume *(49)*. It has been debated whether a coated filament with a larger diameter caused a more complete vessel occlusion with lower residual CBF and better consistency of ischemic lesion volume, without incurring subarachnoid hemorrhage *(45,49)*. Also, in the original Koizumi model, the suture was inserted via the CCA rather than the ECA *(42)*, which was permanently ligated to fix the thread in place and therefore to provide a more homogeneous infarct size (by reducing the variability of the ipsilateral collateral supply). Therefore it has been argued that Koizumi's technique is the method of choice *(50)* in generating a reliably noninvasive proximal MCAO approach with consistent reductions in CBF.

Although there appears to be a tremendous potential in using this model to investigate the deleterious effects of reperfusion injury such as edema and the microvasular alterations that might contribute to such pathology, there are several drawbacks to this model. Even though mechanical damage to the vascular smooth muscle via external compression is obviated, the extent to which the lumen of the vessel (in particular the endothelial cells) is damaged by this procedure has not been fully assessed. In addition to the mechanical consequences of denudation of the endothelial cells *per se*, the loss of the contribution of the endothelium-derived products to cerebrovascular tone may complicate analysis of the results. Although the intraparenchymal vessels would not be affected directly by the mechanical damage from the intraluminal device, there are concerns regarding the influence this damage would have locally on the permeability of the blood–brain barrier and whether it would provide a source of emboli that occlude more distant vessels. The relatively high mortality rate with extended durations of MCAO (70% with 3 h of MCAO) is disappointing and most likely reflects the increased edema and brain swelling; pathology also observed clinically with delayed reperfusion *(7,51–53)*. Nevertheless, this intraluminal thread model has been used to establish correlates between pathological tissue changes observed post mortem and the changes using novel imaging technology with shorter and permanent durations of MCAO with appropriate survival times. Differences between lesion evolution via magnetic resonance imaging (MRI) have also been noted between the Koizumi and Zea Longa approaches, suggesting that infarct expansion may be simulated differentially by these contrasting methods and represent different populations of stroke patients appropriately *(54)*. This provides valuable information to correct clinical use of this technology.

1.4. Functional Consequences of MCAO in Rodents

Although the pathophysiological features of MCAO in rats has been extensively investigated, until recently an examination of the behavioral consequences has received less attention. A description of the functional correlates of ischemic damage is important, as the principle goal of any stroke therapy in humans is the restoration of normal behavioral function of the patient. An appraisal of the behavioral deficits, which are objectively quantified in an animal model of stroke, allows a realistic association between specific pathophysiological mechanisms and specific behavioral impairments.

Most of the behavioral work that has been conducted using the rat pMCAO electrocoagulation technique has focused on simple reflex and motor function during the early phase of infarction, i.e., during the first 24 h *(29)*. Indeed, neurological examinations based on posture and hemiparesis *(29)*, the motor screen test *(55)*, the balance beam test *(55–57)*, the limb placing test *(58)*, and the prehensile-traction test *(59)*, have been widely used to assess outcome because of their simplicity and the fact that they have been developed according to clinical criteria. Although these tests include many parameters or grades to determine total deficits, each parameter tends to be relatively crude. There also tend to be problems such as quantification and objectivity for measurement in the chronic phase, as testing conditions and individuals can vary to different degrees over time. Clearly there is a need to develop fine, objective, and quantitative methods of assessment over time in animal models of MCAO in order to standardize reliable baselines to the severity of ischemia and to be confident in predicting therapeutic efficacy. Few studies have attempted to extend behavioral assessment to determine the chronic consequences of tMCAO after the intraluminal thread technique, even though this surgical approach is commonly used in experimental studies of stroke. In studies that have been extended over time, using alternative surgical procedures for both pMCAO and tMCAO, simple tests of somatosensory and motor function (postural reflex, bilateral sticky label/tactile extinction, beam walking, rotarod) have shown spontaneous recovery *(60–64)*. Indeed, the bilateral sticky label test is an objective test developed by Schallert and Upchurch *(65)* that quantifies the latency to contact and remove sticky tape/labels simultaneously presented on the forepaws of rats. Typically, MCAO produces elevated latencies in contacting and removing tactile stimuli from the contralateral forepaw *(60,61,63,64)* and has similarities to the tactile extinction observed in human stroke patients *(66)* and the contralateral neglect syndrome arising from damage to the parietal cortex in humans *(67)*. Although this test shows reproducible impairments within the acute to subacute phase following MCAO, recovery of variable degrees has been demonstrated within the chronic phase of testing (at 1 mo). Clearly this is a problematic issue, as results from the experimental literature suggest either that rodents have a greater capacity for recovery of function after stroke than humans or that the behavioral tests are insensitive or inappropriate for assessing function over time.

Alternatively, tests of skilled motor function, such as the staircase test, originally developed by Montoya et al. *(68)*, has demonstrated stable and persistent (up to 3 mo) impairments in retrieving food pellets with the contralateral forepaw *(64,69–71)* as well as the ipsilateral forepaw *(64,70,71)*, following MCAO in rats. The staircase test provides a highly objective measurement of independent skilled paw use, requiring the rat to exert precise motor control over each paw in order to grasp and retrieve pellets. Furthermore, the number of attempted but displaced pellets can also be recorded, allowing an assessment of motivation owing to the appetitive nature of this task. Therefore the staircase test provides a good approximation of one of the most pronounced long-term deficits in human stroke patients, finger dexterity.

Therefore a number of behavioral tests have sufficient sensitivity to detect functional impairments, although application on a widespread basis in experimental stroke studies has not been apparent. In several cases relationships between histological outcome and behavioral tests have not been performed; in others, correlations between histological outcome and functional outcome have produced conflicting results *(69,72–74)*. Correlating total lesion size with functional impairment only makes the assumption that the degree or extent of damage is associated with the behavioral deficit. In addition, this assumption does not attempt to identify specific structures that may fulfil the specific functions being probed by the behavioral task. The importance of selective lesion studies in animals has established the crucial relationship between specific or regional brain damage and specific impairments in behavioral performance. The behavioral tasks are designed to yield information concerning distinct components of behaviour that are potentially disrupted by the experimental lesion. Furthermore, Gavrilescu and Kase *(13)* have stated that, depending on the location of the infarct, clinical syndromes vary in stroke patients, implying that functional impairment is a feature of compromised flow within a particular portion of the vascular territory. Moreover, Gavrilescu and Kase *(13)* highlighted the role MRI and computed tomography scans play in improving our understanding of correlations between the anatomical substrates recruited by the infarction process and the neurological status of the individual stroke patient. Therefore, specific anatomical substrates recruited by the infarction process, rather than lesion size *per se*, may be the critical determinants of behavioral impairment following focal cerebral ischemia.

The experimental literature relating to animal studies of stroke has provided an extensive body of evidence to suggest that intact cortical regions surrounding an infarct, as well as contralateral regions, may contribute to the restitution of function following brain injury. Furthermore, changes in axonal outgrowth and synaptogenesis, detected by immunohistochemistry (growth-associated protein-43 and synaptophysin, respectively), have been demonstrated within penumbral and contralateral regions that parallel functional recovery following distal pMCAO in SHRs *(75)*. In addition, recent advances in pharmacological interventions that amplify these cellular events such as amphetamine *(76)*, basic fibroblast factor *(77,78)*, nerve growth factor (NGF) *(79)*, and anti-Nogo-A *(80)*, as well as osteogenic protein-1, that has been postulated to enhance new dendritic sprouting *(81,82)*, have been associated with significantly improved outcome in forelimb and hindlimb tests following focal brain injury in rats. These studies also demonstrate that the enhancement of recovery of function does not necessarily depend on the reduction of total infarct volume but rather on the reorganization of the remaining intact brain. This evidence suggests that the time window for enhancing stroke recovery is potentially much longer than that for reducing infarct volume and implies that recovery-promoting drugs for stroke may have effective time windows of several days or even weeks after the onset of ischemia.

The next sections document appropriate materials and methods to conduct the intraluminal thread MCAO surgical approach (via the modified Koizumi technique) in rats with the aim of assessing animals on objective behavioral tests. This approach can provide useful insights into mechanisms that play a role in spontaneous recovery of function and persistent impairments, ultimately to gauge the efficacy of novel therapeutic strategies.

2. Materials

2.1. Behavioral Testing

1. Parcel tape.
2. Beam-walking apparatus (100-cm horizontal beam).
3. Staircase test apparatus.
4. Coco Pops (Kelloggs, UK).
5. Stopwatches/stop clocks.

2.2. Surgical Procedure

1. Halothane (Concord Pharmaceuticals, UK) in N_2O/O_2.
2. Hibiscrub (Schering Plough Animal Health, UK).
3. Fur shaver.
4. Homeothermic heating blanket with rectal probe (Harvard).
5. Saline.
6. Scalpel.
7. Autoclaved cotton buds.
8. Triangular arrowhead swabs.
9. Watchmaker forceps.
10. Surgical retractors.
11. Aneurysm clips.
12. Microscissors.
13. 3-0 Nylon monofilament (Ethicon, UK).
14. Silicone sealant.
15. Sutures.
16. Operating microscope.
17. Blood gas analyzer.

2.3. Postoperative Care

1. Baby Food (Farleys, UK).
2. Complan (Complan Foods, UK).
3. Weighing boats.
4. Soft bedding.

2.4. Preparation for Brain Removal

1. Halothane (Concord Pharmaceuticals, UK) in N_2O/O_2.
   or Pentoject (Animal Care, UK).

2.5. Perfusion-Fixation for Histopathology

1. Heparin (CP Pharmaceuticals, UK).
2. Saline.
3. Paraformaldehyde (Sigma).
4. Perfusion pump.

2.6. Processing Brains for Histopathological Interrogation

1. Rat Brain Matrix
2. Processing and paraffin-embedding center (Shandon Citadel 1000 processor and Shandon Histocentre 2 embedding center).
3. Paraffin microtome.
4. Microtome blades.
5. Water bath for paraffin sections.
6. Microscope slides.
7. Absolute alcohol.
8. Distilled water.
9. Cresyl violet.
10. Luxol fast blue.
11. Histoclear.
12. Histomount.
13. Cover slips.
14. Image analyzer.

3. Methods

3.1. Animal Housing and Preparation for Behavioral Testing and Postoperative Care

Adult male Sprague-Dawley rats (Charles River, UK; 300–350 g weight at time of surgery) are typically housed singly or in groups of two (*see* **Note 1**) and maintained under a 12-h light/dark cycle with water *ad libitum*.

Food is typically restricted during pretraining and at 7 d post MCAO to facilitate performance on the staircase test, which is an appetitively motivated task. The feeding regimen is controlled so that animals gain weight at a rate of 3–5 g per week, maintaining animals at 85–90 % of their free feeding weight. Animals are provided with food pellets *ad libitum*, together with a mixture of baby food and Complan in appropriate water provided in small weighing boats, on the floor of the cage, from 6 h to 6 d post MCAO to increase postoperative weight and improve recovery. Soft bedding should be introduced to the cage following surgery to keep the animals warm and to facilitate recovery. Rats are initially trained on the following behavioral tasks, prior to surgery: bilateral sticky label test, beam walking and the staircase test.

3.2. Behavioral Training and Testing Following Surgery

3.2.1. Bilateral Sticky Label Test

The bilateral sticky label test *(65)* is used to quantify contralateral neglect/ipsilateral bias and model tactile extinction to double-simultaneous stimulation (DSS), which is observed in human stroke patients *(66)*. Bilateral stimulation of the radial aspect of the forearm is achieved by placing thin strips of brown parcel tape (1.5 × 4 cm) firmly around each animal's wrists, so that they cover the hairless part of the forepaw. In rare instances in which the sticky label comes partially or completely off, without the animal having attempted to remove the tape with its mouth, the trial should be repeated again.

Animals are given three bilateral stimulation trials daily, each lasting up to a maximum of 5 min, and the following parameters are recorded: latency to contact label on left and right forearm (in seconds); latency to remove left and right label on forearm (in seconds); order of contact (total number of times left and right forearms are contacted first); and order of removal (total number of times left and right label on forearms are removed first). Care is taken in each trial to apply each stimulus with equal pressure and to randomize the order of application (left vs right). Latency data are assessed on a daily basis, and order of contact and removal data are represented as the mean ± SEM of two consecutive days, so that a bias analysis can be carried out, i.e., order of contact and removal determined, according to established protocols *(16,64)*.

Animals are trained on this task until a stable baseline for latency and order of contact are established over a period of a week. Final preoperative measurements are carried out over a period of another week prior to surgery to confirm that no bias in order of contact or removal exists prior to surgery, so animals are matched for ability (handedness). This test can be applied daily post surgery, e.g., from 1 to 28 d post MCAO, or at specific time-points post MCAO. Results generated from this test up to 28 d post MCAO can be expressed typically in the format of **Fig. 1** *(64)*, showing a sustained impairment in the latency to contact and remove the contralateral label and an ipsilateral bias following MCAO.

3.2.2. Beam Walking

Beam walking *(58,64,83)* is used as a measure of hindlimb coordination via distance travelled across an elevated 100-cm beam (2.3 cm in diameter, 48 cm off the floor).

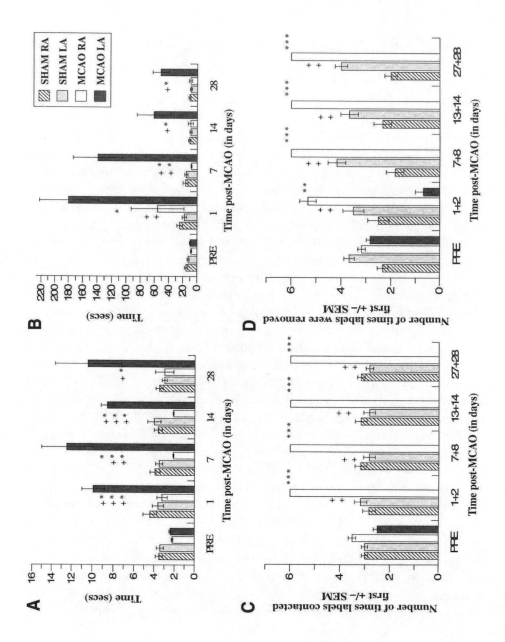

Rats are systematically trained to walk along the elevated beam from start to finish, with the aim of completing the task by 3 min. The total time spent (up to 3 min) on the beam is also recorded as a measure of coordination. A safe/target location, e.g., a flat box, is placed at the end of the beam so that the rat is motivated to cross the beam and complete the task. On occasions rats are "prodded," defined as a gentle tap on the rump delivered with a soft pencil eraser, to facilitate movement across the beam during the training phase (*see* **ref. 83** and **Note 2**).

Each rat is trained twice daily for a maximum of 3 min per trial, and an acquisition curve can be constructed to demonstrate that rats can learn this task to achieve a stable baseline prior to surgery, e.g., typically within 5 d. Following surgery, each testing day can be represented as the mean of two daily sessions (*64*) (**Fig. 2**). This task can be utilized daily post surgery, e.g., from 1 to 28 d post MCAO or can be used at specific time-points post MCAO (*see* **Note 3**). This test identifies a functional impairment relating to the hindlimb within the first 7 d, which has been shown to recover with time (*see* **Fig. 2**) (*64*). Functional recovery on this test at 28 d post MCAO also correlates with the recovery of MRI tissue signatures and tissue salvage within the "penumbral" hindlimb cortex (*see* **Figs. 5** and **6** and **ref. 64**).

3.2.3. Staircase Test

The staircase test (*64,68,70*) is used to measure skilled independent forelimb paw reaching, i.e., pellet recovery is only possible with the left paw from the left stair and with the right paw from the right stair. Pellet recovery is not performed under visual guidance but by using tactile and possibly olfactory cues. Apart from the top two steps of the six-step stairway, from which a few rats use their tongues in the early stages of training, retrieval of a food pellet is only possible with a coordinated grasping action using all digits of the forepaw.

Fig. 1. Mean latency (s ± SD) to contact (**A**) and remove (**B**) adhesive labels placed around the right (ipsilateral) and left (contralateral) forearms by middle cerebral artery occlusion (MCAO) and sham control animals. Total numbers of times that the right (ipsilateral) and left (contralateral) labels were contacted (**C**) and removed (**D**) first by MCAO and then by sham control animals are also depicted (mean ± SEM) in 2-d blocks across the time-course. RA, right forearm; LA, left forearm; *, $p < 0.05$, **, $p < 0.01$, and ***, $p < 0.001$, MCAO (LA) significantly different from respective MCAO ipsilateral forearm (RA). +, $p < 0.05$, ++, $p < 0.01$, and +++, $p < 0.001$, MCAO (LA) significantly different from respective sham control contralateral forearm (LA). Post hoc statistical tests were computed following a significant repeated measures ANOVA (*64*). (Reproduced with permission from **ref. 64**.)

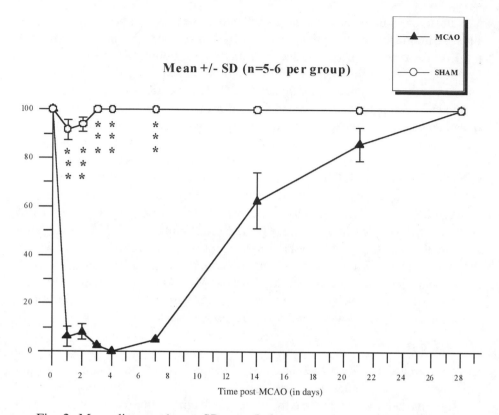

Mean +/- SD (n=5-6 per group)

Fig. 2. Mean distance (cm ± SD; *n* = 5–6 per group) travelled across a 100-cm elevated beam by middle cerebral artery occlusion (MCAO) and sham control animals across the time-course. ***, *p* < 0.001, significantly different from sham control group. Post hoc statistical tests were computed following a significant repeated measures ANOVA. (Reproduced with permission from **ref. *64*.**)

Animals are placed on a mild food-restricted diet during pretraining to provide motivation for food rewards. Animals are then introduced into a Perspex enclosure (300 × 67 × 95 mm, long × wide × high) attached to a holding box. A central plinth (190 × 20 × 48 mm) is positioned between the entry aperture and the front wall. Two removable staircases are positioned, either side of the plinth, by insertion through the front wall. Each stair consists of six steps, each measuring 14 × 17 × 6 mm with a hemispheric cup (11.5 mm in diameter). Between the back stair and the entrance aperture a *V*-shaped barrier is positioned on the floor of the box, so food pellets displaced from the steps can be held without the animal getting hold of them, i.e., pellet recovery is only possible with the left paw from the left stair and with the right paw from the right stair. Animals are given two 5-min sessions daily in the staircase testing box, with approx 10 min between tests. Each step of the stairs is baited with one chocolate food pellet. Each animal had to retrieve as many pellets as possible (maximum of 6 pellets a side per session are typically available, i.e., a total of 12 pellets a side for any one day). At the end of the test period, the stairs are removed, and the animals are returned to their home cage.

Performance is scored as the number of pellets recovered from each stair and the number of pellets displaced but not recovered. Rats are trained over a period of 3 wk on this task, with a criterion of at least six pellets recovered from each side and no more than four pellets displaced per side per day over three consecutive days. Each data point is represented in blocks, i.e., the mean of three consecutive days of testing, from pre- to postoperative blocks, as employed by Marston et al. *(70)* and Virley et al. *(64)*. This test is typically conducted from 7 d post MCAO, in order to give the animals enough time for recovery before food restriction is reintroduced (owing to the appetitive nature of this test). This test demonstrates a significant impairment in both the contralateral and ipsilateral forepaw retrieval following MCAO, without affecting levels of motivation. *See* **Fig. 3** for a description of typical data (from **ref. *64***).

3.3. Surgical Procedures and Confirmation of Successful MCAO

1. After successful training and an overnight fast, animals are randomly assigned to receive either tMCAO (e.g., 90 min) via the intraluminal thread technique by adaption of the method originally described by Koizumi et al. *(39)* or sham surgery, for example. This duration of MCAO is appropriate for assessing long-term function in rats, as opposed to longer durations of, or permanent, occlusion via the intraluminal thread method (*see* **Note 4**).
2. Surgical procedures can be performed under halothane (2/1 mixture of N_2O/O_2) anesthesia.
3. Following exposure of the right CCA, through a midline cervical incision, sterile silk sutures are looped around the common and internal carotid arteries.
4. An aneurysm clip is placed across the common carotid and blood flow in the internal carotid, can be temporarily be arrested using a further clip.
5. The 3/0 nylon monofilament thread, its leading 5-mm end coated with silicone rubber (diameter of 0.30–0.32 mm), is then introduced carefully via a right common carotid arteriotomy and carefully advanced along the ICA, beyond the looped suture so it can be tied followed by removal of the clip on the ICA.
6. The silicone thread is then advanced until its tip is positioned 1 mm beyond the origin of the right MCA (rMCA), as verified by a slight resistance. This is typically 19–22 mm distal to the carotid bifurcation.
7. The filament is tied in place with suture thread, and the aneurysm clip across the CCA is removed.
8. The wound is then sutured closed, and the animal is allowed to recover from anesthesia, typically in an incubator.

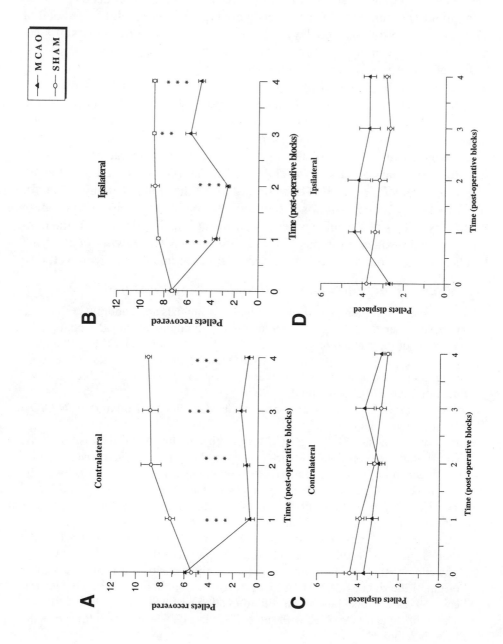

36

9. A saline injection is administered to provide fluid replacement as a result of surgery.
10. Sham surgery is achieved by introducing the thread into the ICA, followed by rapid removal.
11. The rMCA is occluded for e.g., 90 min, after which each rat is briefly reanesthetized with halothane (2/1 mixture of N_2O/O_2), and the thread is withdrawn (not completely removed) to the CCA to permit retrograde blood flow to the rMCA, via the complete circle of Willis, and hence reperfusion.
12. The thread is then cut at the point of the CCA, and the wound is sutured closed.
13. Throughout all surgical procedures, rectal temperature is monitored and maintained at 37 ± 1°C (mean + SD), with a heated electrical blanket.
14. Arterial blood samples can be obtained just prior to inserting the intraluminal thread in both MCAO and sham-operated control animals for assessment of blood gas status (pH, PCO_2, PO_2, HCO_3^-).
15. Successful occlusion is verified by applying the Bederson Neurological Scoring System (grades 0–3) after 60 min of occlusion and on reperfusion. Typically the ischemic deficit can be crudely assessed using the following grading system:

Grade 0 = no deficit.
Grade 1 = failure to extend contralateral forepaw properly.
Grade 2 = decreased grip of contralateral forelimb while tail is gently pulled.
Grade 3 = spontaneous circling or walking to the contralateral side.

This simple neurological scoring system can also be used throughout the extended time-course post MCAO and historically is a commonly used grading system that assesses neurological function in experimental rodent stroke models over time.

3.4. Histological Procedures

1. At the desired end point of the study e.g., 28 d post MCAO, both MCAO and sham animals, for example, are terminally anesthetized and transcardially perfusion-fixed with heparinized saline followed by 4% paraformaldehyde.
2. Brains are then removed (*see* **Fig.** 4 for superficial changes indicative of cavitation at 28 d post MCAO), immersed in fixative, and stored at 4°C for about 7 d.
3. Each brain is then cut into 2-mm-thick coronal blocks for a total of six blocks per animal, using a rat brain matrix.
4. Blocks are then processed for paraffin embedding using a processing and paraffin-embedding center.

Fig. 3. Total number of pellets recovered by contralateral (**A**) and ipsilateral (**B**) forelimbs in preoperative and postoperative blocks. Each block represents the mean ± SEM of six trials over a period of 3 d. Total number of pellets displaced by contralateral (**C**) and ipsilateral (**D**) forelimbs. **, $p < 0.01$, ***, $p < 0.001$, significantly different from the respective sham control groups. Post hoc statistical tests were computed following a significant repeated measures ANOVA. (Reproduced by permission from **ref.** *64*.)

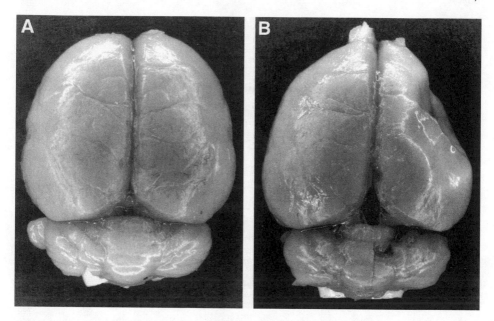

Fig. 4. Dorsal surface profile of (**A**) sham and (**B**) MCAO rat brain at 28 d post-surgery. Note the extensive cavitation on the MCAO rat brain that is attributable to MCA territory infarction.

5. Once coronal blocks have been paraffin-embedded, 10-µm-thick paraffin sections are typically cut using a paraffin microtome and floated out onto a water bath. Once they are flat and crinkles are no longer visible, they are carefully floated onto microscope slides and dried for subsequent staining with Cresyl Violet (Nissl) and Luxol Fast Blue, to delineate regions of gray and white matter loss (**Fig. 5**), respectively, using an appropriate image analysis system. Detailed methods of histological staining and analysis of neuropathology are outlined in other chapters.

3.5. Summary

At the dawn of the new millenium, we are now in a better position to use more sophisticated technology to establish novel treatment strategies that may translate into a clinically realistic benefit. The rigorous assessment of treatment strategies in preclinical animal models of stroke over an extended time-course, in both rodents and nonhuman primates, is a primary objective to predict true therapeutic value before embarking on clinical trials. It is clear that a multimodal approach is of considerable benefit when one is assessing the consequences of central nervous system injury in animal models, as correlations among behavior (**Figs. 1–3**), serial MRI (**Fig. 6**, and Chap. 8), and histological outcome (**Fig. 5**) or genomic information from the same animal can

A Slice 1   **B** Slice 2

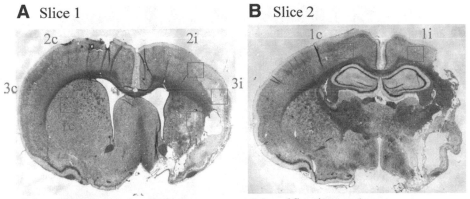

Magnification: x 1   Magnification: x 1

Fig. 5. Photomicrographs of cresyl violet- and Luxol Fast Blue-stained sections depicting ROIs from the same MCAO rat at 28 d post-surgery used for MRI analysis (*see* **Fig. 4**). (**A**) Slice 1 = –0.3 mm from bregma, showing ROI: 1i, ipsilateral caudate putamen (CPU); 1c, contralateral CPU; 2i, ipsilateral forelimb cortex (FLC); 2c, contralateral FLC; 3i, ipsilateral lower parietal cortex (LPC); 3c, contralateral LPC. (**B**) –1.8 mm from bregma, depicting the hindlimb cortex (HLC). 1i, ipsilateral HLC; 1c, contralateral HLC. (Reproduced with permission from **ref. *64*.**)

provide the necessary tools to predict fruitful neuroprotective or restorative treatment strategies. Furthermore, an assessment of pathophysiological mechanisms from tissues harvested following MCAO using novel stroke genomic technologies (*see* Chaps. 6, 11–13, 15) may break new ground in identifying novel targets. If these novel targets have the potential of being chemically tractable, then they too may be assessed in turn in a study design outlined in this chapter to provide evidence for therapeutic efficacy. Therefore this underlying approach is highly relevant to the clinical situation, as monitoring the dynamic pathological and functional changes over time in preclinical animal models of stroke will ultimately provide a robust test of power for predicting the utility of novel treatments on final outcome in patients.

4. Notes

1. Depending on the hypothesis being tested, which has implications for the study design, rats can be either singly housed for drug interventions, i.e., continuous intravenous drug delivery or group housed (typically two or more). If appetitive behavioral tests are being incorporated into the study design, then weight gain and hence food restriction becomes an issue when two or more animals are grouped together.

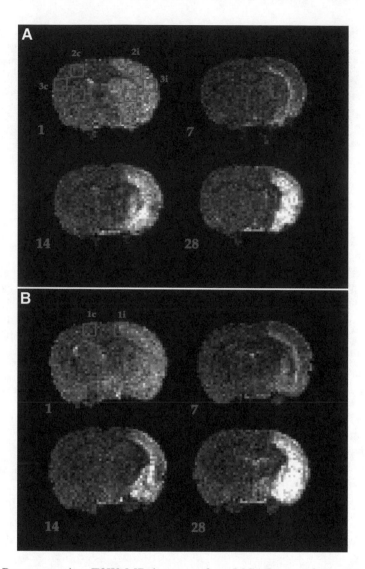

Fig. 6. Representative T2W MR images of an MCAO rat, after surgery. (**A**) Approximately –0.3 mm from bregma, depicting regions of interest (ROI) for analysis of MR tissue signatures at 1, 7, 14 and 28 d post-surgery. 1i, ipsilateral Caudate Putamen (CPU); 1c, contralateral CPU; 2i, ipsilateral ForeLimb Cortex (FLC); 2c, contralateral FLC; 3i, ipsilateral Lower Parietal Cortex (LPC); 3c, contralateral LPC; and (**B**) –1.8 mm from bregma, depicting the HindLimb Cortex (HLC) for analysis across the same time course. 1i ipsilateral HLC; 1c, contralateral HLC. Reproduced by permission of Lippincott Williams and Wilkins (Virley et al., 2000, A temporal MRI assessment of neuropathology following transient MCAO in the rat: Correlations with behaviour. *J. Cereb. Blood Flow Metab.*, **20,** page 566).

2. For pretraining on the beam walking test, on average, animals only require a few gentle "prods" to initiate successful locomotion across the beam. Indeed, "prodded" rats have been shown to attempt to take more steps on the beam, acquiring more task-specific experience, following a sensorimotor cortex lesion, which aided beam walking recovery *(83)*. It must be stressed that "prodding" should *only* be used during the training phase, in order to train the animals successfully on this task. "Prodding" should therefore *not* be used in the postsurgical assessment of rats performing this task.

3. Testing animals daily post surgery raises a number of key issues that need to be addressed. Johansson and Ohlsson *(84)* have demonstrated improvement on the beam walking task using a subjective grading scale, with SHRs subjected to proximal pMCAO and subsequently housed in an enriched environment. The actual compensatory mechanisms that may be stimulated by an enriched environment have not been examined in any great detail, although a study design that assesses MCAO rats tested on a daily basis may induce a stimulating environment to enhance the mechanisms of plasticity in the brain (either ipsilateral or contralateral to the infarct), e.g., to improve hindlimb function. It is important to have appropriate controls in place for any study design that implements a variety of objective behavioral tests to determine whether weekly testing would provide similar stable levels of impairment relative to daily testing. Moreover, recovery of function on the beam walking test needs to be ascertained with appropriate ischemic controls to determine whether recovery was demonstrated as a result of daily testing (familiarization/enrichment), or whether this phenomenon was reproduced with longer intervals between test days.

4. It has been demonstrated that for the intraluminal thread approach in rats, the extent of injury cannot be reduced by reperfusion after 2 h of MCAO *(43,44,85)*, suggesting that occlusion times of 2 h or more yield lesions equivalent in size to those derived from pMCAO *(86,87)* and hence increase the risk of mortality. Alternatively, tMCAO of 90 min, provides a severe enough depression of CBF within the MCA territory to provide ischemic injury as well as allowing the instigation of reperfusion to be beneficial to some regions, consequently improving survival rate *(43,44)*. Therefore 90 min of tMCAO is a rational duration to investigate the long-term effects of focal cerebral ischemia in rats. This model also closely corresponds to human ischemic stroke, in which reperfusion commonly occurs and clinical outcome can be monitored, particularly in the case of cerebral embolism *(7)*. Indeed, an ideal stroke model for neuroprotection and/or regeneration studies has a measurably salvageable ischemic penumbra, high reproducibility, and a low mortality rate.

Acknowledgments

The author would like to thank his colleagues at GlaxoSmithKline and his former colleagues at the Institute of Psychiatry and Queen Mary Westfield College, University of London who are involved with animal models of stroke. The information presented in this chapter has been a shared experience over time through extensive discussions with numerous colleagues, especially Sarah Hadingham, John Beech, Sean Smart, Steve Williams, Andy Parsons, A. Jackie Hunter, Frank Barone, Simon Read, Stuart Smith, Jeffrey Gray, and Helen Hodges, to name a few. I am grateful to all for providing significant insights into such a challenging area with a key objective in mind: using scientific rigor in the preclinical setting to elucidate novel therapeutic strategies for clinical application.

References

1. Sudlow, C. L. M. and Warlow, C. P. (1997) Comparable studies of the incidence of stroke and its pathological subtypes: results from an international collaboration. *Stroke* **28,** 491–499.
2. Warlow, C. P., Dennis, M. S., and van Gijn, J. (1996) Reducing the burden of stroke and improving public health in Warlow, C. P., Dennis M. S., and van Gijn J., (eds.) *Stroke: A practical guide to Management*, Blackwell Scientific, Oxford, pp. 632–649.
3. Lodder, J., Bamford, J. M., Sandercock, P. A. G., Jones, L. N., and Warlow, C. P. (1990). Are hypertension or cardiac embolism likely causes of lacunar infarction? *Stroke* **21,** 375–381.
4. Millikan, C. and Futrell, N. (1990) The fallacy of the lacune hypothesis. *Stroke* **21,** 1251–1257.
5. Chambers, B. R., Norris, J. W., Shurvell, B. L., and Hachinski, V. C. (1987) Prognosis of acute stroke. *Neurology* **137,** 221–225.
6. Saito, I., Segawa, H., Shinokawa, Y., Taniguchi, M., and Tsutsumi, K. (1987) Middle cerebral artery occlusion: correlation of computed tomography with clinical outcome. *Stroke* **18,** 863–868.
7. Ringelstein, E. B., Biniek, R., Weiller, C., Ammeling, B., Nolte, P. N., and Thron, A. (1992) Type and extent of hemispheric brain infarction and clinical outcome in early and delayed middle cerebral artery recanalization. *Neurology* **42,** 289–298.
8. Von Kummer, R. and Forsting, M. (1993). Effects of recanalization and collateral blood supply on infarct extent and brain edema after middle cerebral artery occlusion. *Cerebrovasc. Dis.* **3,** 252–255.
9. Tamura, A., Asano, T., and Sano, K. (1980). Correlation between rCBF and histological changes following temporary middle cerebral occlusion. *Stroke* **11,** 487–493.
10. Tamura, A., Graham, D. I., McCulloch, J., and Teasdale, G. M. (1981). Focal cerebral ischaemia in the rat, 1: Description of technique and early neuropathological consequences following middle cerebral artery occlusion. *J. Cereb. Blood Flow Metab.* **1,** 53–60.
11. Tamura, A., Graham, D. I., McCulloch, J., and Teasdale, G. M. (1981). Focal cerebral ischaemia in the rat. 2. Regional cerebral blood flow determined by [^{14}C]iodoantipyrine autoradiography following middle cerebral artery occlusion. *J. Cereb. Blood Flow Metab.* **1,** 61–69.
12. Siesjo, B. K. (1992). Pathophysiology and treatment of focal cerebral ischaemia. Part 1: Pathophysiology. *J. Neurosurg.* **77,** 169–184.

13. Gavrilescu, T. and Kase, C. S. (1995) Clinical stroke syndromes: clinical-anatomical correlations. *Cerebrovasc. Brain Metab. Rev.* **7,** 218–239.
14. Gresham, G. E., Duncan, P. W., Stason, W. B., et al. (1995) Post-stroke rehabilitation, in *Public Health Service, Agency for Health Care Policy and Research.* PHS, Rockville, MD.
15. Stroke Therapy Academic Industry Roundtable (1999) Recommendations for standards regarding preclinical neuroprotective and restorative drug development. *Stroke* **30,** 2752–2758.
16. Marshall, J. W. B. and Ridley, R. M. (1996) Assessment of functional impairment following permanent middle cerebral artery occlusion in a non-human primate species. *Neurodegeneration* **5,** 275–286.
17. Marshall, J. W. B., Cross, A. J., Jackson, D. M., Green, A. R., Baker, H. F., and Ridley, R. M. (2000) Clomethiazole protects against hemineglect in a primate model of stroke. *Brain Res. Bull.* **52,** 21–29.
18. Virley, D., Hadingham, S. J., Roberts, J. C., et al. (2004) A new primate model of focal stroke: endothelin-1-induced middle cerebral artery occlusion and reperfusion in the common marmoset. *J. Cereb. Blood Flow. Metab.* **24,** 24–41.
19. Marshall, J. W. B., Duffin, K. J., Green, A. R., Ridley, R. M. (2001) NXY-059, a free radical-trapping agent, substantially lessens the functional disability resulting from cerebral ischemia in a primate species. *Stroke* **32,** 190–197.
20. Marshall, J. W. B., Cummings, R. M., Bowes, L. J., Ridley, R. M., and Green, A. R. (2003) Functional and histological evidence for the protective effect of NXY-059 in a primate model of stroke when given 4 hours after occlusion. *Stroke* **34,** 2228–2233
21. Marshall, J. W. B., Green, A. R., and Ridley, R. M. (2003) Comparison of the neuroprotective effect of clomethiazole, AR-R15896AR and NXY-059 in a primate model of stroke using histological and behavioural measures. *Brain Res.* **972,** 119–126.
22. Wolfgang, M. J. and Golos, T. G. (2002) Nonhuman primate transgenesis: progress and prospects. *Trends Biotechnol.* **20, 11:** 479–484.
23. Marvanova, M., Menager, J., Bezard, E., Bontropp, R.E., Pradier, L., Wong, G. (2003) Microarray analysis of non-human primates: validation of experimental models in neurological disorders. *FASEB J.* **17, 3:** 328–346.
24. Robinson, R. G., Shoemake, W. J., Schlumpf, M., Valk, T., and Bloom, F. E. (1975) Effect of experimental cerebral infarction in rat brain on catecholamines and behaviour. *Nature* **225,** 332–334.
25. Yamamoto, M., Tamura, A., Kirino, T., Shimizu, M., and Sano, K. (1988). Behavioural changes after focal cerebral ischaemia by left middle cerebral artery occlusion in rats. *Brain Res.* **452,** 323–328.
26. Bolander, H. G., Persson, L., Hillered, L., d'Argy, R., Ponten, U., and Olsson, Y. (1989) Regional cerebral blood flow and histopathological changes after middle cerebral artery occlusion in rats. *Stroke* **20,** 930—937.
27. Persson, L., Hardemark, H. G., Bolander, H. G., Hillered, L., and Olsson, Y. (1989) Neurologic and neuropathological outcome after middle cerebral artery occlusion in rats. *Stroke* **20,** 641–645.

28. El-Sabban, F., Reid, K. H., Zhang, Y. P., and Edmonds, H. L. (1994). Stability of thrombosis induced by electrocoagulation of rat middle cerebral artery. *Stroke* **25,** 2241–2245.

29. Bederson, J. B., Pitts, L. H., Tsuji, M., Nishimura, M. C., David, R. L., and Bartkowsi, H. (1986) Rat middle cerebral artery occlusion: evaluation of the model and development of a neurological examination. *Stroke* **17,** 472–476.

30. DeLong, W. B. (1973) Anatomy of the middle cerebral artery: the temporal branches. *Stroke* **4,** 412–418.

31. Coyle, P. (1975) Arterial patterns of the rat rhinencephalon and related structures. *Exp. Neurol.* **49,** 671–690.

32. Coyle, P. (1982) Middle cerebral artery occlusion the young rat. *Stroke* **13,** 855–859.

33. Marinkovic, S. V., Milisavljevic, M. M., Kovacevic, M. S., and Stevic, Z. D. (1985). Perforating branches of the middle cerebral artery: microanatomy and clinical significance of their intracerebral segments. *Stroke* **16,** 1022–1029.

34. Rubino, G. J. and Young, W. (1988) Ischemic cortical-lesions after permanent occlusion of individual middle cerebral-artery branches in rats. *Stroke* **19,** 870–877.

35. Zhang, F. and Iadecola, C. (1993) Nitroprusside improves blood flow and reduces brain damage after focal ischaemia. *Neuroreport* **4,** 559–562.

36. Dawson, D. A., Graham, D. I., McCulloch, J., Macrae, J., and Macrae, I. M. (1993) Evolution of ischaemic damage in a new model of focal cerebral ischaemia in the rat. *J. Cereb. Blood Flow Metab.* **13,** S461.

37. Hudgins, W. R. and Garcia, J. H. (1970) Transorbital approach to the middle cerebral artery in the squirrel monkey: a technique for experimental cerebral infarction applicable to ultrastructural studies. *Stroke* **1,** 107–111.

38. Kano, M., Moskowitz, M. A., and Yokata, M. (1991) Parasympathetic denervation of rat pial vessels significantly increases infarct volume following middle cerebral artery occlusion. *J. Cereb. Blood Flow Metab.* **11,** 628–637.

39. Koizumi, J., Yoshida, Y., Nakazawa, T., and Ooneda, G. (1986). Experimental studies of cerebral brain edema. 1. A new experimental model of cerebral embolism in rats in which recirculation can be introduced into the ischaemic area. *Jpn. J. Stroke* **8,** 1–8.

40. Koizumi, J., Yoshida, Y., Nishigaya, K., Kanai, H., and Ooneda, G. (1989). Experimental studies of ischaemic brain edema: effect of recirculation of the blood flow after ischaemia on post-ischaemic brain edema. *Jpn. J. Stroke* **11,** 11–17.

41. Nagasawa, H. and Kogure, K. (1989) Correlation between cerebral blood flow and histologic changes in a new rat model of middle cerebral artery occlusion. *Stroke* **20,** 1037–1043.

42. Zea Longa, Z., Weinstein, P. R., Carlson, S., and Cummins, R. (1989) Reversible middle cerebral artery occlusion without craniectomy in rats. *Stroke* **20,** 84–91.

43. Memezawa, H., Smith, M. L., and Siesjo, B. K. (1992) Penumbral tissues salvaged by reperfusion following middle cerebral artery occlusion in rats. *Stroke* **23,** 552–559.

44. Memezawa, H., Minamisawa, M. L., Smith, M. L., and Siesjo, B. K. (1992). Ischaemic penumbra in a model of reversible middle cerebral artery occlusion in the rat. *Exp. Brain Res.* **89,** 67–78.
45. Kuge, Y., Minematsu, K., Yamaguchi, T., and Miyake, Y. (1995) Nylon monofilament for intraluminal middle cerebral artery occlusion in rats. *Stroke* **26,** 1655–1657.
46. Belayev, L., Alonso, O. F., Busto, R., Zhao, W., and Ginsberg, M. D. (1996) Middle cerebral artery occlusion in the rat by intraluminal suture: neurological and pathological evaluation of an improved method. *Stroke* **27,** 1616–1623.
47. Oliff, H. S., Marek, P., Miyazaki, B., and Weber, E. (1996) The neuroprotective efficacy of MK-801 in focal cerebral ischaemia varies with rat strain and vendor. *Brain Res.* **731,** 208–212.
48. Aspey, B. S., Cohen, S., Patel, M., Terruli, M., and Harrison, M. J. G. (1998) Middle cerebral artery occlusion in the rat: consistent protocol for a model of stroke. *Neuropathol. Appl. Neurobiol.* **24,** 487–497.
49. Takano, K., Tatlisumak, T., Bergmann, A. G., Gibson, D. G., and Fisher, M. (1997) Reproducibility and reliability of middle cerebral artery occlusion using a silicone-coated suture (Koizumi) in rats. *J. Neurol. Sci.* **153,** 8–11.
50. Laing, R. J., Jakubowski, J., and Laing, R. W. (1993) Middle cerebral artery occlusion without craniectomy in rats: which method works best? *Stroke* **24,** 294–298.
51. Hossmann, K. A. and Kleihues, P. (1973) Reversibility of ischaemic brain damage. *Arch. Neurol.* **29,** 375–384.
52. Hallenbeck, J. M. and Dutka, A. J. (1990) Background review and current concepts of reperfusion injury. *Arch. Neurol.* **47,** 1245–1254.
53. Wardlaw, J. M., Dennis, M. S., Lindley, R. I., Warlow, C. P., Sandercock, P. A. G., and Sellar, R. (1993) Does early reperfusion of a cerebral infarct influence cerebral infarct swelling in the acute stage or the final clinical outcome. *Cerebrovasc. Dis.* **3,** 86–93.
54. Parsons, A. A., Irving, E. A., Legos, J. J., et al. (2000) Acute stroke therapy: Issues for translating pre-clinical neuroprotection to therapeutic reality. *Curr. Opin. Invest. Drugs* **1,** 452–463
55. Okada, M., Tamura, A., Urae, A., Nakagomi, T., Kirino, T., Mine, K., and Fujiwara, M. (1995) Long-term spatial cognitive impairment following middle cerebral artery occlusion in rats: a behavioural study. *J. Cereb. Blood Flow Metab.* **15,** 505–512.
56. Feeney, D. M., Gonzalez, A., and Law, W. A. (1982) Amphetamine, haloperidol and experience interact to affect the rate of recovery after motor cortex injuries. *Science* **217,** 855–857.
57. Van Der Staay, F. J., Augstein, K. H., and Horvath, E. (1996) Sensorimotor impairments in Wistar Kyoto rats with cerebral infarction, induced by unilateral occlusion of the middle cerebral artery: recovery of function. *Brain Res.* **715,** 180–188.

58. De Ryck, M., van Reempts, J., Borgers, M., Wauquier, A., and Janssen, P. A. J. (1989). Photochemical stroke model: flunarizine prevents sensorimotor deficits after neocortical infarcts in rats. *Stroke* **20,** 1383–1390.

59. Combs, D. J. and D'Alecy, L. G. (1987) Motor performances in rats exposed to severe forebrain ischaemia: effect of fasting and 1,3-butanediol. *Stroke* **18,** 503–511.

60. Anderson, C. S., Anderson, A. B., and Finger, S. (1991) Neurological correlates of unilateral and bilateral strokes of middle cerebral artery in the rat. *Physiol. Behav.* **50,** 263–269.

61. Markgraf, C. G., Green, E. J., Hurwitz, B. E., et al. (1992) Sensorimotor and cognitive consequences of middle cerebral artery occlusion in rats. *Brain Res.* **575,** 238–246.

62. Markgraf, C. G., Johnson, M. P., Braun, D. L., and Bickers, M. V. (1997) Behavioural recovery patterns in rats receiving the NMDA receptor antagonist MDL 100, 453 immediately post-stroke. *Pharmacol. Biochem. Behav.* **56,** 391–397.

63. Hunter, A. J., Hatcher, J., Virley, D., Nelson, P., Irving, E., Hadingham, S. J. and Parsons, A. A. (2000) Functional assessments in mice and rats after focal stroke. *Neuropharmacology* **39,** 806–816.

64. Virley D., Beech J. S., Smart S. C., Williams S. C. R., Hodges H., and Hunter A. J. (2000)A temporal MRI assessment of neuropathology following transient MCAO in the rat: Correlations with behaviour. *J. Cereb. Blood Flow Metab.* **20,** 563–582

65. Schallert, T., Upchurch, M., Lobaugh, S., and Farrar, S. B. (1982). Tactile extinction: distinguishing between sensorimotor and motor asymmetries in rats with unilateral nigrostriatal damage. *Pharmacol. Biochem. Behav.* **16,** 455–462.

66. Rose, L., Bakel, D. A., Fung, T. S., Farn, P., and Weaver, L. E. (1994) Tactile extinction and functional status after stroke: a preliminary investigation. *Stroke* **25,** 1973–1976.

67. Heilman, K. M., Watson, R. T., and Valenstein, E. (1985) *Neglect and Related Disorders.* Oxford University Press, New York.

68. Montoya, C. P., Campbell-Hope, L. J., Pemberton, K. D., and Dunnett, S. B. (1991) The 'staircase test': a measure of independent forelimb reaching and grasping abilities in rats. *J. Neurosci. Methods* **36,** 219–228.

69. Grabowski, M., Brundin, P., and Johansson, B. B. (1993). Paw-reaching, sensorimotor and rotational behaviour after brain infarction in rats. *Stroke* **24,** 889–895.

70. Marston, H. M., Faber, E. S., Crawford, J. H., Butcher, S. P., and Sharkey, J. (1995) Behavioural assessment of endothelin-1 induced middle cerebral artery occlusion in the rat. *Neuroreport* **6,** 1067–1071.

71. Sharkey, J., Crawford, J. H., Butcher, S. P., and Marston, H. M. (1996) Tacrolimus (FK506) ameliorates skilled motor deficits produced by middle cerebral artery occlusion in rats. *Stroke* **27,** 2282–2286.

72. Yonemori, F., Yamada, H., Yamaguchi, T., Uemura, A., and Tamura, A. (1996) Spatial memory disturbance after focal cerebral ischaemia in rats. *J. Cereb. Blood Flow Metab.* **16,** 973–980.

73. Rogers, D. C., Campbell, C. A., Stretton, J. L., and Mackay, K. B. (1997). Correlation between motor impairment and infarct volume after permanent and transient middle cerebral artery occlusion in the rat. *Stroke* **28,** 2060–2066.
74. Smith, S. E., Hodges, H., Sowinski, P., et al. (1997). Long-term beneficial effects of BW619C89 on neurological deficit, cognitive deficit and brain damage after middle cerebral artery occlusion in the rat. *Neuroscience* **77,** 1123–1135.
75. Stroemer, R. P., Kent, T. A., and Hulsebosch, C. E. (1995) Neocortical neural sprouting, synaptogenesis and behavioural recovery after neocortical infarction in rats. *Stroke* **26,** 2135–2144.
76. Stroemer, R. P., Kent, T. A., and Hulsebosch, C. E. (1998) Enhanced neocortical neural sprouting, synaptogenesis and behavioural recovery with D-amphetamine therapy after neocortical infarction in rats. *Stroke* **29,** 2381–2393.
77. Kawamata, T., Dietrich, W. D., Schallert, T., et al. (1997) Intracisternal basic fibroblast growth factor enhances functional recovery and up-regulates the expression of a molecular marker for neuronal sprouting following focal cerebral infarction. *Proc. Natl. Acad. Sci. USA* **94,** 8179–8184.
78. Li, Q. and Stephenson, D. (2002) Postischemic administration of basic fibroblast growth factor improves sensorimotor function and reduces infarct size following permanent focal cerebral ischemia in the rat. *Exp. Neurol.* **177,** 531–537.
79. Kolb, B., Cote, S., Ribeiro-Da-Silva, A., and Cuello, A. C. (1996) Nerve growth factor treatment prevents dendritic atrophy and promotes recovery of function after cortical injury. *Neuroscience* **76,** (1139–1151).
80. Wiessner, C., Bareyre, F. M., Allegrini, P. R., et al. (2003) Anti-Nogo-A antibody infusion 24 hours after experimental stroke improved behavioural outcome and corticospinal plasticity in normotensive and spontaneously hypertensive rats. *J. Cereb. Blood Flow Metab.* **23,** 154–165
81. Kawamata, T., Ren, J. M., Chan, T. C. K., Charlette, M., and Finklestein, S. P. (1998) Intracisternal osteogenic protein-1 enhances functional recovery following focal stroke. *Neuroreport* **9,** 1441–1445.
82. Ren, J. M., Kaplan, P. L., Charette, M. F., Speller, H., and Finklestein, S. P. (2000). Time window of intracisternal osteogenic protein-1 in enhancing functional recovery after stroke. *Neuropharmacology* **39,** 860–865.
83. Goldstein, L. B. and Davis, J. N. (1990) Beam-walking in rats: studies towards developing an animal model of functional recovery after brain injury. *J. Neurosci. Methods* **31,** 101–107.
84. Johansson, B. B. and Ohlsson, A. L. (1996) Environment, social interaction and physical activity as determinants of functrional outcome after cerebral infarction in the rat. *Exp. Neurol.* **139,** 322–327.
85. Chen, H., Chopp, M., and Welch, K. M. A. (1991) Effect of mild hyperthermia on the ischaemic infarct volume after middle cerebral artery occlusion in the rat. *Neurology* **41,** 1133–1135.
86. Kaplan, B., Brint, S., Tanabe, J., Jacewicz, M., Wang, X. J., and Pulsinelli, W. (1991). Temporal thresholds for neocortical infarction in rats subjected to reversible focal cerebral ischemia. *Stroke* **22,** 1032–1039.

87. Buchan, A. M., Xue, D., and Slivka, A. (1992) A new model of temporary focal neocortical ischaemia in the rat. *Stroke* **23,** 273–279.

3

Mutant Animal Models of Stroke and Gene Expression

The Stroke-Prone Spontaneously Hypertensive Rat

Hilary V. O. Carswell, Martin W. McBride, Delyth Graham,
Anna F. Dominiczak, and I. Mhairi Macrae

Summary

The recent completion of the Human Genome Project provides the potential to advance our knowledge of pathogenesis and identify the gene(s) associated with particular diseases. However, using human DNA to correlate individual genomic variations with particular disorders such as stroke will be extremely challenging because of the large number of variables within an individual, and across different populations. Mutant animal models of stroke such as the stroke-prone spontaneously hypertensive rat (SHRSP) provide the scientist with genetic homogeneity, not possible within a human population, to aid our search for causative genes. This chapter describes the methods our group have employed to study the genetic heritability of stroke sensitivity in the SHRSP. Sections are included on quantitative trait loci, mapping, and congenic strain construction for the identification of genetic determinants of stroke sensitivity in the SHRSP.

KEY WORDS

Genotyping; polymerase chain reaction; microsatellite markers; quantitative trait locus (QTL); genetic mapping; congenic; middle cerebral artery occlusion; infarct volume.

1. Introduction
1.1. Stroke

Stroke is a significant health problem amongst Western societies, being recognized as a major cause of disability, dementia, and death. Its impact on these societies is forecast to increase with growing life expectancy. Despite recogni-

From: *Methods in Molecular Medicine, Vol. 104: Stroke Genomics: Methods and Reviews*
Edited by: S. J. Read and D. Virley © Humana Press Inc., Totowa, NJ

tion of the growing prevalence of this disease in the population, progress in developing effective acute stroke therapies has been very disappointing. Significant advances in our understanding of the mechanisms responsible for stroke-induced brain damage and the efficacy of drugs targeted at these mechanisms have arisen from the experimental stroke literature. However, this knowledge has not translated into effective acute stroke treatments for use in the clinic. In 2001, Kidwell et al. *(1)* reported that from 178 controlled acute stroke trials completed in the 20th century, only 3 trials met conventional criteria for a positive outcome (NINDS tissue plasminogen activator [tPA], PROACT II intraarterial pro-urokinase, and low-molecular-weight heparin). However, the success of the fibrinolytic vascular approach is marred by the limited time window (3 h) and the need for brain scans prior to treatment. From US figures, this currently equates to only 1.5% of stroke patients receiving tPA treatment. Of the >49 neuroprotective agents studied in >114 stroke trials, none has proved to be successful clinically *(1)*. In summary, the increasing prevalence of this disease, combined with the lack of current effective therapies, highlights the urgent need for more research. Identification of the genetic components of stroke will provide greater knowledge of this condition and how it can best be treated.

1.2. Hereditability of Human Stroke

High blood pressure is one of the most important, inheritable, and treatable risk factors for stroke. However, there is comprehensive clinical evidence that genetic inheritance plays a role in susceptibility to stroke, unrelated to high blood pressure. This evidence includes a fivefold increase in the prevalence of stroke among monozygotic compared with dizygotic twins *(2)*. In addition, the Framington Study showed that offspring with paternal and maternal histories of stroke were associated with an increased risk of stroke compared with offspring with no paternal or maternal stroke history *(3)*. Graffagnino et al. *(4)* report that a positive family history for stroke was present in 47% of stroke patients and 24% of nonstroke control subjects. In addition, Diaz et al. *(5)* suggest that living siblings of patients with cerebral infarction and transient ischemic attacks have increased risk of stroke compared with siblings of the patients' spouses.

Genes linked to rare stroke syndromes have been identified. For example, *Notch 3* in cerebral autosomal dominant arteriopathy with subcortical infarctions and leukoencephalopathy (CADASIL) *(6)*, cystatin C in the Icelandic type of hereditary cerebral hemorrhage with amyloidosis *(7)*; the amyloid gene in the Dutch type of hereditary cerebral hemorrhage *(8)*, and the gene encoding KRIT1 (a Krev-1/rap1a binding protein) in hereditary cerebral cavernous malformations *(9,10)*. However, gene linkage to the more common types of stroke (ischaemic, hemorrhagic) and transient ischemic attacks has proved to be a difficult task. This is because genes responsible may be heterogeneous among the different types of common stroke, and, within one patient, mulitple genes may contribute to the stroke. In addition, using human DNA to correlate individual genomic variations with particular disorders such as stroke will be extremely challenging because of the large number of variables within an individual, and across different populations *(11)*. With the aid of new genealogical techniques and with the use of a relatively homogeneous population, the first major locus for common stroke has recently been successfully mapped to human chromosome 5 *(12)*. The next challenge is to find this particular association in a different human population.

1.3. Mutant Animal Models for Study of the Genetics of Stroke: The SHRSP

Genetic dissection of stroke has been hindered by the late onset and heterogeneity of the condition in humans. However, inbred animal models with a predisposition and increased sensitivity to stroke, such as the stroke-prone spontaneously hypertensive rat (SHRSP), have brought stroke genetics within our grasp. Stroke in the SHRSP has a unique similarity to the human condition, with the occurrence of stroke being a complex, polygenic, and multifactorial disorder in which both hypertension and salt diet increase stroke proneness. The SHRSP is regarded as a good pathogenetic model for human stroke, with major predilection sites of ischemic damage being the anteromedial cortex, the occipital cortex, and the basal ganglia *(13)*.

The SHRSP was derived by Okamoto et al. in 1974 *(14)* by selective breeding of Wistar Kyoto (WKY) rats that had developed spontaneous hypertension when fed the Japanese diet (high in sodium, low in potassium, and low in protein, compared with Western chow). Over many generations of selective breeding, the strain known as the spontaneously hypertensive rat (SHR) was established. Further selective breeding of SHRs that developed stroke symptoms led to the generation of the SHRSP, which has genetically determined hypertension and stroke susceptibility. The SHRSPs exhibit a high frequency of spontaneous strokes, when fed a high-salt diet *(14)*, and increased sensitivity to experimental stroke *(15–17)* compared with WKY rats. Stroke sensitivity in these animals is not secondary to hypertension: young SHRSP rats whose hypertension has not yet developed exhibit the same increased sensitivity to stroke as adult SHRSPs *(15)*; SHRSPs with blood pressure controlled with antihypertensive agents continue to exhibit increased sensitivity to stroke compared with WKY rats *(18)*, and WKY rats made hypertensive using deoxycorticosterone acetate (DOCA)-salt are not as sensitive to stroke as SHRSP *(19)*.

Here we describe quantitative trait loci (QTL) mapping and congenic strain construction for the identification of the genetic determinants of stroke sensitivity in the SHRSP. QTL mapping is a phenotype-driven approach that does not require prior knowledge of either causative genes or their function and can lead to the identification of novel genes involved in disease. A well-character-ized stroke phenotype that has a bimodal distribution between SHRSP and WKY rats, is volume of infarction 24 h after a distal occlusion of the middle cerebral artery (MCAO) (**Fig. 1**; *16,17*). The genomic resources available for the rat are considerable and include a large number of microsatellite markers, which allow high-resolution genetic linkage mapping of the rat genome *(20,21)*. Phenotypes can be linked to a QTL using these genetic markers, spread evenly throughout the genome, and tested on a large cosegregating population that has been phenotypically assessed using computer software.

The identification of a large chromosomal region containing a QTL is only the first step in the ultimate goal of gene identification. The next step is the construction of congenic strains by which the existence of a QTL can be veri-fied. Congenic strain construction involves the transfer of defined chromo-somal segments from a donor to a recipient strain. The traditional breeding strategy takes approx 3–4 yr *(22)*. However, this time can be shortened by utilizing a marker-assisted "speed" congenic breeding strategy, previously tested in the mouse *(23)*. Jeffs et al. *(24)* were the first to show that this strategy can be successfully applied in the rat. Lander and Schork *(25)* proposed that screening of polymorphic genetic markers covering the entire background of the genome could be used to select male offspring with the least donor alleles in their background. This rapid congenic approach reduces the average time required to produce strains to approx 2 yr *(24)*. To ensure transfer of the entire QTL, original congenic strains tend to have large chromosomal regions introgressed. Therefore smaller substrains (minimal congenics) are required to reduce the size, and hence the implicated region.

2. Materials
2.1. High-Throughput Fluorescent Genotyping

2.1.1. Genomic DNA Isolation (Animal Tail Tip)

1. 0.5 *M* EDTA (pH 8.0): sterilize by autoclaving.
2. Nuclei Lysis Solution (cat. no. A7943, Promega).
3. Protein Precipitation Solution (cat. no. A7953, Promega)j
4. Proteinase K (20 mg/mL).
5. Isopropanol.
6. 70% Ethanol.
7. Benchtop microcentrifuge (Eppendorf).
8. Whirli Mixer (FSA Laboratory Supplies).
9. Hybridization oven.

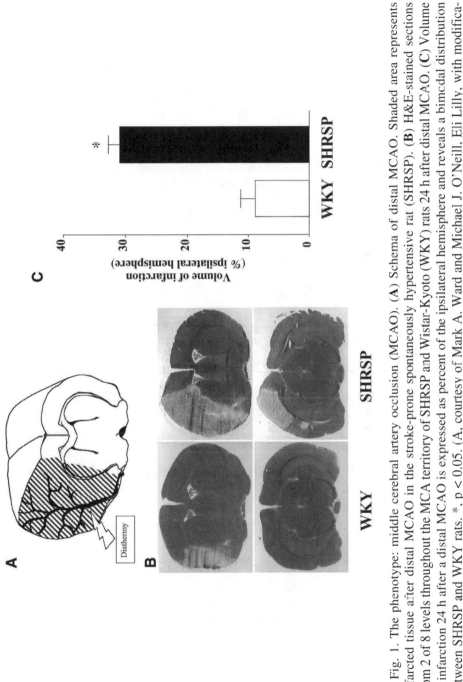

Fig. 1. The phenotype: middle cerebral artery occlusion (MCAO). (**A**) Schema of distal MCAO. Shaded area represents infarcted tissue after distal MCAO in the stroke-prone spontaneously hypertensive rat (SHRSP). (**B**) H&E-stained sections from 2 of 8 levels throughout the MCA territory of SHRSP and Wistar-Kyoto (WKY) rats 24 h after distal MCAO. (**C**) Volume of infarction 24 h after a distal MCAO is expressed as percent of the ipsilateral hemisphere and reveals a bimodal distribution between SHRSP and WKY rats. *, $p < 0.05$. (A, courtesy of Mark A. Ward and Michael J. O'Neill, Eli Lilly, with modifications.)

10. P1000, P200, and P20 pipets (Gilson).
11. Sterile pipet tips.
12. 1.5-mL vol Sterile Eppendorf tubes.

2.1.2. Quantification of Genomic DNA

1. UV/Visible Spectrophotometer (Pharmacia Biotech Ultrospec 2000).
2. Quartz cuvets.
3. Sterile water.
4. Parafilm.

2.1.3. Polymorphic Fluorescent Microsatellite Markers

1. TET, FAM, or HEX (Applied Biosystems, Inc.).
2. Color fluorophore.

2.1.4. Polymerase Chain Reaction (PCR)

1. Deep 96-well plate.
2. Sterile water.
3. 96-Well skirted PCR plates (cat. no. AB-0800, ABgene Thermo-fast).
4. PCR plate sticky lids.
5. 1 mM deoxynucleotide triphosphates (dNPTs) stocks.
6. 10X Hotstar Taq DNA polymerase buffer (Qiagen, UK).
7. Forward and reverse primers (20 pmol/µL concentration).
8. W1 detergent.
9. Hotstar Taq DNA polymerase (Qiagen, UK).
10. Peltier Thermal Cycler hot lid PCR machine (MJ Research, cat no. PTC-225)
11. Multichannel electric pipet.

2.1.5. PCR Pooling

1. 96-Well skirted PCR plates (cat. no. AB-0800, ABgene Thermo-fast).
2. Multichannel electric pipet.
3. Sterile water.

2.1.6. Casting Gels for the ABI Prism 377XL

1. P5000, P1000, and P40 tips.
2. 100-mL beaker.
3. 50% Long Ranger gel solution (Cambrex).
4. Ultrapure urea.
5. Deionizing resin (Bio-Rad AG 501-X8 [D].
6. Small stirrer bar.
7. Deionized water.
8. 50-mL Measuring cylinder.
9. Ammonium persulfate (APS).
10. 1.5-mL Eppendorf tube.

11. Whatman cellulose filters.
12. 10X TBE.
13. TEMED (*N,N,N',N'*-tetramethylenediamine; Sigma, cat. no. 7024).

2.2. The Phenotype: Distal Middle Cerebral Artery Occlusion

1. Halothane (Concord Pharmaceuticals, Dunmow, Essex, UK) in N_2O/O_2 or other suitable anesthetic.
2. Intubation tube for artificial respiration (Baxter Healthcare Quick-cath 16G, Deerfield, MA).
3. Rodent Ventilator (Harvard).
4. Hibiscrub (Schering Plough Animal Health, UK).
5. Fur shaver (Sterling 2plus, by Whal Clipper, Sterling, IL).
6. Homeothermic heating blanket (Harvard) or cork pad (RA Lamb, London, UK) and heating lamp to maintain body temperature under anesthesia.
7. Surgical retractors (not commercially available; cut to size and shape from copper sheet).
8. Scalpel, size 22 A blade (Maersk Medical, Sheffield, UK).
9. Cotton buds and gauze pads.
10. Triangular arrowhead swabs (John Weiss & Son cat. no. 30761).
11. Watchmakers forceps (Downs Surgical, Sheffield , UK).
12. Dental drill and handpiece (Volvere Vmax NSR, Nakanishi, Japan).
13. Dental needle, 30 gage (Terumo, Tokyo, Japan).
14. Bipolar diathermy forceps and unit (Aesculap, part of Downs Surgical, Eschmann TBD60, Lancing, UK).
15. Vannas microscissors (Downs Surgical).
16. Absorbable sutures (USS DG Tyco Healthcare, Gosport, UK).
17. Laser Doppler flowmetry equipment (Moor Instrument).

2.3. Histology for Measurement of Infarct

2.3.1. Harvest of Brains

1. Isopentane (BDH VWR, Leicestershire, UK).
2. Dry ice (BOC).
3. Thermometer capable of measuring temperatures down to –50°C.
4. Tall, thin beaker able to withstand temperature of dry ice (made in lab from copper).
5. Lipshaw (M1 embedding matrix Thermo Shandon PA 15275, USA).
6. Mounting agent (Cryomatrix, cat. no. 6769006, Life Sciences, Cheshire, UK).
7. Chuck and cryostat (Bright Instruments, Huntingdon, UK).
8. Microscope slides with frosted end for pencil mark-ups (BDH, VWR).

2.3.2. Staining of Tissue With H&E

1. Microscope slide racks (RA Lamb, London, UK).
2. Square glass beakers to hold the slide racks (RA Lamb, London, UK).

3. Absolute alcohol.
4. Hematoxylin (Surgipath, Peterbourgh, Cambridgeshire, UK).
5. Acidic methylated alcohol.
6. Scot's tap water substitute.
7. Eosin (Surgipath).
8. Histoclear (National Diagnostics, Hull, UK).
9. Histomount (RA Lamb).

2.3.3 Quantification of Volume of Infarct

1. Image analyzer (MCID, M4, Imaging Research, St. Catherine's, Ontario, Canada).

3. Methods
3.1. Production of an F_2 Population

Two reciprocal F_1 genetic crosses can be generated by crossing one male SHRSP with two WKY females (cross 1) and one male WKY with two SHRSP females (cross 2). From the F_1 rats of each cross, three males and six females are brother × sister mated to generate F_2 rats. All rats are housed under controlled conditions of temperature (21°C) and light (12-h light/dark cycle; 7 AM to 7 PM) and are maintained on normal rat chow (rat and mouse No. 1 maintenance diet, Special Diet Services) and water *ad libitum*. Litters are weaned and sexed after 3 wk and maintained by sibling group and sex thereafter.

3.2. High-Throughput Fluorescent Genotyping

3.2.1. Genomic DNA Isolation From Animal Tail Tip

1. Briefly anesthetize the animals with halothane/oxygen, remove a 4-mm tip from the tails, and immediately cauterize the wound (*see* **Note 1**).
2. Prepare the tail tip samples for digestion; add 120 µL of 0.5 M EDTA to 500 µL of Nuclei Lysis Solution for each sample into an appropriate sized tube.
3. Chill the mix on ice for 5 min.
4. Add 600 µL of the prepared mix to each tail tip sample, and add 17.5 µL of 20mg/ml Proteinase K solution. Incubate the samples in a rotating hybridization oven at 37°C overnight (*see* **Note 2**).
5. Allow the samples to cool to room temperature.
6. Add 200 µL of Protein Precipitation Solution to each of the samples and mix by vortex for 20 s.
7. Chill the samples on ice for 5 min.
8. Centrifuge samples for 4 min at 13,000–16,000g at room temperature (*see* **Note 3**).
9. Transfer the supernatant to a fresh 1.5-mL Eppendorf tube.
10. Add 600 µL of room temperature isopropanol and mix by inversion (*see* **Note 4**).
11. Centrifuge the sample for 2 min at 13,000–16,000g at room temperature.

12. Carefully decant the supernatant and ensure that a white pellet is visible and it is not disturbed.
13. Wash the DNA pellet by adding 1 mL of 70% ethanol.
14. Centrifuge the sample for 1 min at 13,000–16,000g at room temperature.
15. Carefully remove the supernatant by pipet and invert the tube on a piece of absorbent paper.
16. Allow the DNA to air-dry for 5 min
17. Add 100 µL of sterile water to rehydrate the DNA pellet, and then store the samples in a 4°C fridge overnight.

3.2.2. Quantification of Genomic DNA

1. Allow the spectrophotometer to self-calibrate.
2. Blank-correct by adding 1 mL of sterile water to a clean quartz cuvet, and insert it into the main reading holder.
3. Replace the cuvet into the main reading holder. The absorbance reading should be zero.
4. Empty the cuvet, and wash thoroughly with distilled water.
5. Add 995 µL of sterile water to the cuvet.
6. Add 5 µL of sample DNA to the cuvet, and mix well by inversion. (Cover the cuvet opening with parafilm and mix.) This reading will give a reading for the 1:200 dilution factor for the DNA.
7. Insert the cuvet into the main reading holder and press Run.
8. Record the OD_{260} concentration value given on the machine.
9. To obtain the 260/280 ratio value, press the down arrow key (*see* **Note 5**).
10. Add a further 5 µL of the sample DNA to the cuvet (mix as before) and repeat as before. This value will give a reading for the 1:100 dilution factor for the DNA.
11. Empty the contents, wash the cuvet with distilled water, and repeat the procedure for each sample.
12. Make DNA stocks of 20 ng/µL.

3.2.3. Polymorphic Fluorescent Microsatellite Markers

Polymorphic microsatellite markers are synthesized with a fluorescent molecule (either TET, FAM, and HEX, available from ABI) at the 5' end of the forward primer. Microsatellite markers of a similar size are synthesized with different color fluorophores. This allows the greatest flexibility when considering pooling strategies (*see* **Subheading 3.2.5.**).

3.2.4. Polymerase Chain Reaction

1. Prepare a deep-well plate of 20 ng/µL DNA stock: add the calculated volume of DNA to a sterile deep-well plate and make the volume up to a total of 500 µL with sterile water. This will provide the template DNA for future PCR procedures.
2. Label thermo-fast 96-well plates with the appropriate primer used for PCR and date.

3. Transfer 5 μL of the previously prepared 20 ng/μL DNA template with a multichannel pipet. This will give the PCR reaction DNA concentration of 100 ng.
4. Prepare a PCR master mix (*see* **Note 6**), using the following quantities per sample: 2.0 μL 10X Hotstar Taq Buffer, 1.0 μL W1 detergent, 4.0 μL dNTP's (1 m*M* stock), 0.5 μL of each forward and reverse primer (20 pmol/μL stock), 6.96 μL sterile water, and 0.04 μL Hotstar *Taq* enzyme (*see* **Note 7**).
5. Add 15 μL of the prepared master mix to each of the samples.
6. Firmly place a sticky lid on the prepared PCR plate and seal tightly.
7. Place the plate in a PCR machine and tighten the lid (*see* **Note 8**).
8. Amplify under the following conditions: 95°C 10 min, 55°C 1 min; 72°C 2 min, 94°C 1 min for 35 cycles; then 55°C 1 min, 72°C 10 min, and 12°C thereafter.

3.2.5. PCR Pooling

Pooling of PCR products prior to loading samples onto the acrylamide gel reduces the number of lanes required for genotyping and is a key step for highthroughput genotyping. Pooling involves mixing between 5 and 10 PCR amplicons from a number of microsatellite amplification reactions into a single tube. Dilutions of this pooled mixture are then loaded onto the acrylamide gel. The only requirement is that microsatellite markers with similar PCR product sizes be either synthesized with different fluorophores or placed in different pools.

3.2.6. Casting Polyacrylamide Gels for the ABI Prism 377XL

1. To a 100-mL beaker, add 18.0 g of urea, 5.0 mL of Long Ranger 50% Stock Gel Solution, a spatula-full of Bio-Rad AG 501-X8 (D) resin, a stirrer bar, and deionized water up to approx 45 mL. Leave the gel solution to stir for 15 min.
2. Set the cassette on a flat surface with the laser shutter, and therefore the bottom of the gel, closest to you, and place the back plate into the cassette so that the etched writing is on the bottom surface (back to front as you look at it). Take care to avoid touching the bottom 3 inches of the plate, as this is the read region. Pull the plate toward you until it rests on the metal pins.
3. Position the spacers so that the cutouts are at the top of the gel and facing inward.
4. Position the front plate on top of the spacers with the straight edge at the bottom of the gel cassette.
5. Verify that the plates and spacers are aligned and that the cassette is square, and then secure all but the bottom and top clips on each side of the cassette.
6. Position the clear plastic top clamp at the top of the plates, and clip it into place.
7. Take the gel injection device, pull the black handles up, and position it on the bottom of the gel plates. Push the handles down, thus pulling the gel injector onto the bottom of the plates. Twist the bottom clips on the cassette so they fit over the gel injector.
8. Take the assembled cassette through to the sequencer along with the comb.
9. Pour the gel solution into a clean 50-mL measuring cylinder, and make up the volume to 45 mL with deionized water.

10. Assemble the filter unit with a cellulose filter in place. The filters are packed with the filtration surface uppermost. Connect to the water pump and turn on the water. Check that the tap water is not flashing back into the filter unit!

11. Pipette 5 mL of 10X TBE into the filter, allow this to go through, and then add the gel solution in the order stated to prevent deionization of the TBE. Once the gel solution has been added, screw the lid onto the unit and stand it on the shelf next to the dH₂O still. Leave to de-gas.

12. Make up at least 260 µL of 10% w/v ammonium persulfate (APS).

13. Pour the de-gassed gel solution back into the beaker, and rinse the filter unit. Add 250 µL of 10% APS, 25 µL TEMED and stir carefully with a 60-mL syringe. Draw the solution into the syringe, screw the syringe into the gel injection device, and push the solution into the plates, tapping them to maintain a smooth interface as you do.

14. Any bubbles greater than 2 cm from the spacers will affect the loading of samples. These should be removed using a bubble catcher.

15. Wet the comb with gel solution and push it into the top of the plates. Check that no bubbles have been caught, and clamp together with three bulldog clips.

16. Leave the gel for at least 1 h to polymerize.

3.2.7. Washing Gel Plates

1. Lay the cassette flat, and unclip the top buffer tank and plates from the cassette. Wash the cassette.

2. Use a plastic wedge to split the plates, and remove the gel by laying a sheet of tissue on the plate and lifting it off (*see* **Note 9**)

3. Wash all components with hot tap water only (*see* **Note 10**), using the sponge and plastic tray. Plates should only be washed in the plastic tray.

4. Rinse all components with distilled water and leave to dry on a paper towel or rack. Stand the plates upside down so that all residual water runs away from the read region.

3.2.8. Genotyping

An ABI 377 DNA sequencer is used with the appropriate Genotyper 2.1 software to identify DNA fragments of differing sizes with laser signals. Polyacrylamide gels are run and software facilitates semiautomated calling of genotypes.

3.3. Genetic Mapping Software

Software such as Mapmaker maps microsatellite markers relative to one another based on recombination fractions. Putative QTLs are then mapped to the molecular markers using multipoint linkage analysis with linkage being determined as a logarithm of odds (LOD) score.

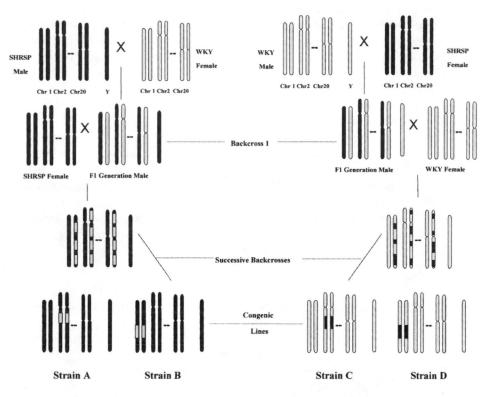

Fig. 2. Construction of reciprocal congenic strains using a marker-assisted strategy with stroke-prone spontaneously hypertensive rat (SHRSP) and Wistar-Kyoto (WKY) parental strains.

3.4. Marker-Assisted "Speed" Congenic Strategy

1. Congenic strains are constructed using a speed congenic (or marker-assisted) strategy whereby various segments of the rat chromosome of interest are introgressed from the normotensive WKY strain to the genetic background of the SHRSP, and in the reciprocal direction from SHRSP to the genetic background of the WKY (**Fig. 2**).
2. Reciprocal F_1 generations are produced by crossing WKY and SHRSP. Resulting male F_1 hybrids are mated to the desired recipient strain (WKY or SHRSP) females (*see* **Notes 11** and **12**).
3. Microsatellite markers throughout the chromosomal region of interest, and additional markers broadly spanning the remaining genome, are genotyped (*see* **Subheading 3.2.3.**) in the offspring from this first backcross (*see* **Note 13**). Those males identified as heterozygous for marker alleles within the chromosomal

region of interest, but with most homozygosity for recipient alleles throughout the remaining genome, are selected as the "best" males for breeding. These males are backcrossed to recipient strain females to produce a second backcross generation and the offspring are genotyped.

4. Repeat procedure by backcrossing "best" male offspring until all donor alleles in the genetic background (indicated by the background markers) are eradicated (*see* **Note 14**). Approximately four to five backcross generations are required to achieve recipient strain homozygosity in all background markers. The differential chromosomal region is then fixed and made homozygous by crossing appropriate males and females.

5. Fixed congenic strains are maintained by brother–sister mating.

6. Confirmation of successful QTL capture by phenotypic measurement within a congenic strain is followed by congenic substrain (or minimal congenic) production with smaller donor regions. This strategy dissects the introgressed region, allowing further localization of the QTL.

7. Congenic substrain production is undertaken by backcrossing congenic males to recipient females to yield rats heterozygous within the original introgressed segment, while maintaining recipient strain homozygosity in the remaining genome.

8. Resulting heterozygous F1 rats are intercrossed and the offspring genotyped to identify appropriate males and females with smaller regions of donor allele homozygosity. These smaller regions are fixed, and the new substrains are maintained by brother–sister mating.

3.5. The Phenotype: Distal Middle Cerebral Artery Occlusion

Genetics of stroke studies in humans are impeded by the heterogeneity of human stroke (e.g., occlusive and hemorrhagic strokes, which can occur in any part of the brain; the possibility of secondary insults; or the potential for interference from other disease processes that may also be present). In animal models, experimental stroke can be induced by occlusion of one specific cerebral artery (normally the middle cerebral artery [MCA], the blood vessel most commonly affected in human occlusive stroke) with control over the physiological variables that can influence stroke severity (e.g., blood pressure, body temperature, plasma glucose). Many different MCA occlusion models have been developed in rodents and have been reviewed elsewhere *(26)*. Permanent distal occlusion of the MCA *(16)* was found to provide the most reproducible infarct and the best bimodal distribution between the SHRSP and WKY strains in studies from our laboratory. Infarct volume 24 h after distal MCA occlusion has therefore been used as a phenotype for stroke sensitivity in the SHRSP (**Fig. 1C**) (*see* **Notes 12** and **15–29**).

1. Autoclave all surgical instruments prior to surgery. Use and adopt as clean an environment and technique as possible with the use of Hibiscrub. Shave, swab, and cut incision site between the left eye and ear.

2. Retract the temporal muscle overlying the site of the left MCA using copper retractors

3. Expose the MCA via a subtemporal craniectomy (*see* **Notes 20–23**) and remove the dura (*see* **Note 21**).
4. Electrocoagulate a 2-mm segment of the MCA, distal to the inferior cerebral vein (**Fig. 1A**), using a pair of bipolar forceps (*see* **Note 22**), and section the MCA using Vannas spring-loaded microscissors to ensure occlusion.
5. Suture the muscle and then the skin wound.
6. Sham-operated animals undergo the same procedure to expose the MCA, but the artery is not occluded.
7. Changes in local cerebral blood flow can be monitored in real time with laser Doppler flowmetry using a probe placed on the surface of the skull. The bone of the skull must be thinned at the site where the laser Doppler probe is placed (usually over the cortex of the MCA territory), and care is taken to avoid placing the probe near any large blood vessels on the surface of the cortex. Alternatively, fiberoptic laser Doppler probes can be stereotaxically implanted into the brain for monitoring of changes in cerebral blood flow after MCA occlusion (*see* **Notes 24–26**). Postoperative care and adverse effects, *see* **Notes 27–29**.

3.6. Histology for Quantification of Infarct

3.6.1. Harvest of Brains

Under deep surgical anesthesia, decapitate and then dissect the brain from the skull. Drop into isopentane that has been cooled on dry ice to –42°C for 10 min (rat brain) or 2 min (mouse brain) (*see* **Note 30**). Remove, immediately coat with Lipshaw, mount in mounting agent, and cut coronal sections (20 μm) on a cryostat (*see* **Note 31**).

3.6.2. H&E Staining of Tissue

Hematoxylin is a basophilic stain that stains the nucleus of a cell blue and is used in conjunction with eosin which stains the cytoplasm pink or red (*see* **Note 32**).

1. Fix cryostat sections in 10% formal saline for 5–10 min. Wash well for 3 min.
2. Dehydrate sections in graded alcohols (70% - 100% -70%) and rinse in water, before submersion in haematoxylin for a period of between 1–2 min.
3. Rinse the sections in water and differentiate in acidic methylated alcohol (1% HCl).
4. Following another rinse in water, place the sections in Scot's tap water substitute (STWS) for 1 min, rinse again and place in eosin (aqueous) for 3 min (*see* **Note 33**)
5. After a final wash in water, dehydrate the sections through a graded series of alcohols, clear in histoclear and apply coverslips with histomount.

3.6.3. Quantification of Infarct

Infarct areas can be measured directly from H&E-stained sections selected at eight predetermined coronal planes covering the territory of the MCA (*27*). The infarct volume is then derived from integration of areas of infarction over the eight planes with end points of 12.5 mm anterior and 0.05 mm posterior to the interaural line. A fundamental problem with accurate infarct measurement is the complication of cerebral edema, which causes brain swelling within the infarct and adjacent tissue within the first 24–72 h following the stroke. Brain swelling can vary from experiment to experiment and can result in overestimated of infarct size. The H&E-stained sections in Fig. 1B demonstrate the presence of cerebral edema in the ipsilateral hemisphere. To correct for brain swelling, the infarct volume can be expressed as a percentage of the ipsilateral hemisphere (*see* **Notes 34** and **35**).

3.7. Results and Discussion

Using the linkage analysis approach, our group has discovered that there is at least one gene or more in an area of chromosome 5 in the F_2 population that contributes to stroke sensitivity in response to a focal ischemic insult (**Fig. 3**). We have located a QTL with a LOD score of 16.6 on rat chromosome 5, independent of blood pressure *(16)*. Rubattu et al. *(36)*, using latency to spontaneous stroke on high salt, have located three rat QTLs on chromosomes 1, 4, and 5 for susceptibility to stroke, again independent of blood pressure. Interestingly, the QTL on rat chromosome 5 located by Jeffs et al. *(16)* and Rubattu et al. *(36)* overlap each other, but in the former study the QTL is linked to increased sensitivity to stroke and in the latter study it is linked to protection against susceptibility to stroke. Ikeda et al. *(37)* used reduction in brain weight as a measure of development of stroke in the SHRSPs and found a linkage to rat chromosome 4, again independent of high blood pressure. SHRSP exhibit lower serum cholesterol than WKY rats, and this genetic trait has been significantly linked to three QTLs on rat chromosomes 5, 7, and 15, although no direct connection with stroke etiology has been made *(38)*. The authors show that the QTL on chromosome 5 is found in male rats only, and thus identical genes for that phenotype and for infarct volume after MCAO are unlikely.

Few studies have been published so far confirming the existence of stroke QTL by the use of congenic strains. Preliminary results published by Rubattu et al. *(39)* showed that congenic strains carrying a chromosome 1 QTL for stroke proneness exhibited a 50% incidence of stroke vs 0% in SHR and 100% in SHRSP parental strains. Nonstroke-related studies using congenics to confirm QTL are numerous and include our own studies whereby chromosome 2 blood pressure QTL regions were successfully transferred from WKY into an SHRSP background, significantly reducing blood pressure (and vice versa using reciprocal congenic strains) *(24)*. Further examples include the use of Dahl and SHR rat congenic strains to prove the existence of blood pressure QTL on various rat chromosomes *(22,40–42)*. Jiang et al. *(43)* have successfully transferred the salt-resistant renin allele into Dahl salt-sensitive rats to prove the

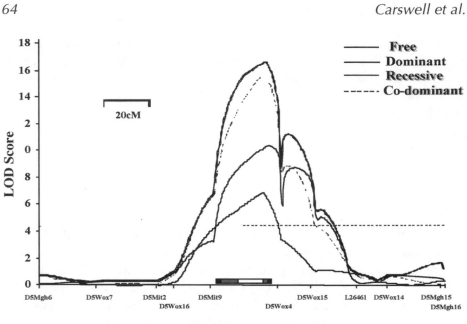

Rat Chromosome 5

Fig. 3. Multipoint sex-adjusted lod score for the locus affecting infarct volume expressed as a percentage of ipsilateral hemisphere on rat chromosome 5. The 1-, 3-, and 4.3-lod support intervals are shown as black, gray and open bars, respectively *(16)*.

role of the renin gene in the development of hypertension in Dahl salt-sensitive rats. By carrying out differential expression studies using DNA microarrays on congenic strains created by transferring a region of the chromosome 4 QTL from the Brown Norway genome to the SHR, Aitman et al. *(44)* have identified one causative gene (*Cd36*) underlying a QTL responsible for insulin resistance.

Construction of congenic strains takes many years, and a commonly implemented strategy for causative gene identification in the meantime is to test differential gene expression or candidate gene expression between SHRSPs and WKY rats. Using differential gene expression techniques, Kirsch et al. *(45)* showed altered gene expression within cerebral capillaries in SHRSPs compared to SHRs. The sulfonylurea receptor 2B was shown to be upregulated, and the rat homologue for the mouse G protein signalling 5 regulator was shown to be downregulated in the SHRSP cerebral capillaries *(45)*. However, the disadvantage of this approach is that there is not always a match with known proteins. The candidate gene approach is more straightforward, studying abnormal expression, structure, function, and/or regulation of candidate genes in the SHRSP compared with the control strain, WKY or SHR. If gene expression is measured after experimental stroke, there is, however, a difficulty in defining whether the SHRSP-specific gene expression is caused by genetically determined factors (i.e., contributory/causative genes) or an increased degree of brain damage after stroke (i.e., consequential genes). Candidate genes can be selected for analysis by either their presence within a QTL or by their known association with stroke.

3.7.1. Candidate Genes Located Within Identified QTLs

Of all the candidate genes for stroke in the SHRSP, the most investigated is atrial natriuretic peptide (ANP). The loci identified by Rabutta et al. *(36)* and Jeffs et al. *(16)* both colocalize with the genes encoding ANP. Although SHRSP/Gla showed no significant differences in either structure or expression of ANP compared with the WKY rat and no association of ANP with increased sensitivity to MCAO in the SHRSP was found *(46)*, SHRSP/Hedelberg showed alterations in structure, expression, and in vitro function of ANP in SHRSPs compared with SHRs as well as a downregulation of brain ANP in the stroke-prone phenotype *(47)*. The conflicting results may be explained partly by technical differences such as different quantification methods and partly by substrains of SHRSP. Other examples of candidate genes for stroke that have been investigated in the SHRSP include the disheveled-1 gene, which is involved in the Notch signaling pathway implicated in CADASIL in humans. It mapped to the homologous region of the rat chromosome 5 QTL for sensitivity to stroke *(16)* but had a similar sequence in SHRSPs and WKY rats and is thus excluded as a causative gene for sensitivity to stroke *(48)*.

3.7.2. Candidate Genes With a Known Association With Stroke

Following the initial ischemic/hypoxic episode, there is a cascade of events involving upregulation and downregulation of many different genes and proteins, some of which are recognized to have a role in stroke (for recent reviews on gene expression during cerebral ischemia, *see* **rcf.** *49*) and that can be tested for differential expression in SHRSPs and WKY rats. The advantage of this analysis is that genetic linkage analysis is not a prerequisite and may more speedily provide information on mechanisms that cause the increased sensitivity to MCAO or stroke proneness in SHRSPs. For example, brain CuZn SOD and Mn SOD protein content and activity have been shown to be lower in SHRSPs compared with WKY rats *(50)*, which may explain the higher lipid peroxidation content in SHRSPs compared with WKY rats *(51)*. SHRSPs exhibit ineffective dilator reserve in collateral vessels to the MCA *(52)* and a greater deficit in blood flow after MCAO in SHRSPs compared with WKY rats *(53)*. Rubattu et al. *(54)* tested several candidate genes for vascular dysfunction but found no evidence for a role of renin, angiotensinogen, angiotensin-converting enzyme, angiotensin II AT1b receptor, ANP, brain natruristic protein, ANP GC-A receptor, kallikrein, and endothelial nitric oxide synthase. Wang et al. *(55)* observed a downregulation of angiopoietin-1 (which is shown to promote neovascularisation after hypoxia) post MCAO in SHRSPs compared with SHRs and WKY rats. Angiotensin II type I receptor expression is increased in brain microvessels in SHRSP compared with WKY rats. Antagonism of these receptors in the SHRSP reduced endothelial injury, possibly revealing a role for angiotensin II in hypertensive microvascular injury *(56)*. Gene expression for redox regulatory proteins and proteins involved in energy metabolism were shown to be attenuated after hypoxia in SHRSPs compared with WKY in isolated cortical neurons exposed to hypoxia and reoxygenation *(57)*. Decreased neuronal bcl-2 expression after hypoxia in vitro in SHRSPs compared with WKY rats *(57)* may explain why neuronal apoptosis after cerebral ischaemia was apparent in the SHRSP but not in WKY rats *(58)*. Finally, a putative role for the influence of the SHRSP Y chromosome on stroke sensitivity in the SHRSP, indicated using F_1 hybrids *(17)*, was not proved using Y consomic strains *(59)*.

3.8. Summary

This chapter focuses on how to localize the gene(s) responsible for the increased sensitivity to focal cerebral ischemia in SHRSPs. The need for information on stroke genes is emphasized by the fact that no effective stroke therapy exists for most stroke patients despite the ever increasing elderly population. The methodology we have detailed describes QTL mapping, congenic strain construction, and a well-characterized stroke phenotype. Using the linkage analysis approach, we have described how we discovered the QTL on chromosome 5 for stroke sensitivity in response to a focal ischemic insult, independent of blood pressure. This information has added to the findings on susceptibility to spontaneous strokes (stroke proneness), again independent of blood pressure. The ultimate aim of this work is to provide important information on the pathophysiology of stroke and the potential for new therapies to limit stroke-induced damage. Furthermore, if equivalent genetic mutations are found in humans, this information could be used to identify people at greater risk of sustaining a significant stroke and provide an opportunity for prophylactic treatment. Rat models such as the SHRSP have helped to overcome some of the enormity of the challenge of stroke genetics by providing an inbred genetic model where by experimental stroke can be induced reproducibly and confounding factors controlled. However, the task ahead remains challenging. Genes and their functions must be identified, bearing in mind that the existence of multiple interacting genes (epistasis) may be required for a functional role in complex diseases such as stroke. Since rats do not necessarily share the same stroke-related genes as humans, once identified, the rat genes must be carefully extrapolated to humans and evaluated in human stroke.

4. Notes

1. Do not remove any more than 4 mm, as this can affect DNA yields and quality.
2. Alternatively, incubation may be done for 3 h at 55°C with hourly vortexing.
3. The white tight pellet will contain the precipitated protein. If no pellet is seen at this stage, then the sample was not cooled to room temperature before the addition of the Protein Precipitation Solution. As a solution, cool the sample to room temperature, vortex briefly for 20 s, spin again, and proceed as normal.
4. At this stage white DNA thread-like strands should be visible.
5. The ratio value given on the spectrophotometer gives an indication of the quality of the DNA extracted. The normal value for quality DNA is 1:800.
6. The PCR preparation should be done on ice. Once the dNTPs have been defrosted, use and discard the remainder. Repeated freeze-thawing of dNTPs is not good for successful PCR.
7. We use a hot start enzyme that reduces the number of stutter peaks aiding accurate genotyping. This is one-tenth the recommended amount of *Taq* polymerase, but in our hands this is sufficient for successful PCR, and costs are kept down.
8. The PCR reaction is mineral oil-free. To ensure that the PCR sample does not evaporate, the hotlid must be tightened firmly.
9. Don't pull the spacers out before splitting the plates, as it tears them, and don't twist the wedge to split the plates—just push it in until the plates split.
10. Wash plates with hot water only; never use any detergent, as this increases background fluorescence.
11. To ensure the correct origin of the Y chromosome in the fixed congenic strain, male F_1 hybrids having a SHRSP father should be backcrossed to SHRSP recipient strain females, and likewise male F_1 hybrids having an WKY father should be backcrossed to WKY recipient strain females.
12. Male rodents are more commonly used in ischemia research than females to avoid the influence of fluctuating estrogens and progesterone on infarct size *(30,31)*. If females are used for ischemia research, hormonal influences can be controlled, either by monitoring the estrous cycle or by using ovariectomized animals.
13. To reduce costs and minimize rat numbers, all females generated by backcrossing can be culled at weaning.
14. Computer simulations have indicated that a relatively modest selection effort (60 background markers, 25 c*M* marker spacing, 16 males per generation) would typically reduce unlinked donor genome contamination to <1% by four backcross generations *(28,29)*.
15. Always occlude the MCA on the same side of the brain for each study.
16. Analgesics for postoperative pain relief are seldom used in ischemia research, as they may influence the amount of ischemic damage induced by MCA occlusion and therefore may compromise results.

17. Halothane offers fast onset and recovery from anesthesia and is inexpensive relative to other volatile anesthetics. Care must be taken in the choice of alternative anesthetic regimes, e.g., ketamine is not recommended as it is an N-methyl-D aspartate receptor antagonist. Propofol (bolus plus top-up injections or infusion), which is short acting and has a fast onset and recovery time, could be considered as an alternative.

18. Intubation and artificial respiration should be considered even if injectable rather than volatile anesthesia is used, as some anesthetics can cause apnea. An intubation tube size of 16 gage is recommended for rats of 250–350 g. A good tip for intubating rats is to angle the tube upward with the animal in a supine position, to help avoid insertion into the esophagus. It should be noted that the tips of intubation tubes can become blocked with mucus secretions during lengthy procedures. A length of fine polythene tubing attached to a 5- or 10-mL syringe can be used to aspirate and withdraw any mucus buildup at the tube tip that may compromise ventilation. At the end of the procedure, allow animals to regain spontaneous respiration before removing the intubation tube.

19. Body temperature should always be monitored and maintained within normal limits throughout anesthesia. Hypothermia of a couple of degrees C can reduce infarct size, and conversely hyperthermia can increase infarct size. Physiological monitoring of blood gases and blood pressure are not essential for short periods of anesthesia but can help reproducibility in infarct size if physiological variables are maintained within normal limits throughout the study. Arterial CO_2 ($PaCO_2$) should be maintained between 36 and 42 mmHg. Cerebral blood flow (CBF) is sensitive to changes in $PaCO_2$, and levels outside these limits can influence the severity of ischemia by influencing CBF. Similarly, blood pressures that fall below the lower limit of cerebral autoregulation (~50–60 mmHg in a normotensive rat) can compromise CBF, exacerbate ischemia, and thereby increase infarct size.

20. Thin the bone of the skull overlying the MCA. Avoid drilling continuously in one particular spot, as this transfers heat from the drill burr to the bone and underlying tissue, which can result in tissue damage and artifacts in stained tissue sections. By thinning the bone, you can locate the position of the underlying MCA; when it is thin enough, the bone can be peeled away at the appropriate site

21. When removing the dura, bend the tip of a dental needle and use this to pierce and peel back the dura.

22. Use copious amounts of sterile saline to cool the drill tip while making the craniectomy. When drilling and using diathermy forceps, a fine brain temperature probe (type IT-21 tissue-implantable thermocouple microprobe, Physiotemp Instruments, Clifton, NJ) can be inserted into the temporal muscle to give an assessment of brain temperature. Ensure that the temperature does not rise above 38°C, and flush the area regularly with sterile saline. Sterile saline, flushed over the MCA, also helps to prevent the bipolar forceps from sticking to tissues when one is electrocoagulating the artery. Replace the saline continuously. If the saline heats up, this can contribute to the brain temperature during the electrocoagulation procedure.

23. Use triangular arrowhead swabs to mop up and stem bleeding from bone, as well as blood and cerebrospinal fluid from the craniectomy site. Bone wax can also be used to control any bleeding from bone.

24. When monitoring changes in local CBF with laser Doppler flowmetry, the probe remains in the same position during basal and ischemic measurements and should not be moved

25. Place the probe tip on the thinned bone, and apply some sterile saline to improve the signal

26. Probe placement onto thinned bone is preferable to drilling a craniotomy with probe placement directly onto the dura of the exposed brain.

27. Recovery and postoperative care. Rats are monitored closely during recovery from anesthesia, given 2 mL subcut sterile saline to prevent dehydration, and kept in a warm environment (e.g., 28°C). They are initially provided with water and softened rat chow and housed individually to allow wound healing. Thereafter group housing is recommended, and animals can be maintained under normal controlled temperature (21°C) and light (12-h dark/light cycle) conditions and provided with normal rat chow (*see* **Note 29**) and water *ad libitum*.

28. Recovery rate is good with this model of MCAO. In the experience of our laboratory less than 5% of animals die during surgery and less than 1% from anesthetic accident. Risk of infection is rare in SHRSP and WKY (less than 1%). However, our experience with Lister hooded rats indicates a higher rate of infection, which can be prevented and controlled for by use of antibiotics before and after MCAO. A transient (24-h) disruption of feeding can be expected in 100% of animals with prolonged (72-h) disruption of feeding in about 20%. About 5% of animals may have prolonged disruption of feeding for more than 72 h. If body weight drops by 10%, then rehydration therapy is recommended (0.5 mL sterile physiological saline subcutaneously). If body weight drops by 20%, then food supplementation such as Complan (by gavage if required) is recommended. A higher than 20% drop in body weight is beyond the accepted severity limit for this model, and euthanasia should be determined by the inability of animals to eat and drink. Softened food or food treats such as fruit puree and fluid support therapy should be offered. If failure to drink persists beyond 48 h or failure to eat persists beyond 72 h, the animal should be euthanized.

29. In terms of behavioral changes, it can be expected that up to 100% of animals exhibit circling and transient mild hemiparesis affecting the contralateral forelimb, although these symptoms are less apparent in the distal model compared with the proximal MCA occlusion model. Lethargy, altered consciousness, hunching, and piloerection may be exhibited up to 24 h postoperatively in up to 100% of animals. The intracranial temporal approach often affects the jaw alignment. Animals that have undergone the intracranial temporal approach may need to be continually supplied with softened diet throughout the survival period. In addition, teeth clipping may need to be carried out up to twice a week if the animals do not eat dry diet or use chew blocks. Teeth clipping can be done under brief anesthesia if required. Chromodacryorrhea with unilateral loss of blink

reflex and dry eye (up to 100%) is often associated with intracranial MCAO and can be prevented and controlled by taping or suturing the ipsilateral eye shut during surgery. Eye ointment (e.g., chloromycetin, orbenin) and/or drops (e.g., Viscotears, Lacri lube) can be applied to the eye to limit this adverse effect.

30. Check that the temperature of the isopentane is –42°C throughout the solution by mixing frequently while it is placed on dry ice. Higher temperatures at the top of the solution can introduce ice crystals within the frozen brain. The recommended height and volume of the cylinder to use for the isopentane are approx 15 cm and approx 450 mL, resepectively.

31. As an alternative to freezing the fresh brain, animals can be perfusion-fixed using 4% paraformaldehyde in 50 mM phosphate buffer (e.g., if brain sections are also required for immunohistochemistry). To dissolve paraformaldehyde, heat to no higher than 60°C in phosphate buffer. Following perfusion fixation, postfix the brain in the skull for 24 h, then remove the brain, and postfix for another 24 h in fresh fixative. The brain can be stored for up to 4–5 d in phosphate buffer before being processed for paraffin embedding. Once embedded in paraffin wax, coronal brain sections can be cut (6 μm is ideal for microscopy) using a microtome. Note that brains can shrink because of the processing for paraffin embedding. This shrinkage can be corrected for (*see* **Note 35**).

32. As an alternative to H&E staining, 2.3,5-triphenyltetrazolium chloride (TTC) can be employed. It is a rapid, convenient, inexpensive, and reliable method for detection and quantification of cerebral infarction in rats 24 h after occlusion *(32)*. TTC is oxidized to formazan (red) by mitochondrial enzyme systems when mitochondria are viable. Either the brain can be sliced and immersed in TTC solution (2% TTC in physiological saline) for 30–60 min (a heat lamp may be placed over it) or the animals can be perfused with TTC (2% TTC in physiological saline or phosphate buffer) after perfusion with saline to remove blood. However, there are at least two limitations to the use of TTC for quantification of infarct size. First, the reliability of TTC for measuring infarct volume at early time-points (<24 h) post occlusion is questionable since it has been shown that at 4 h post occlusion, the TTC immersion technique grossly underestimates the area of ischemic damage *(33)*. Second, at later time-points, such as 36 h post occlusion, macrophages and microglia (containing viable mitochondria) will invade the damaged tissue to remove dead cells and debris and would be stained by TTC and confound identification of the infarct boundary *(34)*.

33. Check differentiation under the microscope before using eosin. Nuclei should be stained relative to the rest of the tissue. If not, then resubmerge in hematoxylin. If background staining is too dark, then resubmerge in acid alcohol. This is a subjective stage in the staining.

34. Swanson et al. *(35)* describes an alternative method to minimize the error owing to brain swelling by measuring the volume of surviving normal gray matter rather than the volume of tissue that has undergone infarction.

35. To correct for brain swelling and shrinkage owing to fixation and processing, distribution of the infarct can be mapped onto line diagrams of the eight coronal

planes throughout the MCA territory *(27)*. Areas of infarct can be measured from the diagrams by image analysis with the volume of infarct derived from integration of areas of infarction over the eight planes with end points of 12.5 mm anterior and 0.05 mm posterior to the interaural line *(27)*.

References

1. Kidwell, C. S., Liebeskind, D. S., Starkman, S., and Saver, J. L. (2001) Trends in acute ischemic stroke trials through the 20th century. *Stroke* **32,** 1349–1359.
2. Brass, L. M., Isaacsohn, J. L., Merikangas, K. R., Robinette, C. D. (1992) A study of twins and stroke. *Stroke* **23,** 221–223.
3. Kiely, D. K., Wolf, P. A., Cupples, L. A., Beiser, A. S., Myers, R. H. (1993) Familial aggregation of stroke. The Framingham Study. *Stroke* **24,** 1366–1371.
4. Graffagnino, C., Gasecki, A. P., Doig, G. S., Hachinski, V. C. (1994) The importance of family history in cerebrovascular disease. *Stroke* **25,** 1599–1604.
5. Diaz, J. F., Hachinski, V. C., Pederson, L. L., Donald, A. (1986) Aggregation of multiple risk factors for stroke in siblings of patients with brain infarction and transient ischemic attacks. *Stroke* **17,** 1239–1242.
6. Joutel, A., Corpechot, C., Ducros, A., et al. (1996) Notch3 mutations in CADASIL, a hereditary adult-onset condition causing stroke and dementia. *Nature* **24,** 707–110.
7. Palsdottir, A., Abrahamson, M., Thorsteinsson, L., et al. (1988) Mutation in cystatin C gene causes hereditary brain haemorrhage. *Lancet* **2,** 603–604.
8. Levy, E., Carman, M. D., Fernandez-Madrid, I. J., et al. (1990) Mutation of the Alzheimer's disease amyloid gene in hereditary cerebral hemorrhage, Dutch type. *Science* **248,** 1124–1126.
9. Laberge-le Couteulx, S., Jung, H. H., Labauge, P., et al. (1999) Truncating mutations in CCM1, encoding KRIT1, cause hereditary cavernous angiomas. *Nat. Genet.* **23,** 189–193.
10. Sahoo, T., Johnson, E. W., Thomas, J. W., et al. (1999) Mutations in the gene encoding KRIT1, a Krev-1/rap1a binding protein, cause cerebral cavernous malformations (CCM1). *Hum. Mol. Genet.* **8,** 2325–2333.
11. Alberts, M. J. (2001) Genetic update. Impact of the human genome projects and identification of a stroke gene. *Stroke* **32,** 1239–1241.
12. Gretarsdottir, S., Sveinbjornsdottir, S., Jonsson, H.H., et al. (2002) Localization of a susceptibility gene for common forms of stroke to 5q12. *Am. J. Hum. Genet.* **70,** 593–603.
13. Yamori, Y., Horie, R., Handa, H., Sato, M., Fukase, M. (1976) Pathogenetic similarity of strokes in stroke-prone spontaneously hypertensive rats and humans. *Stroke* **7,** 46–53.
14. Okamoto, K., Yamori, Y., and Nagaoka, A. (1974) Establishment of the SHRSPs (SHR). *Circ. Res.* **34 (suppl. I),** I143—I153
15. Coyle, P. and Jokelainen, P. T. (1983) Differential outcome to middle cerebral artery occlusion in spontaneously hypertensive stroke-prone rats (SHRSP) and Wistar Kyoto (WKY) rats. *Stroke* **14,** 605–611.

16. Jeffs, B., Clark, J. S., Anderson, N. H., et al. (1997) Sensitivity to cerebral ischaemic insult in a rat model of stroke is determined by a single genetic locus. *Nat. Genet.* **16,** 364–367.

17. Carswell, H. V., Anderson, N. H., Clark, J. S., et al. (1999) Genetic and gender influences on sensitivity to focal cerebral ischemia in the stroke-prone spontaneously hypertensive rat. *Hypertension* **33,** 681–685.

18. Fujii, K., Weno, B. L., Baumbach, G. L., Heistad, D. D. (1992) Effect of antihypertensive treatment on focal cerebral infarction. *Hypertension* **19,** 713–716.

19. Coyle, P. (1984) Outcomes to middle cerebral artery occlusion in hypertensive and normotensive rats. Hypertension 6(2 Pt 2),I69-74.

20. Bihoreau, M.T., Gauguier, D., Kato, N., et al. (1997) A linkage map of the rat genome derived from three F2 crosses. *Genome Res.* **7,** 434–440.

21. Steen, R. G., Kwitek-Black, A. E., Glenn, C., et al. (1999). A high-density integrated genetic linkage and radiation hybrid map of the laboratory rat. *Genome Res.* **9,** AP1–AP8.

22. Frantz, S. A., Kaiser, M., Gardiner, S., et al. (1998) Successful isolation of a rat chromosome 1 blood pressure QTL in reciprocal congenic strain. *Hypertension* **32,** 639–646.

23. Morel, L., Yu, Y., Blenman, K. R., Caldwell, R. A., Wakeland, E. K. (1996) Production of congenic mouse strains carrying genomic intervals containing SLE-susceptibility genes derived from the SLE-prone NZM2410 strain. *Mammalian Genome* **7,** 335–339.

24. Jeffs, B., Negrin, C. D., Graham, D., et al. (2000). Applicability of a speed congenic strategy to dissect blood pressure QTL on rat chromosome 2. *Hypertension* **35,** 179–187.

25. Lander, E. S. and Schork, N. J. (1994) Genetic dissection of complex traits. *Science* **265,** 2037–2048.

26. Macrae, I. M. (1992) New models of focal cerebral ischaemia [Review]. *Br. J. Clin. Pharmacol.* **34,** 302–308.

27. Osborne, K. A., Shigeno, T., Balarsky, A. M., et al. (1987) Quantitative assessment of early brain damage in a rat model of focal cerebral ischaemia. *J. Neurol. Neurosurg. Psychiatry* **50,** 402–410.

28. Wakeland, E., Morel, L., Achey, K., Yui, M., and Longmate, J. (1997) Speed congenics: a classic technique in the fast lane (relatively speaking). *Immunol. Today* **18,** 472–477.

29. Markel, P., Shu, P., Ebeling, C., et al. (1997) Theoretical and empirical issues for marker-assisted breeding of congenic mouse strains. *Nat. Genet.* **17,** 280–284.

30. Hurn, P. D. and Macrae, I. M. (2000) Estrogen as a neuroprotectant in stroke. *J. Cereb. Blood Flow Metab.* **20,** 631–652.

31. Carswell, H. V., Dominiczak, A. F., Macrae, I. M. (2000a) Estrogen status affects sensitivity to focal cerebral ischemia in stroke-prone spontaneously hypertensive rats. *Am. J. Physiol. Heart Circ. Physiol.* **278,** H290–H294

32. Bederson, J. B., Pitts, L. H., Germano, S. M., Nishimura, M. C., Davis, R. L., Bartkowski, H. M. (1986) Evaluation of 2,3,5-triphenyltetrazolium chloride as a

stain for detection and quantification of experimental cerebral infarction in rats. *Stroke* **17,** 1304–1308.

33. Hatfield, R. H., Mendelow, A. D., Perry, R. H., Alvarez, L. M., Modha, P. (1991) Triphenyltetrazolium chloride (TTC) as a marker for ischaemic changes in rat brain following permanent middle cerebral artery occlusion. *Neuropathol. Appl. Neurobiol.* **117,** 61–67.

34. Liszczak, T. M., Hedley-Whyte, E. T., Adams, J. F., et al. (1984) Limitations of tetrazolium salts in delineating infarcted brain. *Acta. Neuropathol. (Berl)* **65,** 150–157.

35. Swanson, R. A., Morton, M. T., Tsao-Wu, G., Savalos, R. A., Davidson, C., Sharp, F. R. (1990) A semiautomated method for measuring brain infarct volume. *J. Cereb. Blood Flow Metab.* **10,** 290–293.

36. Rubattu, S., Volpe, M., Kreutz, R., Ganten, U., Ganten, D., Lindpaintner, K. (1996) Chromosomal mapping of quantitative trait loci contributing to stroke in a rat model of complex human disease. *Nat. Genet.* **13,** 429–434.

37. Ikeda, K., Nara, Y., Matumoto, C., et al. (1996) The region responsible for stroke on chromosome 4 in the stroke-prone spontaneously hypertensive rat. *Biochem. Biophys. Res. Commun.* **229,** 658–662.

38. Kato, N., Tamada, T., Nabika, T., et al. (2000) Identification of quantitative trait loci for serum cholesterol levels in stroke-prone spontaneously hypertensive rats. *Arterioscler. Thromb. Vasc. Biol.* **20,** 223–229.

39. Rubattu, S., Ganten, U., Volpe, M., and Lindpainter, K. (1999) Increased incidence of stroke in congenic rats carrying the SHRSP-derived stroke-related locus, STR-1. *Circulation* **100,** Suppl 1.

40. Garrett, M. R., Dene, H., Walder, R., et al. (1998) Genome scan and congenic strains for blood pressure QTL using Dahl salt-sensitive rats. *Genome Res.* **8,** 711–723.

41. Deng, A. Y., Dene, H., Rapp, J. P. (1997) Congenic strains for the blood pressure quantitative trait locus on rat chromosome 2. *Hypertension* **30,** 199–202.

42. Alemayehu, A., Breen, L., Krenova, D., Printz, M. P. (2002) Reciprocal rat chromosome 2 congenic strains reveal contrasting blood pressure and heart rate QTL. *Physiol. Genomics* **10,** 199–210.

43. Jiang, J., Stec, D. E., Drummond, H., et al. (1997) Transfer of a salt-resistant renin allele raises blood pressure in Dahl salt-sensitive rats. *Hypertension* **29,** 619–627.

44. Aitman, T. J., Glazier, A. M., Wallace, C. A., et al. (1999) Identification of Cd36 (Fat) as an insulin-resistance gene causing defective fatty acid and glucose metabolism in hypertensive rats. *Nat. Genet.* **21,** 76–83.

45. Kirsch, T., Wellner, M., Luft, F. C., Haller, H., Lippoldt, A. (2001) Altered gene expression in cerebral capillaries of stroke-prone spontaneously hypertensive rats. *Brain Res.* **910,** 106–115.

46. Brosnan, M. J., Clark, J. S., Jeffs, B., et al. (1999) Genes encoding atrial and brain natriuretic peptides as candidates for sensitivity to brain ischemia in stroke-prone hypertensive rats. *Hypertension* **33,** 290–297.

47. Rubattu, S., Lee-Kirsch, M. A., DePaolis, P., et al. (1999) Altered structure, regulation, and function of the gene encoding the atrial natriuretic peptide in the stroke-prone spontaneously hypertensive rat. *Circ. Res.* **85,** 900–905.

48. De Lange, R. P., Burr, K., Clark, J. S., et al. (2001) Mapping and sequencing rat dishevelled-1: a candidate gene for cerebral ischaemic insult in a rat model of stroke. *Neurogenetics* **3**, 99–106.
49. Sharp, F.R., Lu, A., Tang, Y., Millhorn, D.E. (2000) Multiple molecular penumbras after focal cerebral ischemia. J. Cereb. Blood Flow Metab. 20, 1011-32.
50. Kimoto, S., Nishida, S., Funasaka, K., Nakano, T., Teramoto, K., and Tomura, T. T. (1995) Regional distribution of superoxide dismutase in the brain and myocardium of the stroke-prone spontaneously hypertensive rat. *Clin. Exp. Pharmacol. Physiol. Suppl.* **22**, S160–S161.
51. Ito, H., Torii, M., Suzuki, T. (1993) A comparative study on lipid peroxidation in cerebral cortex of stroke-prone spontaneously hypertensive and normotensive rats. *Int. J. Biochem.* **25**, 1801–1805.
52. Coyle, P. and Heistad, D. D. (1991) Development of collaterals in the cerebral circulation. *Blood Vessels* **28**, 183–189.
53. Carswell, H. V., Anderson, N. H., Morton, J. J., McCulloch, J., Dominiczak, A. F., Macrae, I. M. (2000b) Investigation of estrogen status and increased stroke sensitivity on cerebral blood flow after a focal ischemic insult. *J. Cereb. Blood Flow Metab.* **20**, 931–936.
54. Rubattu, S., Giliberti, R., Russo, R., Gigante, B., Ganten, U., Volpe, M. (2000) Analysis of the genetic basis of the endothelium-dependent impaired vasorelaxation in the stroke-prone spontaneously hypertensive rat: a candidate gene approach. *J. Hypertens.* **18**, 161–165.
55. Wang, M. M., Klaus, J. A., Joh, H. D., Traystman, R. J., Hurn, P. D. (2002) Postischemic angiogenic factor expression in stroke-prone rats. *Exp. Neurol.* **173**, 283–288.
56. Takemori, K., Ito, H., Suzuki, T. (2000) Effects of the AT1 receptor antagonist on adhesion molecule expression in leukocytes and brain microvessels of stroke-prone spontaneously hypertensive rats. *Am. J. Hypertens.* **13**, 1233–1241.
57. Yamagata, K., Tagami, M., Ikeda, K., Yamori, Y., Nara, Y. (2000) Altered gene expressions during hypoxia and reoxygenation in cortical neurons isolated from stroke-prone spontaneously hypertensive rats. *Neurosci. Lett.* **284**, 131–134.
58. Tagami, M., Ikeda, K., Nara, Y., et al. (1997) Insulin-like growth factor-1 attenuates apoptosis in hippocampal neurons caused by cerebral ischemia and reperfusion in stroke-pronespontaneously hypertensive rats. *Lab. Invest.* **76**, 613–617.
59. Negrin, C. D., McBride, M. W., Carswell, H. V., et al. (2001) Reciprocal consomic strains to evaluate Y chromosome effects. *Hypertension* **37**, 391–397.

4

Gene Therapy in Neurological Disease

Midori A. Yenari and Robert M. Sapolsky

Summary

Advances in the area of stroke and other neurodegenerative disorders have identified a variety of molecular targets for potential therapeutic intervention. The use of modified viral vectors has now made it possible to introduce foreign DNA into central nervous system cells, permitting overexpression of the protein of interest. A particular advantage of the herpes simplex system is that the herpes virus is neurotropic and is therefore suited for gene therapy to the nervous system. The vectors used by our group to date utilize an amplicon-based bipromoter system, which permits expression of both the gene of interest and a reporter gene. Using this strategy, we have been successful in transferring potentially neuroprotective genes to individual central nervous system cells. Using this approach, it is possible to show that gene therapy both before and after insult is feasible. Some limitations of this technique exist, the main one being delivery and extent of transfection. Although application to clinical stroke is probably remote, viral vector-mediated gene therapy provides a unique and powerful tool in the study of molecular mechanisms involved in brain injury.

KEY WORDS

Gene therapy; rat; stroke; cerebral ischemia; herpes simplex; neuroprotection.

1. Introduction

Several studies have shown that cerebral ischemia alters gene expression and that some of the induced genes may determine whether the injured neuron will live or die. Among the many genes that have been identified to participate in ischemia, those that possess neuroprotective properties may be excellent candidates for gene therapy *(1,2)*. The use of transgenic animals which either overexpress or fail to express gene products of interest has already greatly enhanced our knowledge of the molecular biology of ischemia. Gene transfer provides another means of studying this area as well. Some advantages of gene

From: *Methods in Molecular Medicine, Vol. 104, Stroke Genomics: Methods and Reviews*
Edited by: S. J. Read and D. Virley © Humana Press Inc., Totowa, NJ

transfer over transgenic animals use is that transgene expression can be altered in genetically normal animals, and application of candidate genes can also be studied post insult. Although there are a number of technical obstacles to be overcome before this approach can be used at the clinical level, we believe this approach is novel and provides another means of better understanding the complex molecular biology underlying ischemic injury. We describe our strategy for studying such questions through viral vector-mediated gene therapy to cerebral neurons using defective herpes simplex viral vectors.

Viral vectors provide a rational approach to gene transfer therapy, as these agents naturally transfer genetic material into host cells. To use viruses as vectors for delivery of foreign genetic material, they must retain their ability to infect target cells but not cause damage. Following viral infection, injury to the host cell generally results when there is replication and subsequent cell lysis. Several viral strains have been used for gene therapy that contain deletions in the genes controling replication and cell lysis. A few of these strains are appropriate for gene transfer into the central nervous system (3). Herpes simplex virus (HSV) is a natural choice for gene transfer into the adult central nervous system as the virus is neurotropic (4,5). It is taken up almost exclusively by neurons after direct parenchymal injection; however, ependymal cells will take up the viral vector after intraventricular injection. It is a DNA virus and remains episomal after transfection, making it ideal for use in postmitotic cells such as neurons. Its insert size is also large (36 kb), and it can be purified to high titers. Although herpes is known to cause devastating human illnesses, replication-incompetent strains have been developed that contain mutations within genes responsible for cytotoxicity (6). The d120 strain contains a 4.1-kb deletion mutation in α4, an immediate early gene essential for early and late gene induction (6). Without α4, the virus is incapable of performing replication, capsid synthesis, and assembly (7). Prior studies in our laboratories failed to demonstrate increased cytotoxicity up to 5 d after injection into rodent striata and hippocampus (8). In addition, control vectors (containing only the reporter gene, lacZ) do not induce the stress protein Hsp72 (9). Here we outline how viral vectors are generated and how they are applied in rodent models of ischemic brain injury to determine whether potentially beneficial genes improve neuron survival.

2. Materials

2.1. Generation of Viral Vectors

1. Vero cells (African green monkey kidney cells; ATCC CCL81, American Type Culture Collection, Rockville, MD).
2. E5 cells and d120 helper virus (Dr. N.A. DeLuca, University of Pittsburgh, Pittsburgh, PA) (6).

3. Bipromoter plasmid (R. Sapolsky, Stanford University).
4. Dulbecco's modified Eagle's medium (DMEM) (cat. no. 11965-092, Gibco-BRL).
5. Opti-MEM (cat. no. 31985-070, Gibco-BRL).
6. NuSerum IV (cat. no. CB55004, Gibco-BRL).
7. Penicillin and streptomycin (Gibco).
8. Lipofectamine (cat. no. 18324-012, Gibco-BRL). Store at 4°C).
9. X-gal (5'-bromo-4-chloro-3-inodyl-β-D-galactopyranoside; Molecular Probes, Eugene, OR).

2.2. Ischemia Models

1. Hamilton syringe (Hamilton, Reno, NV).
2. Microsyringe pump (Micro4, World Precision Instruments, Sarasota, FL).
3. 3-0 Surgical nylon sutures (Ethicon, Somerville, NJ).
4. Paraformaldehyde, sucrose, dimethylsulfoxide (DMSO), and cresyl violet (Sigma).

3. Methods

The vectors used in our studies are referred to as "bipromoters," in that two separate promoters drive expression of two different genes *(10)*. An advantage of this system is that both a reporter gene (e.g., the *E. coli lacZ* gene) and the gene of interest are coexpressed within targeted cells. As the gene of interest is often induced within cells following various insults, transgene expression owing to transfection and endogenous gene expression can be difficult to differentiate. With these vectors, transfected cells can overexpress the gene of interest and can also be identified by staining for β-galactosidase (β-gal), the gene product of *lacZ*. We have adapted this system for gene transfer studies using an in vivo models of cerebral ischemic injury.

3.1. Vector Construction

The method used by our group for vector generation is an amplicon ("cloning amplifying") based system (**Fig. 1**). This in contrast to a recombinant method where the gene of interest is spliced into the viral vector and components of the viral genome are deleted to reduce cytotoxicity. In the amplicon system, a plasmid containing the gene(s) of interest and a replication-incompetent helper virus are cotransfected into a cell line that contains all the necessary components for virus replication and assembly. The transfected cell line produces progeny virion containing either the plasmid of interest or helper virus. Progeny virion are purified from cellular components, titered, and saved in aliquots until use in the ischemia experiments.

The construction of amplicon plasmids and vectors has been described in detail elsewhere *(10–14)*. The amplicon plasmid pα22βgalα4<gene of inter-

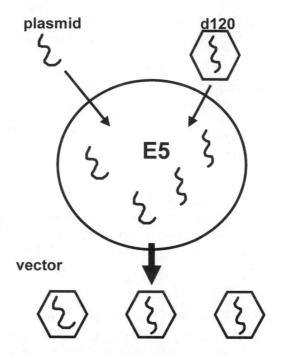

Fig. 1. Amplicon (cloning amplifying vector) system for viral vector generation. The plasmid of interest is cotransfected with a replication-defective helper virus (d120) in E5 cells, which are stably transformed with α4. Since the d120 strain has a mutation in the immediate early gene, α4, this system provides the defective herpes virus with all of the necessary tools to replicate, assemble, and cause cytotoxicity. The resulting vectors are concatemers of the helper virus and plasmid of interest, which are then purified from cellular debris. Used with permission from **ref**. *5*.

est> contains the gene of interest and the *Escherichia coli lacZ* gene under the control of the HSV α4 and α22 promoters, respectively (**Fig. 2A**). In some instances, it may be desirable to generate two constructs, one with the transgene under the α4 promoter, and the reporter under α22, and another with the promoters switched to control for any potential differences. The HSV oriS and the "α" sequence were also included to provide the necessary *cis*-signals for replication and packaging of the amplicon DNA. Control plasmids are identical in every way, except that a stop codon is inserted immediately after or shortly downstream of the α4 promoter (pα22βgalα4s or α4s) (**Fig. 2B**; *see* **Note 1**). In the former construct, no transgene is produced downstream of α4, and in the latter, a truncated, presumably nonfunctioning peptide is produced. In both control plasmids, the reporter gene is used to identify transfected cells. Bicistronic vectors have been constructed in the past by our group with an internal riboso-

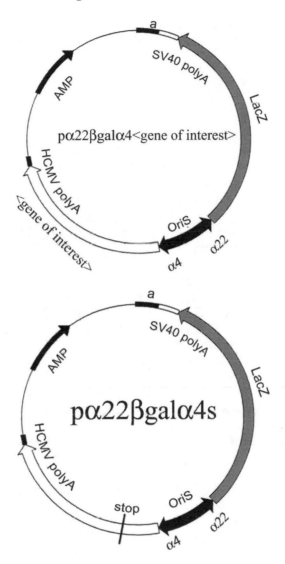

Fig. 2. Diagrams of the bipromoter plasmids. **(Top)** Plasmid that contains the gene of interest and reporter gene (*lacZ*), under control of the α4 and α22 promoters, respectively. The two promoters are separated by the Oris, which is necessary for replication. **(Bottom)** Control plasmid identical to plasmid (A), except that there is a stop codon after the α4 promoter. Used with permission from **ref. 5**.

mal entry site (IRES) connecting the gene of interest followed by the reporter; however, these constructs led to variable expression of the reporter. In contrast, the bipromoter system led to more consistent protein expression of both genes, with similar, although not necessarily identical, kinetics. Coexpression of both gene products was found in over 98% of all transfected cells *(11)*.

3.1.1. Transfecting Cell Lines With Plasmid and Helper Virus

1. This process requires the use of two cell lines, Vero (African green monkey kidney cells; ATCC CCL81) and E5 (provided by Dr. N. A. DeLuca, University of Pittsburgh, Pittsburgh, PA) *(6)*, both of which are derived from African green monkey kidney fibroblasts.
2. Vectors are generated by cotransfection of the bipromoter plasmid and the *d120* helper virus (HSV-1 strain KOS, also provided by Dr. N. A. DeLuca) *(6)* into E5 cells. The E5 cell line contains the missing α4 gene; therefore, the d120 virus can replicate in the same manner as the wild type.
3. E5 cells are grown in DMEM to approx 70% confluence.
4. Plasmids are first transfected into the E5 cells using lipofectamine (Gibco) according to the manufacturer's directions.
5. For each T-25 flask, 4 µg of DNA plus 14 µL of lipofectamine are suspended in 2.2 mL Opti-MEM (Gibco) prior to applying to cells.
6. Cells are incubated in lipofectamine for 6–8 h at 37°C in a 5% CO_2 incubator, after which cultures are diluted with an additional 2.8 mL DMEM plus 20% NuSerum and incubated for an additional 18 h.
7. After plasmid transfection, cells are infected with the *d120* helper virus. Stocks containing approx 1.1×10^8 virion particles/mL are used and are stored at –80°C and thawed to room temperature prior to use.
8. Dilute vector stocks in DMEM containing 10% NuSerum containing penicillin and streptomycin in different concentrations to generate vector stocks with at least two multiplicities of infection (MOI), or the proportion of viral particles per cell, to ensure optimal cell line transfection. (One MOI may prove more efficient than the other depending on variables such as culture confluence, helper virus concentration, and so on). MOIs should be approx 0.01, 0.03, and 0.1, with lower MOI if the cells are not confluent. Therefore, for a MOI of 0.1, 10^5 virus particles are needed for every 10^6 cells.
9. Media containing lipofectamine and plasmid should be discarded and replaced with media containing helper virus and incubated for 2–3 d.
10. Once cells reach 100% cytopathicity (cells appear ballooned), flasks should be frozen at –80°C.

3.1.2. Harvesting and Purifying Amplicons

1. Thaw flasks partially, as ice crystals will be used to scrape cells off the flask bottoms by shaking.
2. Collect media into sterile test tubes and keep on ice.
3. Sonicate for three 5-s bursts (for T-25 flasks) to release the virus from the cells.
4. Purify viral stocks from cellular components using a sucrose gradient.
5. Transfer the virus into 15-mL centrifuge tubes and spin at 1845*g* for 5 min at 4°C.
6. Overlay supernatants on 2 mL of 25% sucrose placed in ultracentrifuge tubes, and then spin at 75,000*g* for 15 h.

7. Collect the pellet and resuspend in sterile phosphate-buffered saline (PBS; approx 100–200 µL in order to have sufficiently high titers for injection into brain).
8. Aliquot virus into appropriate tubes for future use (usually 20 µL/tube for in vivo ischemia studies), and store at –70°C for up to 1–2 yr. Some vector should be reserved for titering.

3.1.3. Titering Virus

1. Titers of helper virus are determined using a standard plaque assay on E5 cells, whereas titers of amplicons are determined by quantifying the number of β-gal-expressing cells.
2. For helper virus titers, vector samples are plated on E5 cells at dilutions of 10^{-2} and 10^{-3} into DMEM/10% NuSerum/Pen-Strep for each MOI.
3. Incubate vector samples for 1 h, and then add fresh media containing 0.1% human serum to bind excess virus in the media.
4. After 2–3 d, plaques should be visible in the cultures and can be counted.
5. Amplicon titers are determined by incubating Vero cells in dilutions of 10^{-5} to 10^{-6} for 16–24 h.
6. Vero cells are fixed in 0.5% gluteraldehyde and stained with X-gal (a chromogenic substrate of β-galactosidase) to identify cells successfully transfected with the bipromoter vector.
7. Optimal titers are on the order of $1 \times 10^{6-7}$ particles/mL, with ratios of 1:2 to 1:7 for the plasmid and helper virus, respectively (*see* **Note 2**).

3.2 Ischemia Protocols

3.2.1. Vector Delivery

1. Rats are anesthetized with 1–2% isofluorane by face mask and placed in stereotaxic frames.
2. Burr holes are carefully drilled into the skull at prespecified sites, depending on the sterotactic coordinates for the model used.
3. For the focal cerebral ischemia model (described below in **Subheading 3.2.2.**), vector is injected into the striata where infarcts are produced reproducibly. Two injections per striata are made with 2.5–3 µL of vector instilled at each site using a Hamilton syringe (Hamilton, Reno, NV) at a rate of 0.5–1 µL/min either by hand or using a microsyringe pump (Micro4, World Precision Instruments).
4. Stereotactic coordinates from Bregma are: AP 0, ML 3.5 mm, and DV 5 and 6 mm. Either the vector of interest or control vectors are directly injected bilaterally into the striata of rats in order that remaining reporter gene-positive cells in the ipsilateral ischemic striatum can be compared with the contralateral nonischemic striatum, which serves as an internal control.
5. For the global cerebral ischemia model, vectors are targeted to the CA1 region of hippocampi. Coordinates are AP –3.8 mm, ML 1.7 mm, and DV 1.8 mm.
6. First, 3 µL of vector are injected using a microinfusion pump at a rate of 0.5 µL/min (*see* **Note 3**).

7. Vectors can be injected at various time-points before or after ischemia onset, but only one set of injections per animal has been performed to date.
8. Post insult, viral vector delivery showing significant neuroprotection has been carried out by our group up to 2 h after ischemia onset with the 70-kDa heat shock protein (Hsp70) *(15)*. Viral vector delivery at later time-points (5 h) has not been associated with improved neuroprotection using vectors expressing Hsp70(15) or Bcl-2 *(16)*.

3.2.2. Transient Focal Ischemia Model

1. Male Sprague-Dawley rats weighing 290–310 g are anesthetized by face mask with 3% halothane plus oxygen and air, which is then decreased to 1–2% throughout the remainder of the procedure.
2. Ischemia is induced using an occluding intraluminal suture as described in detail previously *(17,18)*.
3. The occluding suture is kept in place for 60 min, resulting in a mild but reproducible striatal insult with variable involvement of the cortex allowing for transcription and translation of the transgenes. It should be noted that longer time periods have not been studied by our group.
4. At the end of the ischemic period, the suture is removed and the animal was allowed to recover.
5. Forty-eight hours later (or at other selected time-points as described previously), the animal is euthanized, and brain sections are prepared for histological analysis.

3.2.3. Global Cerebral Ischemia Model

1. Rats weighing 350–450 g are anesthetized as described above in **Subheading 3.2.2.** and subjected to global cerebral ischemia as described previously *(19)*.
2. The femoral artery and vein are cannulated with PE-50 tubing in order to monitor mean arterial blood pressure (MABP) and to infuse heparin and blood, respectively.
3. The common carotid arteries (CCAs) are exposed via a midline incision in the neck.
4. Heparin (500 IU/kg in normal saline, 100 IU/mL) is infused into the femoral venous catheter.
5. Hemorrhagic hypotension is induced by withdrawal of venous blood until the MABP falls to 30 mmHg.
6. CCAs are temporarily occluded for 8 min using aneurysm clips, and MABP is maintained at 30 mmHg.
7. After 8 min, the clips are removed and the blood returned to the animal.
8. Sham-operated (control) rats undergo similar exposure to anesthesia and surgical manipulation, but the CCAs are not occluded and MABP is not altered by blood withdrawal.
9. After 72 h, animals are euthanized, and the brains are prepared for histological analysis.

3.2.4. Histopathology/Cell Counts

1. Brains are postfixed in a 3% paraformaldehyde/20% sucrose solution for 1–2 d, and 25-µm frozen sections in the coronal plane are taken at 100 mm increments 0.5 mm anterior and 0.5 mm posterior to the infusion sites. This ensures that approx 90% of all transfected cells are collected for counting.
2. Sections are costained with X-gal (to identify vector-infected neurons expressing β-gal) and cresyl violet (to delineate regions of infarction and cell death).
3. The numbers of transfected neurons are counted at 40x magnification using previously established criteria *(17)*.
4. For the focal cerebral ischemia model, the numbers of surviving neurons are expressed as the ratio of X-gal-positive blue neurons in the ischemic striatum compared with the contralateral nonischemic striatum. These ratios can then be used to compare survival in active and control vector treated animals (*see* **Note 4**).

As the extent of vector-mediated infection is limited to only a few hundred striatal neurons within 500 µm of the injection site (Corresponding to an estimated efficiency of transfection on the order of 10%), it was not expected that overall infarct size would be affected (*see* **Note 5**). On the other hand, improvement in vector-infected striatal neuron survival could be explained by smaller infarcts in one group compared with the other. To confirm that improved striatal neuron survival was not caused by an imbalance between groups with regard to the severity of ischemia, the cresyl violet-stained sections were inspected for injury by both gross visual inspection and examination of brain sections at high power (40x objective). Brains were assigned a numeric score on a scale of 0–3, with 0 representing no damage and 3 representing severe damage with involvement of the striatum and surrounding cortex. Animals with no visible damage (score of 0) were excluded from the analysis. For all the studies conducted by our group to date, we have not detected any significant imbalances with regard to infarct scores between groups.

For the global cerebral ischemia model, a parallel set of sham-operated, vector-injected animals are included for comparison. The number of X-gal-positive pyramidal neurons in hippocampus CA1 are counted, as described for the focal model. The numbers of surviving, X-gal-positive neurons in ischemic rats are normalized to the mean number of X-gal-positive neurons in vector-injected, sham-operated rats. The ratios of surviving vector-transfected neurons can then be compared between animals receiving active vector.

3.3. Conclusions

We have shown that gene transfer therapy in experimental stroke is feasible, but it is currently limited by the regions that these viral vectors can infect and the route of administration. Nevertheless, these techniques can provide insights into the pathophysiology of cerebral ischemia that are complementary to studies utilizing transgenic animals. Moreover, these models have been instrumental in establishing that some of the genes induced in response to cerebral injury are in fact neuroprotective when expressed at high levels. Methods of expressing transgene throughout the brain or expressing genes of a diffusible nature can be explored to determine whether gene transfer might reduce overall infarct size and improve neurological recovery. These latter strategies, employed by a few groups, have already shown that this is possible *(28,29)*.

4. Notes

1. *Control vectors.* Two different kinds of control vectors have been used by our group. The first control vector uses the plasmid pα22βgal, which contains only *lacZ*, the reporter gene. This type of control was used for the experiments examining the stress protein Hsp72. The second type of control utilizes the same bipromoter system as the plasmid of interest, except that a stop codon is inserted in the middle of the gene of interest. In this fashion, a truncated, and therefore nonfunctioning, gene product is produced along with the reporter gene. However, this latter control vector has the potential for encoding peptide fragments that may have unanticipated biological activity and for this reason is used less by our group.

2. *Vector generation.* Multiple freeze–thaw cycles will decrease viral titers by approx 10% or more; therefore, it is important to aliquot stocks appropriately. When collecting vector samples for titering, it is advisable to use vector that has been through the same number of freeze–thaw cycles as the vector that will be used in future experiments. (Usually stocks should be subjected to no more than one freeze–thaw cycle.)

3. *Vector delivery.* For optimal dispersal of vector, in order that transfected cells do not "clump" near the injection site, it is helpful to instill vector slowly at rates no faster than those suggested above. Furthermore, it is critical that a pump be used for delivering vector to confined areas such as the hippocampus in order to contain transfecton to the CA1 region. Otherwise, vector will travel along white matter tracts in the corpus callosum or migrate to the dentate. In addition, it is useful to allow the syringe needle to remain in the tissue for an additional minute to allow vector to diffuse, and to avoid tracking vector along the needle tract.

4. *Transgene expression and function.* Transgene expression following intracerebral vector injection is observed as early as 4 h and peaks at about 9–24 h, with persistent expression of rare cells by 7 d *(5,11,16,20)*. Expression decreases shortly after 24 h, but counts are still within 25–30 % of peak counts from d 2–4. Optimal time-points for counting surviving transfected cells for the ischemia models are 1–3 d, taking into consideration that ischemic lesions may not mature for a few days and CA1 neuron loss is not typically seen in the global model until 3 d. The reasons for transient expression from viral vectors are not well known but are thought to be owing to promoter inactivation or, in the case of herpes simplex, viral latency. We previously found that even before viral DNA degrades, vectors peripheralized to the nucleus at the time transcription stopped, suggesting that shortened gene expression might be owing to extrusion of the vectors

from the nucleus *(21)*. It should also be noted that the tissue half-life of β-gal is particularly long and may persist well after transcription and protein translation has ceased *(22,23)*. For instance, vectors expressing Hsp70 begin to express at the same time as the *lacZ* reporter, with a similar peak expression. However, vector-mediated Hsp70 protein is no longer observed after approx 48 h, whereas β-gal is observed for several days more *(9)*. Therefore, it is important to confirm expression not only of the reporter but also the gene product of interest as well. Reporter gene expression in ischemic brain appears to follow a similar time-course as that in uninjured brain.

Although we show that the number of transfected neurons is greater in the groups receiving the treatment vector, an important issue is whether these neurons are physiologically normal. Prior studies indicate that transfection with a vector expressing the glucose transporter gene (*glut-1*) increases glucose uptake compared with control vectors *(24)*. Furthermore, gene transfer of the calcium binding protein calbindin D28K leads to improved neuronal function *(25)* and behavior *(26)* in rats subjected to kainic acid toxicity. Overexpression of the antioxidant genes catalase, glutathione peroxidase and superoxide dismutase using these vectors led to increased enzymatic activity of each protein using standard assays and was associated with less neuronal damage in culture (R. Sapolsky, personal observations).

5. *Vector delivery and extent of transfection.* A significant problem common to several areas of research is that of vector delivery and extent of transfection. In our hands, intravascular injection resulted in uptake by only endothelial cells, and did not penetrate the blood–brain barrier (BBB) (R. Sapolsky, personal observations). Doran et al. *(27)* combined adenoviral vectors with the osmotic agent mannitol to open the BBB transiently but only saw expression within pericapillary astrocytes. Intraventricular injection resulted in uptake only by ependymal cells; therefore, for the candidate genes studied by our group to date, direct cerebral injection has been the only reliable method to transfect neurons. It should also be pointed out that the vectors only transfect a limited number of neurons (a few hundred per nonischemic striatum); therefore, overall infarct size is not altered, and certainly behavioral scores in the cerebral ischemia models may be meaningless.

Acknowledgments

The authors would like to thank J. MClaughlin, H. Zhao, and S. Kelly for expert technical assistance and advice and Elizabeth Hoyte for preparation of figures. This work was supported by NIH NINDS grants RO1 NS40516 (to M.A.Y.) and P01 NS37520 (to M.A.Y. and R.M.S.) and the Adler Foundation (to R.M.S.).

References

1. Sapolsky, R. M. and Steinberg, G. K. (1999) Gene therapy using viral vectors for acute neurologic insults. *Neurology* **53,** 1922–1931.

2. Yenari, M. A., Dumas, T. C., Sapolsky, R. M., and Steinberg, G. K. (2001) Gene therapy for treatment of cerebral ischemia using defective herpes simplex viral vectors. *Neurol. Res.* **23,** 543–552.

3. Karpati, G., Lochmuller, H., Nalbantoglu, J., Durham, H. (1996) The principles of gene therapy for the nervous system. *Trends Neurosci.* **19,** 49–54.

4. Glorioso, J. C., Goins, W. F., Meaney, C. A., Fink, D. J., DeLuca, N. A. (1994) Gene transfer to brain using herpes simplex virus vectors. *Ann. Neurol.* **35 (suppl.),** S28–S34.

5. Yenari, M., Fink, S., Lawrence, M., et al. (1998) Gene transfer therapy for cerebral ischemia, in *Pharmacology of Cerebral Ischemia* (Krieglstein, J., ed.), Medpharm Scientific Publishers, Stuttgart, pp. 453–465.

6. DeLuca, N. A., McCarthy, A. M., Schaffer, P. A. (1985) Isolation and characterization of deletion mutants of herpes simplex virus type 1 in the gene encoding immediate-early regulatory protein ICP4. *J. Virol.* **56,** 558–570.

7. Fink, D. J., DeLuca, N. A., Goins, W. F., Glorioso, J. C. (1996) Gene transfer to neurons using herpes simplex virus based vectors. *Annu. Rev. Neurosci.* **19,** 265–287.

8. Ho, D. Y., Fink, S. L., Lawrence, M. S., et al. (1995) Herpes simplex virus vector system: analysis of its in vivo and in vitro cytopathic effects. *J. Neurosci. Methods* **57,** 205–215.

9. Yenari, M. A., Fink, S. L., Sun, G. H., et al. (1998) Gene therapy with HSP72 is neuroprotective in rat models of stroke and epilepsy. *Ann. Neurol.* **44,** 584–591.

10. Ho, D. Y. (1994) Amplicon-based herpes simplex virus vectors. *Methods Cell Biol.* **43,** 191–219.

11. Fink, S. L., Chang, L. K., Ho, D. Y., Sapolsky, R. M. (1997) Defective herpes simplex virus vectors expressing the rat brain stress-inducible heat shock protein 72 protect cultured neurons from severe heat shock. *J. Neurochem.* **68,** 961–969.

12. Lawrence, M. S., Ho, D. Y., Sun, G. H., Steinberg, G. K., Sapolsky, R. M. (1996) Overexpression of Bcl-2 with herpes simplex virus vectors protects CNS neurons against neurological insults in vitro and in vivo. *J. Neurosci.* **16,** 486–496.

13. Ho, D. Y., Lawrence, M. S., Meier, T. J., et al. (1995) Using of herpes virus vectors for protection from necrotic neuron death, in *Viral Vectors* (Kaplitt, M. G. and Loewy, A. D., eds.), Academic, New York, pp. 133–155.

14. Ho, D. Y., Saydam, T. C., Fink, S. L., Lawrence, M. S., and Sapolsky, R. M. (1995) Defective herpes simplex virus vectors expressing the rat brain glucose transporter protect cultured neurons from necrotic insults. *J. Neurochem.* **65,** 842–850.

15. Hoehn, B., Ringer, T. M., Xu, L., et al. (2001) Overexpression of HSP72 after induction of experimental stroke protects neurons from ischemic damage. *J. Cereb. Blood Flow Metab.* **21,** 1303–1309.

16. Lawrence, M. S., Sun, G. H., Ho, D. Y., Sapolsky, R. M., Steinberg, G. K. (1997) Herpes simplex viral vectors expressing Bcl-2 are neuroprotective when delivered following a stroke. *J. Cereb. Blood Flow Metab.* **17,** 740–744.

17. Yenari, M. A., Minami, M., Sun, G. H., Meier, T. J., Kunis, D. M., McLaughlin, J. R., et al. (2001) Calbindin d28k overexpression protects striatal neurons from transient focal cerebral ischemia. *Stroke* **32,** 1028–1035.

18. Yenari, M., Palmer, J., Sun, G., de Crespigny, A., Moseley, M., Steinberg, G. (1996) Time-course and treatment response with SNX-111, an N-type calcium channel blocker, in a rodent model of focal cerebral ischemia using diffusion-weighted MRI. *Brain Res.* **739**, 36–45.
19. Kelly, S., Zhang, Z. J., Zhao, H., et al. (2002) Gene transfer of HSP72 protects cornu ammonis 1 region of the hippocampus neurons from global ischemia: influence of Bcl-2. *Ann. Neurol.* **52**, 160–167.
20. McLaughlin, J. R., Ho, D. Y., and Sapolsky, R. M. (1997) An adenovirus vector expressing lacZ for gene transfer into the hippocampus and striatum. *Soc. Neurosci. Abs.* **23**, 1622.
21. Tsai, D. J., Ho, J. J., Ozawa, C. R., Sapolsky, R. M. (2000) Long-term expression driven by herpes simplex virus type-1 amplicons may fail due to eventual degradation or extrusion of introduced transgenes. *Exp. Neurol.* **165**, 58–65.
22. Schafer, H., Schafer, A., Kiderlen, A. F., Masihi, K. N., Burger, R. (1997) A highly sensitive cytotoxicity assay based on the release of reporter enzymes, from stably transfected cell lines. *J. Immunol. Methods* **204**, 89–98.
23. Neve, R. L. (1993) Adenovirus vectors enter the brain. *Trends Neurosci.* **16**, 251–253.
24. Ho, D. Y., Mocarski, E. S., Sapolsky, R. M. (1993) Altering central nervous system physiology with a defective herpes simplex virus vector expressing the glucose transporter gene. *Proc. Natl. Acad. Sci. USA* **90**, 3655–3659.
25. Dumas, T. C., McLaughlin, J. R., Ho, D. Y., Lawrence, M. S., Sapolsky, R. M. (2000) Gene therapies that enhance hippocampal neuron survival after an excitotoxic insult are not equivalent in their ability to maintain synaptic transmission. *Exp. Neurol.* **166**, 180–189.
26. Phillips, R. G., Monje, M. L., Giuli, L. C., et al. (2001) Gene therapy effectiveness differs for neuronal survival and behavioral performance. *Gene Ther.* **8**, 579–585.
27. Doran, S. E., Ren, X. D., Betz, A. L., et al. (1995) Gene expression from recombinant viral vectors in the central nervous system after blood-brain barrier disruption. *Neurosurgery* **36**, 965–970.
28. Zhang, W. R., Sato, K., Iwai, M., Nagano, I., Manabe, Y., Abe, K. (2002) Therapeutic time window of adenovirus-mediated GDNF gene transfer after transient middle cerebral artery occlusion in rat. *Brain Res.* **947**, 140–145.
29. Betz, A. L., Yang, G. Y., Davidson, B. L. (1995) Attenuation of stroke size in rats using an adenoviral vector to induce overexpression of interleukin-1 receptor antagonist in brain. *J. Cereb. Blood Flow Metab.* **15**, 547–551.

5

Stem Cell Transplantation After Middle Cerebral Artery Occlusion

Paul Stroemer and Helen Hodges

Summary

Stem cell lines have been and are being developed to treat damage in the central nervous system after stroke. Stem cells are able to migrate to areas of damage and to differentiate into neurons and glia. Grafts of murine stem cells have been shown to promote recovery from behavioral dysfunction after stroke. We have developed protocols to optimize behavioral testing, animal recovery, and stem cell delivery after middle cerebral artery occlusion. In this chapter we discuss study protocols aimed at integrating in vitro preparation of cells, small animal surgery, behavioral testing batteries, and histological analysis.

KEY WORDS:

Stem cell; cell grafting; middle cerebral artery occlusion; behavioral analysis.

1. Introduction

Treatment of stroke has focused primarily on stopping or reducing the primary pathologies after onset. Much time and energy have been spent with agents to limit ischemic damage by excitotoxic, oxidative, and apoptotic mechanisms. Unfortunately, these various approaches have met with limited success in the clinic, with the only current effective stroke treatment being the use of tissue plasminogen activators to eliminate thrombosis and restore blood supply (1). Even with the advent of effective therapies to limit ischemic damage, the core of the stroke may not be salvageable. The resulting cell death will almost certainly lead to behavioral dysfunction. Poststroke treatment will attempt to promote behavioral recovery and stimulate neuronal plasticity (2,3).

From: *Methods in Molecular Medicine, Vol. 104: Stroke Genomics: Methods and Reviews*
Edited by: S. J. Read and D. Virley © Humana Press Inc., Totowa, NJ

Restoration of damaged brain tissue has been attempted by a variety of methods. Blocks of solid tissue have been implanted into animals following occlusion in an attempt to "fill the hole" left by the lesion. Solid tissue transplants have been used since 1905 *(4)*; however, the problems of integrating the graft into the parenchyma have not been resolved. Cortical grafts have been placed into areas of damaged cortex, and tissue survival has been observed. There is some integration of the tissue with afferent and efferent connections with the host brain *(5)*. Recovery of function with tissue blocks implanted into stroked animals was greatly enhanced by enriched environmental conditions, suggesting that the enriched environment was of a greater value than the graft *(5)*. Hippocampal cell suspension grafts have been shown to replace damaged tissue and promote behavioral recovery after four-vessel occlusion (4VO). For the grafts to be successful, they have to be homologous to the damaged regions *(6,7)*.

The use of stem cells to promote behavioral recovery after stroke has several advantages over primary solid tissue or cell suspension grafts. First, cells can be grown in adequate amounts for large-scale clinical use. This supply also increases the reliability of cell supply compared with arranging delivery of donor tissue. Stem cell populations can be expanded almost indefinitely in vitro, but cells will differentiate in vivo. Clonal cell lines can be thoroughly characterized, increasing reliability in differentiation. Primary fetal tissue grafts can have several cell types in its populations that may have unwanted growth characteristics *(8)* and may carry pathogens unless they are highly screened. Preparation of cells from the original tissue may also damage cells, reducing their viability for grafting. The stem cells are multipotential, turning into neurons and glia *(6)*. The cells migrate, allowing flexibility of placement, and once they have "settled" achieve an appropriate phenotype *(9)*, whereas primary grafts, either in suspension or solid, have difficulty integrating into the brain.

We have been developing conditionally immortalized stem cell lines for the treatment of neurodegenerative conditions. Stem cells that have been isolated from primary tissue enter a period of senescence where they stop dividing and die. Some cells may pass this period of senescence, become "immortal," and continue to divide. Cell lines can be "immortalized" by the insertion of oncogenes that will allow persistent replication. Although they provide expansive numbers for grafting, these immortal cells may be tumorigenic when implanted into the brain. The MHP36 cell line is conditionally immortalized in that it is temperature-sensitive, meaning it will divide when grown at 33°C, but once cells are maintained at a temperature of 37°C they stop dividing and will differentiate into neuronal or glial phenotypes. This switch allows for the expansion and growth of the cell line in vitro and safety in vivo *(7)*. The hope of the use of stem cells in treating stroke or other central nervous system (CNS) diseases is that new cells can replace or augment damaged tissue. This replacement of tissue should lay the anatomical foundation for improved behavioral recovery or the replenishment of lost neurotransmitters.

The location of the graft site itself may be of utmost importance in determining cell survival *(9)*. The hemisphere ipsilateral to the lesion is undergoing massive remodeling after the onset of damage caused by occlusion. Dead or damaged tissue will be removed by the neuroimmune system, resulting in deformation of the striatum and cortex and expanded cerebral ventricles. Severe infarcts may leave little viable tissue around the infarction to host the grafts, and the tissue when sectioned may bear little relation to brain atlases. However, the surrounding tissue has axonal growth and synaptogenesis following damage that should promote cell integration into the surviving tissue.

Grafting of cells contralateral to the lesion places them in an environment that is more conducive to cell survival than the immediate lesion. Cells placed into the contralateral hemisphere are able to migrate across the brain to the damaged hemisphere. It is possible that the grafted cells are attracted toward growth factors in the damaged areas of the brain. It should be noted, however, that the "healthy" hemisphere also has remodeling owing to the loss of input from the damage with synaptogenesis in the weeks after damage *(10)*. This remodeling may provide growth factors as well because cells grafted ipsilateral to the damage migrate across to the healthy hemisphere. Grafting of cells into the contralateral hemisphere resulted in a higher rate of survival compared to the ipsilateral grafts *(9,11)*.

Grafting stem cells into stroke animals requires quite involved protocols and the interaction of varied specialties. An example of the elements involved is shown in **Fig. 1**. Combining these areas will place heavy burdens on time for animal and histological groups. Extra technical staff may be needed to complete all tasks involved. The key element for all involved is the strict adherence to timelines and the consistent time and grafting of cells. An example of a study timeline is shown in **Fig. 2**.

2. Materials

1. Coco Pops (Kellogg's).
2. Staircase test apparatus.
3. Parcel tape.
4. *d*-Amphetamine sulfate (Sigma).
5. Rotameter (TSE, Germany).
6. 3-0, 4-0 prolene and 4-0 vicryl (Ethicon).
7. Clear silicone sealant (Halfords, UK).
8. Halothane (Concord Pharmaceuticals, UK).
9. Duphalyte (Fort Dodge Animal Health, UK).
10. Saline (Baxter, UK).
11. Complan (Complan Foods, UK).

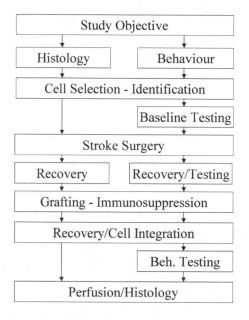

Fig. 1. Flowchart for determination of requirements for a stroke stem cell study. Behavioral testing needs to be integrated into the study design during planning stages.

12. *N*-acetyl-L-cysteine (Sigma).
13. Medrone (Pharmacia).
14. Cyclosporin A (CSA; Sandoz).
15. Cremaphore El (Sigma).
16. Hemacytometer (Hausser Scientific).
17. Trypan blue (Sigma).
18. Water maze (HVS Image, UK).
19. Pentoject (Animal Care, UK).
20. Heparin (CP Pharmaceuticals, UK).
21. Paraformaldehyde (Sigma).
22. Primary antibodies (Chemicon).
23. Secondary antibodies (Molecular Probes).
24. Confocal microscope (Leica).
25. Image Pro Plus (Media Cybernetics).

3. Methods

The methods described here outline the following procedures: (1) prelesion animal assessment, (2) middle cerebral artery occlusion, postocclusion care, and testing; (3) cell grafting and immunosuppresion; (4) timing of a behavioral testing battery; and (5) histological analysis of lesion volume and cell survival and differentiation.

Study Weeks

1	2	3	4	5	6	7-10	11	12	13	14	15

Week	Activity
1	Animals into unit
2	Base line testing
3	Surgery
4-5	Recovery
5	Testing
6	Grafting
7-10	Recovery/Integration
10-14	Sensori-motor Testing
13-14	Morris Water Maze
15	Rotameter Testing
	Perfusion

Fig. 2. Typical time-course for a 14–15-wk study. Histological studies may have varied timings to evaluate cell survival and differentiation at different time-points.

3.1. Preocclusion Activity

3.1.1. Cell Line Selection

Cell line candidates should be chosen before the study begins and sufficient quantities of cells grown to ensure that grafting occurs on time. Rodent cell lines will need to either be genetically modified to express a protein marker (e.g., βgal) or be labeled with a fluorescent marker (e.g., PKH26) to allow cell identification. Cell preparation techniques should be optimized before animals are occluded to ensure there is good cell viability and consistency in delivery at the time of grafting.

3.1.2. Animal Arrival Quarantine

Animals arrive from the supplier (Charles River) and are housed in quarantine for 5 d. Animals are examined daily for any nasal discharges or signs of respiratory distress.

3.1.3. Pretraining Behavioral Battery

After release from quarantine, animals are tested twice on the Bilateral Asymmetry Test before occlusion to establish baseline behavior.
1. Rotational bias is tested in a rotameter with injection of *d*-amphetamine sulphate (2.5 mg/kg sc).
2. Rotations are recorded from 30 to 60 min after injection.
3. If animals demonstrate spontaneous bias with *d*-amphetamine sulphate (>70% of

rotations in either a clockwise or counterclockwise direction), they are removed from the study.
4. Training in paw reaching using a staircase task is conducted for a total of 10 working days.

3.1.4. Bilateral Asymmetry Test

This test assesses sensorimotor neglect arising from cortical involvement in stroke damage.
1. A strip of tape approx 1 × 5 cm is placed around each front paw of the rat at the wrist, alternating the paw first taped.
2. The rat is then placed in a box for 5 min, and a timer is started.
3. The experimenter records the following measures:

   a. Latency to contact a tape (defined as a bite or a scratch) for each paw.
   b. Latency to remove a tape (defined as complete removal) for each paw.

4. If the animal does not contact or remove one or both tapes after 5 min, a latency of 5 min (300 s) is recorded.
5. Each rat is tested for two consecutive 5-min trials *(11)*.

3.1.5. Rotation

Biased rotation is a reliable measure of motor asymmetry (striatal damage) after middle cerebral artery occlusion (MCAO), which persists indefinitely. The rotameter systems can pick up both spontaneous rotation bias (typically contralateral to the damage, possibly a weakened forepaw) and rotation induced by amphetamine, which results from increased dopamine stimulation on the intact side.
1. The rotameter consists of a set of eight bowls, each with a harness and swivel connected to a computer, which records the direction of the rats' movements.
2. Drugs or vehicle are given, and the animals are placed in the bowls and tethered for 30 min prior to recording of movements.
3. The movements are recorded automatically for 30 min.
4. We use a dose of 2.5 mg/kg of *d*-amphetamine sulfate in a volume of 1.0 mL/kg sc. This dose increases rotation activity to a similar extent in control and stroke animals, but with a marked ipsilateral bias in stroked rats *(11)*.

3.1.6. Paw Reaching

Fine motor skills are tested using a staircase apparatus. This task requires substantial pretraining with a minimum of 10 training runs prior to the occlusion.
1. Rats are placed in a testing chamber consisting of a cage with a platform ledge having steps on either side. Each step has a small depression, allowing the placement of a food reward. Our practice has found breakfast cereal (Cocoa Puffs) to be a reliable reward.
2. Two rewards are placed on each step.
3. The rat is introduced into the cage and allowed to reach for the rewards for 5 min.

4. After the testing period, the steps are withdrawn and the numbers of disturbed and consumed pellets are recorded for each side.
5. The food rewards are introduced to the home cages prior to the testing days, allowing the animals to become accustomed to the reward (*see* **Note 1**).

3.2. Occlusion and Group Selection

3.2.1. Middle Cerebral Artery Occlusion

1. Animals (male Sprague-Dawley rats 300–350 g) are subjected to 70 min of MCAO using the filament insertion model *(12)* under halothane (in 70%/30% NO_2/O_2) anesthesia.
2. The filament is coated with silicone at the tip to a diameter of 0.38 mm to ensure that the base of the middle cerebral artery (MCA) is blocked at the circle of Willis.
3. The filament is inserted 19–21 mm via the common carotid trunk caudal to the bifurcation of the internal and external branches. The external branch is not ligated or cauterized to prevent back blood flow.
4. This approach to insertion results in the permanent ligation of the ipsilateral common carotid artery (CCA), which in our experience results in better consistency in lesion placement and volume with resulting behavioral dysfunction.
5. Surgery is performed with animals placed on homoeothermic blankets to maintain the temperature at $36 \pm 1°C$.
6. After insertion of the filament, animals are removed from halothane anesthesia and allowed to recover.
7. Animals are kept in a heated chamber during the awake occlusion period and for 2 h after filament removal.
8. Torso twist and circling activity are recorded at 60 min of occlusion to establish neurological deficit.
9. If animals fail to demonstrate these abnormal behaviors, they are culled from the study.
10. At the end of the occlusion period, animals are reanesthetized, the filament is retracted, the excess is trimmed off, and the wound is closed.
11. Sham-operated animals have the filament inserted into the CCA to a depth of 10 mm and then retracted.

3.2.2. Postoperative Care and Neurological Scoring

1. The first 2 d after MCAO animals receive a combination of Duphalyte and 4% glucose/saline (50:50, 5 mL, sc, bid).
2. Animals are fed a combination of Complan and wet mash until body weight recovers to preocclusion levels *(13)*.
3. Regular food pellets are placed in the cage hoppers starting at 2 d post-stroke.
4. Animals are housed singly on paper cage liners and scored daily for neurological dysfunction using a 12-point examination protocol for a minimum of 7 d (Table 1).
5. Animals that do not exhibit neurological dysfunction may be marked for removal from the study.

Table 1
Postoperative Care Sheet[a]

Day

	1	2	3	4	5	6	7	8	9	10	11	12	13	14
Grasping														
Placing reaction														
Visual placing														
Righting reflex														
Tilted cage top														
Horizontal bar test														
Spontaneous motility														
Circling														
Tail lifting														
Weight (g)														
Food intake														
Feces														
Injections														
Initials														

[a]Neurological scoring is performed daily, for a minimum of 7 d and then until animals are placed into group housing. Weight, food intake, and feces monitoring indicate the general health of animals.

6. Eating behavior, body weight, urine, and fecal production are also recorded to indicate general health status.
7. Once animals recover preoperative body weight, they are housed on sawdust for 2 d, and the Complan/mash diet is withdrawn.
8. If body weight is maintained over this transition period, the animal is then group housed.
9. Further information concerning post-MCAO animal care can be found in Modo et al. *(13)*.

3.2.3. Postocclusion Behavioral Testing

Once animals reach preoperative body weight, they have a round of testing for bilateral asymmetry, paw reaching, and rotameter behavior. Rotameter testing must be conducted with a minimum 10-d delay after occlusion to prevent animals from rotating contralateral to the expected direction.

3.2.3.1. Balancing Groups for Dysfunction

1. Data from rotational bias on the rotameter are used to balance dysfunction between experimental groups, as they have the least confounding variables.
2. Groups are developed to include a sham-operated group, an occluded group grafted with vehicle, and occluded groups grafted with cells as necessary.
3. Available animals are ranked according to bias and treatment groups of equivalent bias selected.

3.2.3.2. Removal of Nonimpaired Rats

Any stroked animal that does not display dysfunction in the bilateral asymmetry test or rotational bias in the expected direction is removed from the study. Similarly, any sham-operated animals displaying rotational bias are removed from the study.

3.2.4. Pregraft Immunosuppresion With CSA

Animals are injected with CSA suspended in cremaphore El (CSA 10 mg/kg sc) 24 h before grafting.

3.3. Grafting of Stem Cells

3.3.1. Graft and Stereotaxic Placement of Cells

1. Animals are anesthetized with halothane or injectable anesthetic, prepared for surgery, and placed into a stereotaxic frame (*see* **Note 2**).
2. The skin is incised, and the skull is exposed.
3. Bregma is located, and holes are drilled in the desired anterior and lateral stereotactic coordinates *(16)* for exposure of the dura.
4. Cells suspended in *N*-acetyl-L-cysteine are drawn into a Hamilton syringe (10 µL).
5. The syringe is lowered into the brain to the desired ventral coordinates, cells are injected (2 µL/2 min), and the needle is left in place for another 2 min (*see* **Note 3**).

6. Once all injections are made, the wound is closed with subcutaneous vicryl sutures *(9)*.
7. *See* **ref.** *13* for typical ipsilateral, contralateral, and intraventricular coordinates.

3.3.2. Immunosuppresion

1. Once animals are grafted, they receive injection of CSA (10 mg/kg sc) and medrone (20 mg/animal sc).
2. To avoid dehydration, animals are also injected with 4% glucose saline (5 mL sc).
3. If an injectable anesthetic is used, a reversing agent is given.

3.3.3. Cell Viability

1. At the end of the grafting session, cell viability is checked using trypan blue exclusion, and the percentage of live cells is recorded (*see* **Note 4**).

3.3.4. Post Grafting Care

1. Animals are housed singly on paper liners and injected with Duphalyte/glucose saline (5 mL sc) for 2 d after surgery.
2. Animals are again fed on a diet of mash and Complan until preoperative body weight is reached.
3. The immunosuppresion regime continues with medrone (10 mg/animal sc for 14 d) and CSA (10 mg/kg sc) in cremaphore every 3 d for the rest of the study.
4. Once animals reach preoperative body weight, they are housed on sawdust, and the complan/mash diet is removed.
5. If animals maintain body weight for 2 d they can then be group housed.

3.4. Behavioral Testing

3.4.1. Animal "Holiday"

Animals are not tested for 6 wk after grafting to allow for the migration, differentiation, and integration of cells into the parenchyma and to avoid "task learning".

3.4.2. Sensorimotor Testing

At 6–12 wk, post graft, animals begin bilateral asymmetry testing with two trials a week. Paw reaching with the staircase task is tested once or twice a week.

3.4.3. Cognitive Testing

1. At 11 and 12 wk post graft, animals are tested on the Morris Water Maze with two trials a day, maximum swim time 60 s over 10 working days. This procedure assesses spatial learning and memory, which are adversely affected by both hippocampal and parietal cortex damage after stroke.

2. The apparatus consists of a circular, black plastic tank (2 m diameter, 500 mm high), filled with water (at $26 \pm 2°C$) to a depth of 250 mm.
3. The rat's task is to find a 100-mm platform made of Plexiglas mounted 20 mm below the water surface.
4. The maze is conceptually divided into three concentric circles (annuli A, B, and C, with A in the center) and four quadrants (1–4). The platform is located in the middle of one of the quadrants within annulus B. Four start positions designated N, S, E, and W are used in quasirandom order.
5. Rats are placed in the pool, facing the wall, and allowed to swim for 1 min. If they find the platform within this time they are allowed to remain there for 10 s; if they fail to find the platform, they are guided to and placed on it for 10 s.
6. They are then removed either to a holding cage, or, after being towelled dry, to a home cage.
7. The swim path is recorded by an image analyzing system (HVS Ltd.). Latency to reach the platform, distance swum, speed, and time spent in each quadrant/annulus are calculated and saved on disk.
8. For standard acquisition, two trials are given each day, with a 10-min intertrial interval.
9. At the end of training (10 d), rats are given a probe trial, with the platform removed, to assess memory for its location by the percentage of time spent in the training quadrant (**9**) and time spent in the platform area.

3.4.4. Rotameter Testing

At 13 and 14 wk post graft, animals are tested in the rotameter with three trials of *d*-amphetamine sulfate (2.5 mg/kg, sc) and three trials of saline.

3.4.5. Tissue Retrieval

1. At 15 wk post graft, animals are sacrificed for histological examination.
2. Animals are exsanguinated with heparinized saline (5000 U/L saline) followed by perfusion with fixative (4% paraformaldehyde in phosphate buffer).

3.5. Histological Processing

3.5.1. Tissue Preparation

1. After 24 h in 4% paraformaldehyde fixative, the brains are placed in two changes of water to remove excess fixative solution, allowing for safer handling during processing.
2. The brains are then placed in a solution of 30% sucrose (w/v) in phosphate-buffered saline for cryoprotection. Once brains sink in the sucrose, they are ready for freezing and sectioning.
3. It should be noted that chronically stroked brains may be quite fragile, with sizable lesions, and may need extra time in paraformaldehyde before cutting (**Fig. 3**).

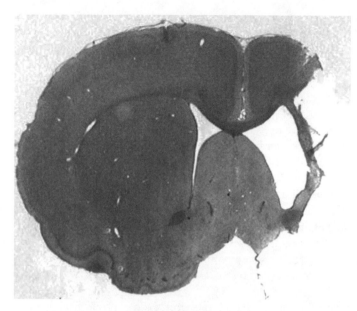

Fig. 3. Brain section taken from animal 4 mo after MCAO (Cresyl violet staining, 50-mm section). The striatum and cortex ipsilateral to the infarction have degenerated, and the ventricles have enlarged.

3.5.2. Cutting

1. Sections are cut at 50 µm on a sledge microtome with every 20th section mounted on gelatine-coated slides for determining lesion volume.
2. The remaining sections are collected serially and placed in 30% sucrose to allow for freezing until further use.

3.5.3. Lesion Volume Determination

1. Lesion volume is assessed in images of unstained whole brain sections taken from a Leica microscope by a digital video camera using Image Pro Plus software.
2. Ten sections are used to estimate lesion volume using Simpson's rule by subtracting the area of the damaged hemisphere from the area of the intact hemisphere.
3. All images are taken at the same magnification (×1.6) and light-intensity settings in a single session.
4. Computerized system allows for calculation of missing or removed tissue.

3.5.4. Cell Identification

1. Transplanted human cells are identified using immunohistochemistry.
2. Primary antibodies raised against human cell nuclei are incubated overnight at room temperature before a secondary antibody is applied for 1 h.

3. Rodent cells labeled with fluorescent markers (e.g., PKH26) can be identified directly using fluorescence microscopy.
4. To determine graft survival semiquantitatively, transplanted cells in the same focal plane are counted (×400 magnification) bilaterally in three regions of interest (ROIs): somatosensory cortex, striatum, and septum at the level of implantation in a rectangular field (0.202 × 0.252 mm) to provide an appreciation of graft survival within areas equivalent relative to implantation tracts in each ROI for each brain.
5. Since it is difficult to measure thinly scattered grafted cells, the purpose of standardized ROI cell counts is not to estimate the total number of grafted cells, but to obtain an accurate count within selected ROIs across groups (9).
6. If more accurate cell counts are required, stereological counting methods should be used.

3.5.5. Phenotyping

Once cell survival has been quantified, the transplanted cells can be phenotyped to evaluate differentiation; examples of identified cells are shown in **Fig. 4**. The remaining sections may be run through immunohistochemical protocols to double-label human nuclei and cellular markers. Exact immunohistochemical protocols can differ between laboratories and depend on the antibodies used in the study and the supplier's recommendations. Suggested markers are listed in Table 2 (9).

3.6. Data Analysis

3.6.1. Bilateral Asymmetry, Paw Reaching, Morris Water Maze

The bilateral asymmetry, paw reaching, and Morris Water Maze data are analyzed using repeated-measures ANOVA with Bonferroni's *post hoc* tests to compare group means.

3.6.2. Lesion Volumes and Rotameter Data

Lesion volumes and rotameter data are compared between groups by a one-way ANOVA followed by Bonferroni tests.

4. Notes

1. Some animals might not eat the pellets owing to damage in the gustatory cortex and may lose the motivation to eat "treats." Similarly, it may not be appropriate to compare performance between sides within animals, as ipsilateral performance may be affected after MCAO. If an animal is not eating pellets from either side of the apparatus, it may be necessary to assess its status outside of the testing period.
2. Anesthesia selection for grafting surgery must be made to avoid impairing behavioral recovery, as some compounds (e.g., benzodiazepines) will cause a regression in neurological recovery after stroke (14).

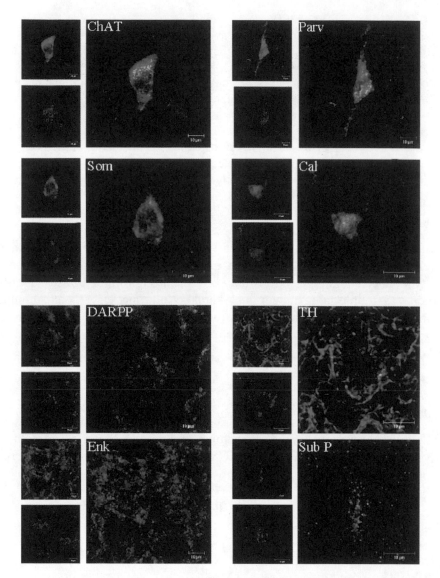

Fig. 4. Confocal images of neuronal phenotypes of transplanted cells in the striatum. Transplanted cells (PKH26 in red) appropriately differentiated into interneuronal phenotypes expressing choline actyl transferease (ChAT), parvalbumin (Parv), somatostatin (Som), and calretinin (Cal). Transplanted cells also differentiated into striatal GABAergic output neurons by DARPP-32 (a dopamine and cyclic-AMP-regulated phosphoprotein), and were inclosed by markers for substance P (sub P), tyrosine hydroxylase (TH), and (ENK) enkephalin. The staining shown indicates that transplanted cells differentiate into site-appropriate phenotypes and integrate into the host neuropil. (Reproduced with permission from **ref. 9.**)

Table 2
Primary Antibodies[a]

Cell type	Marker	Supplier
Nondifferentiated stem cells	Nestin	Chemicon
Neurons	NeuN	Chemicon
Newly differentiated neurons	Doublecortin	Chemicon
Astrocytes	GFAP	DAKO
Dividing cells	KI67	Novacastra
Cholinergic cells	ChAT	Chemicon
GABAergic output neurons	DARPP-32	Santa Cruz
Dopaminergic fibers	Tyrosine hydroxylase	Chemicon
Human cells	Human nuclear	Chemicon

[a]Suggested markers for phenotyping differentiated cells using immunohistochemistry and antibody suppliers.

ChAT, choline acetyltransferase; DARPP-32, dopamine and cAMP-regulated phosphoprotein-32; GABA, γ-aminobutyric acid; GFAP, glial fibrillary acidic protein.

3. **Caution:** any work with genetically modified cells *must* be approved by the local Genetic Modification Safety Committee. Procedures for cell handling, codes of practice, and facilities must be in place before grafting of modified cells.
4. Cell viability must be determined at the time of delivery by the supplying party. Grafting of damaged or dead cells can exaggerate behavioral deficits and increase lesion volume after grafting *(15)*.

References

1. Hofmiejer J., Van Der Worp H., and Kappelle J. (2003) Treatment of space-occupying cerebral infarction. *Crit. Care Med.* **31,** 617–625.
2. Stroemer R. P., Kent T., and Hulsebosch C., (1998) Enhanced neocortical neural sprouting, synaptogenesis, and behavioural recovery with D-amphetamine therapy after neocortical infarction in rats. *Stroke* **29,** 2381–2393.
3. Chen P., Goldberg D., Kolb B., Lanser M., Benowitz L. (2002) Inosine induces axonal rewiring and improves behavioral outcome after stroke. *Proc Natl Acad Sci USA* **99,** 9031–9036.
4. Saltykow, S. (1905) Versuche uber Gehirnreplantation. *Arch. Psychatr. Nervenkrankheiten* **40,** H11.
5. Grabowski M., Sorensen J., Mattsson B., Zimmer J., Johansson B. (1995) Influence of an enriched environment and cortical grafting in functional outcome in brain infarcts of adult rats. *Exp. Neurol.* **133,** 96–102.
6. Gray J. A., Grigoryan G., Virley D., Patel S., Sinden J. D., Hodges H. (2000) Conditionally immortalized multipotential and multifunctional neural stem cell lines as an approach to clinical transplantation. *Cell Transplant.* **9,** 153–168.

7. Gray J. A., Hodges H., and Sinden J. D. (1999) Prospects for the clinical application of neural transplantation with the use of conditionally immortalized neruoepithelial stem cells. *Philos. Trans. R. Soc. Lond.* **354,** 1407–1421

8. Freed C., Greene P., Breeze R., et al. (2001) Transplantation of dopamine neurons for severe Parkinson's disease. *N. Engl. J. Med.* **344,** 701–709.

9. Modo M., Stroemer R. P., Tang E., Patel S., Hodges H. (2002) Effects of implantation site of stem cell grafts on behavioural recovery from stroke damage. *Stroke* **33,** 2270–2278.

10. Jones T. and Schallert T. (1992) Overgrowth and pruning of dendrites in adult rats recovering from neocortical damage. *Brain Res.* **581,** 156–160.

11. Veizovic T., Beech J., Stroemer R., Watson W., Hodges H. (2001) Resolution of stroke deficits following contralateral grafts of conditionally immortal neuroepithelial stem cells. *Stroke* **32,** 1012–1019.

12. Longa E., Weinstein P., Carlson S., Cummins R. (1989) Reversible middle cerebral artery occlusion without craniectomy in rats. *Stroke* **20,** 84–91.

13. Modo M., Stroemer R. P., Tang E., Veizovic T., Sowinski P., Hodges H. (2000) Neurological sequelae and long-term behavioural assessment of rats with transient middle cerebral artery occlusion. *J. Neurosci. Meth.* **104,** 99–109.

14. Goldstein L. (1993) Basic and clinical studies of pharmacologic effects on recovery from brain injury. *J. Neural. Transplant. Plast.* **4,** 175–192.

15. Modo M., Stroemer R. P., Tang E., Patel S., Hodges H. (2003) Effects of implantation site of dead stem cells in rats with stroke damage. *Neuroreport* **14,** 39–42.

16. Paxinos, Watson. (1999) The *Rat Brain in Stereotoxic Coordinates*. Academic (Elsevier).

6

Endogenous Brain Protection

Models, Gene Expression, and Mechanisms

Frank C. Barone

Summary

Almost all injurious stimuli, when applied below the threshold of producing injury, activate endogenous protective mechanisms that significantly decrease the degree of injury after subsequent injurious stimuli. For example, a short duration of ischemia (i.e., ischemic preconditioning [PC]) can provide significant brain protection to subsequent long-duration ischemia (i.e., ischemic tolerance [IT]). PC/IT has recently been shown in human brain, suggesting that learning more about these endogenous neuroprotective mechanisms could help identify new approaches to treat patients with stroke and other central nervous system disorders/injury. This chapter provides a brief overview of PC/IT research, illustrates the types of data that can be generated from in vivo and in vitro models to help us understand gene and protein expression related to induced neuroprotective mechanisms, and emphasizes the importance of future research on this phenomenon to help discover new mechanisms and targets for the medical treatment of brain and other end-organ injuries.

KEY WORDS

Ischemic preconditioning; ischemic tolerance; middle cerebral artery occlusion; ischemic stroke; neuroprotection; neurological deficits; gene expression; heat shock proteins; in vitro brain protection model; signaling pathways; endogenous neuroprotective mechanisms; caspases; caspase 3; heat shock protein 70.

1. Introduction

1.1. Ischemic Preconditioning/Ischemic Tolerance

Most stimuli that can cause injury will protect the brain against subsequent severely injuring stimuli if applied at a low intensity (i.e., subthreshold for

From: *Methods in Molecular Medicine, Vol. 104: Stroke Genomics: Methods and Reviews*
Edited by: S. J. Read and D. Virley © Humana Press Inc., Totowa, NJ

inury) prior to that severe injury. For ischemic stimuli, this phenomenon has been termed ischemic preconditioning (PC) or ischemic tolerance (IT). More generally, PC/IT is the reaction to a potentially noxious stimulus, such as hypoxia, ischemia, or inflammation. For example, a short ischemic event (i.e., PC), can result in a subsequent resistance to severe ischemic tissue injury (i.e., IT). This phenomenon has been described in several organs (i.e., especially the brain and heart) and by various experimental paradigms and may represent a fundamental cell/organ response to certain types or levels of injury *(1–9)*. An early, short-lasting IT was observed initially in the heart *(10,11)*. Myocardium protection was observed starting between 1 and 60 min of PC and lasting for less than 3 h *(10,12)*. A later, delayed tolerance (i.e., a "second window of protection") also was identified in the heart beginning about 24 h after ischemic preconditioning *(10,13)*. No studies on brain tolerance have identified an early, short-lasting PC protective effect, such as that observed in the heart. Ischemic tolerance in the brain has been reviewed recently *(1–5)*. PC/IT has many triggers, with the list including global *(6,7)* and focal *(8,9)* cerebral ischemia, hyperbaric oxygenation *(14)*, inflammation *(15)*, epilepsy *(16)*, metabolic inhibition and oxygen free radicals *(17)*, both hypothermia *(18)* and hyperthermia *(19)*, cortical spreading depression *(20)*, and cerebellar stimulation *(21)*.

The PC stimuli need to follow certain intensity and duration parameters to result in IT. For example, although brief PC ischemia can produce IT for global brain ischemia *(2,4,7)*, repetitive ischemic episodes of 2–5-min durations (spaced at 1-h intervals) are not protective and result in more severe neuronal damage than a single episode of the same duration *(22,23)*. In terms of PC and subsequent later protection from focal stroke injury, several reports have established this principle in the brain. For example, a mild global or focal ischemic insult significantly decreased the infarct size after transient or permanent middle cerebral artery occlusion (pMCAO) *(6,8,24–26)*. However, little is known of the molecular underpinning of PC-induced IT, although certain heat shock proteins have been implicated *(1,2,4,24,26,27)*.

1.2. Brain Ischemia Involves Neuroprotective and Neurodestructive Processes

The brain changes associated with brain ischemia involve a progression of both injurious and protective processes as brain injury evolves and then is repaired following an insult such as a stroke. These have been described in detail previously *(1,4,28–30)*. Focal cerebral ischemia induces a complex series of mechanisms that damage brain cells. In infarcted tissue, neurodestruction has overwhelmed neuroprotection. The major pathophysiological mechanisms of tissue destruction in stroke involve acute mechanisms of excitotoxicity, delayed mechanisms of inflammation, and apoptosis. Corresponding protective tissue responses include upregulation of neurotrophic factors and protective/stress genes/proteins. Later regenerative/restorative processes also allow recovery of some brain function post stroke. Because PC/IT paradigms can probe the mechanisms of endogenous brain neuroprotection in the absence of brain destructive processes, it is believed that an understanding of the signaling, processes, and mechanisms involved in PC-induced IT can reveal new targets and approaches to protect the brain and other end organs from injury/disease. Approaches in focal ischemia have identified many gene changes that might be associated with cell survival or death *(29–33)*. Since PC/IT has been demonstrated in the human brain (e.g., transient ischemic attacks [TIAs], which by definition do not cause structural damage but do induce protection against a subsequent stroke *[34,35]*), this strategy appears to have significant probability for success.

1.3. PC/IT: Profile, Stimuli, and Signaling

PC/IT in the brain is associated with a protected state that develops over hours and persists for days *(7,8)* and involves *de novo* protein synthesis *(8)*. Many PC/IT mechanisms appear similar or identical to highly conserved responses to oxygen deprivation; however, preischemic increases in cerebral blood flow or energy reserves are not mechanisms of brain PC/IT *(1,8)*.

Several endogenous neuroprotective mechanisms depend on glutamate-receptor activation and Ca^{2+} signaling and can affect the composition of receptors, and increase inhibitory neurotransmitters. The breakdown products of ATP (e.g., adenosine), the adenosine A1 receptor and ATP-sensitive K^+ (K_{ATP}) channels *(36)* are involved in the early events leading to ischemic cell death.

The proinflammatory stimulus lipopolysaccharide (LPS) can precondition the brain against subsequent ischemia. LPS-induced brain protection involves increases in activity of superoxide dismutase (SOD) and endothelial nitric oxide synthase *(37)*. LPS significantly increases interleukin-1 (IL-1) and tumor necrosis factor (TNF). Interestingly, preconditioning can be produced by IL-1 and protects CA1 hippocampal neurons from global ischemia in the gerbil, which is blocked by IL-1 receptor antagonist *(38)*.

Interleukin-1 has been implicated as a mediator of neuronal injury after cerebral ischemia. IL-1β mRNA is elevated soon after focal ischemia *(39–41)*, and IL-1β protein has been demonstrated to increase in the ischemic brain *(42)*. IL-1β may be detrimental to neuronal survival after ischemic injury since IL-1β administration markedly increases brain edema and the degree of infarct injury *(43)* whereas neutralizing antibodies to IL-1β delivered 1 h post MCAO or inhibition of IL-1β-converting enzyme by pharmacologic or genetic manipulations reduce ischemic brain damage following MCAO *(43–45)*. IL-1 receptor antagonist (IL-1ra), an endogenous antagonist of the IL-1 receptor

(46,47), has demonstrated neuroprotective activity under many conditions *(48–51)*. Recently, a putative role for the IL-1 system, the neuroprotective protein IL-1ra in particular, in PC-induced IT has been studied *(8)*.

Preconditioning with the inflammatory cytokine TNF induces manganese SOD (MnSOD) *(52)*. Preclinical in vivo and in vitro stroke models using cortical neurons, cortical astrocytes, and microvessel endothelial cells have now demonstrated that TNF and its downstream mediator ceramide is involved in brain tolerance signaling *(53–55)*. TNF and ceramide block recruitment of the Ca^{2+}-binding protein p300/CBP to the p65 nuclear factor (NFB) binding site on DNA, which modifies NFB-driven gene expression, inhibits expression of proinflammatory intercellular adhesion molecule 1 mRNA, and preserves expression of MnSOD, which apparently contribute to brain tolerance *(56)*. TNF-converting enzyme (TACE) upregulation has been demonstrated after ischemic brain damage and also appears to be involved in TNF release and its role in IT. Western blot analysis has shown that TACE expression is increased after PC and is blocked by the selective TACE inhibitor BB-1101 *(57)*.

1.4. PC/IT Mechanisms: Gene and Protein Expression

Potential IT mechanisms may be divided into two categories, as described previously *(4)*:

1. A cellular defense function against ischemia that may be increased by neuronal mechanisms (e.g., arise by posttranslational modification of proteins or by expression of new proteins via a signal transduction system to the nucleus). These mechanisms appear to strengthen the influence of survival factors or inhibit apoptosis.
2. A cellular "stress response" with synthesis of new stress proteins that produce an increased capacity for maintenance of cellular integrity. These are cellular "chaperone proteins" that work by unfolding misfolded cellular proteins and disposing of unneeded denatured cellular proteins. The processing of unfolded proteins has been demonstrated to be an important factor in cell survival and cell death.
   The brain may be protected from ischemia by using multiple mechanisms that are available for cellular survival. If tolerance induction can be manipulated and accelerated by a drug treatment that is safe and effective enough, it could greatly improve the treatment of stroke.

Ischemic injury induces elevated expression of heat shock proteins (Hsps) in the brain *(58–63)*. Hsps contribute to the cellular repair processes by refolding denatured proteins and acting as molecular chaperones in normal processes such as protein translocation and folding *(64–66)*. The highly inducible member of the 70-kDa family of Hsps, Hsp70, has been associated with resistance (i.e., tolerance) to ischemic injury in the heart *(67–69)* and brain *(24,70–74)*. Brain tolerance to ischemic injury has been observed following heat shock treatment *(75)* and several types of PC treatments. Mild global ischemic insults can reduce the extent of brain/neuronal damage after subsequent more severe ischemic insults *(1–9,27,71–79)*. These situations were related to Hsp70 accumulation. However, brief repetitive middle cerebral artery occlusion can decrease infarct size following a subsequent 100-min occlusion, but Hsp70 accumulation and degradation did not always match the acquisition and decay of ischemic tolerance *(2)*.

Other heat shock proteins have also been implicated in cellular resistance to injury. The 27-kDa heat shock protein (Hsp27) has been shown to increase cell resistance to oxidative injury *(80)* or thermal stress *(81)*. Transgenic mice that express increased Hsp27 suffer reduced seizures and reduced hippocampal cell damage following kainate treatment *(82)*. Recently, Hsp27 has been shown to be highly inducible in the neocortex of rats after seizure activity *(83)*, photothrombotic injury *(84)*, and cortical spreading depression *(85)*. In fact, Hsp27 is expressed following cerebral ischemia *(86)* and following PC in IT *(24,87)*. Although the cellular stress response can be a major component of ischemic tolerance in the brain, neurons seem to be protected by divergent mechanisms of tolerance induction. Even a brief period of ischemia can trigger an enormous complexity of gene expression in neurons *(88–91)*. Protein kinase signaling, immediate early genes, growth factors, and antioxidant defence enzymes are examples of such a response *(1,4)*. Thus, PC-induced ischemic tolerance can be expected to have multiple and important overlapping mechanisms.

Much of the acquisition of ischemic tolerance might be explained by the stress response of the brain tissue. When exposed to a noxious environment, the stress signal in the cell is transferred to heat shock factors to convey information to the heat shock elements that exist upstream to stress-inducible Hsp genes. Then Hsp gene expression is greatly enhanced, and the cell acquires transient tolerance to further stresses. These characteristics are common to a wide variety of cells, and neurons may not be the exception. The stress response is essential for cell survival because cells cannot cope with a sudden accumulation of denatured proteins or protein aggregation that arises from various stresses such as thermal, anoxic or oxidative stresses. Protein aggregates do not accumulate in unstressed cells because they are continuously degraded by cellular health control machinery. This mechanism consists of chaperoning unfolded proteins and subsequent proteolysis by the ubiquitin–proteasome pathway *(4)*.

The family of stress proteins or heat shock proteins is large, diverse, and abundant. They are expressed on cellular stress, but some members exist under unstressed conditions. They are subdivided according to their molecular size and location within the cell *(92,93)*. Protein molecules are one of the most sensitive sensors of stressful conditions. On ischemic stress, increased intracellular Ca^{2+}

and nitric oxide produce damage and unfold the normal conformation of proteins within the cytoplasm and endoplasmic reticulum. These denatured protein molecules are partially renatured by chaperones. If the chaperone system cannot counter increased unfolded proteins, heat shock factor 1 (HSF1) is activated. HSF1 then enters the nucleus and greatly enhances expression of Hsp genes. When an excess amount of Hsps is produced by the stress response, it is believed to confer tolerance to subsequent stresses. Protein aggregates do not accumulate in unstressed cells because they are continuously degraded by cellular health control machinery. This mechanism consists of chaperoning unfolded proteins and subsequent proteolysis by the ubiquitin–proteasome pathway *(94)*. Once this system fails, cells undergo apoptosis. When cells are subjected to stressful conditions, protein molecules in the cytoplasm and in the endoplasmic reticulum are denatured and unfolded. If they are not refolded with the aid of molecular chaperones such as Hsp70, this ubiquitin–proteasome pathway usually degrades them. The function of this pathway is believed to be essential for cell survival.

When ischemic tolerance is induced by preceding global brain ischemia, Hsp70 increases in the hippocampal CA1 pyramidal cells *(71,72,78,95)*. The amount of Hsp70 was also shown to correlate well with tolerance acquisition in a gerbil model of chronic hypoperfusion of the brain *(95)*. Increased amounts of Hsp70 appear to be necessary for ischemic tolerance, since experimental manipulation that inhibits Hsp70 function abolishes ischemic tolerance. Increased synthesis of Hsp70 was blocked by MK801 (an N-methyl-D-aspartate [NMDA] antagonist) or anisomycin (a protein synthesis inhibitor) *(96)*. MK801 inhibited Hsp70 synthesis and abolished ischemic tolerance after 2 min of preconditioning ischemia. In contrast, anisomycin also inhibited Hsp70 synthesis but did not inhibit tolerance induction. This experiment could be interpreted as signifying that augmented NMDA neurotransmission is necessary for tolerance induction *(97)* but Hsp70 is not. Inhibition of protein synthesis using anisomycin, however, might itself be a stressful stimulus to cell survival under these conditions and thus might confound this aspect of the study *(98)*. An increase in Hsp27 was also demonstrated in the brain after tolerance induction *(24,87)*. The direct role of Hsp27 in neuronal ischemic tolerance remains to be explained since Hsp27 expression is observed mainly in glial cells *(24)*, as will be described in this chapter. Induction of the Hsp110/105 family was studied in rats that had acquired ischemic tolerance by transient ischemia *(99)*. Ischemic tolerance also facilitates the recovery of SOD synthesis, and, this presumably contributes to protecting neurons from ischemia.

1.5. PC/IT: Types of Models and Monitoring Approaches

Brain protection is most typically monitored by measuring the degree of injury/infarct. However, other methods are available. Small animal magnetic resonance imaging (MRI) has recently become available and has been applied to study the tolerance phenomenon in focal ischemia. MRI has demonstrated the evolution of infarction and its reduction by tolerance induction *(100,101)*. Once ischemic tolerance is induced, it attenuates not only neuronal injury but also the extent of ischemic brain edema by protecting vascular cells *(102)*. Functional/motor effects are clearly protected by PC/IT *(8)*. It is interesting that spontaneously hypertensive stroke-prone rats (SHRSPs) that are more sensitive to and exhibit greater brain injury after cerebral ischemia and also exhibit a greater degree of neurological deficits with dramatically less spontaneous recovery of neurological deficits, also exhibit a significantly reduced degree of IT to PC *(101,103,104)*. Memory impairment can be another way to monitor PC/IT. Clear protective effects of IT on the impairment of memory acquisition following brain ischemia have been demonstrated *(105–107)*.

1.6. PC/IT: Signalling and Cellular Mechanisms

Brief ischemia in the brain triggers a complex pattern of gene expression, which includes the immediate-early genes (IEGs) *fos*, *jun*, and *Krox (4)*. Preconditioning ischemia causes postischemic increase in c-JUN protein expression but not in the other IEGs *(4)*. IT after PC can be associated with the activation of specific transcription factors such as c-JUN *(108)*. Expression of c-fos has been implicated in the protection of the contralateral hippocampus after unilateral PC *(109)*. The role of early response genes in regulating genetic responses by acting as transcription factors is critical to tissue changes induced by a variety of different stimuli *(110)*. Following ischemia, IEG transcription factors are markedly increased, apparently associated with their activation of diverse target genes *(111–113)*. Although the broad implications of altered early response gene expression on cellular/tissue responses to stress are recognized, the role of PC in altering early response gene expression in PC-induced IT is unknown. More importantly, signals, which might promote processes that can add to brain injury, may be affected very early, as shown by changes in the response of these early response genes following severe focal stroke in previously PC animals.

Severe ischemia activates protein phosphorylation *(114)*, and this phosphorylated state persists for a few days, as was confirmed with calcium/calmodulin-dependent protein kinase II *(115)* and mitogen-activated protein kinases *(116)*. PC blocks this ischemia-induced phosphorylation. Since the cascade of phosphorylation may work as an amplifier of neuronal injury, preconditioning could cancel this detrimental cascade and normalize the intracellular signal transduction. Postischemic activation of Akt/protein kinase B may contribute to the induction of ischemic tolerance *(117)*. Akt was activated after sublethal ischemia that could induce tolerance, and inhibition of Akt activity resulted in attenuation of ischemic tolerance. Possible molecules lying down-stream to Akt are BAD, caspase-9, Bcl-2, and trophic factors such as brain-derived neurotrophic factor (BDNF), and Hsp70.

As outlined previously *(1)*, ischemia induces genes that regulate apoptosis, including *bax* (found in cells programmed to die) and *bcl2* (found in cells pro-grammed to survive). In some PC/IT models, BCL2 protein induction parallels initiation and cessation of tolerance, and *bcl2* antisense oligonucleotides block tolerance *(118)*. The transcription factors that drive the induction of *bcl2* expression during tolerance include the cyclic adenosine monophosphate (cAMP) response element binding protein (CREB), as the *bcl2* promoter con-tains a camp response element (CRE) *(119)*. Multiple protein kinases can acti-vate transcription via the response element through phosphorylation of CREB *(119,120)*. *bcl2* appears to act at the mitochondria, where the upregulation of base-excision repair capability distinguishes tolerance from ischemic injury *(121)*. However, tolerance to hypoxia can occur in the absence of *bcl2* induc-tion, thus indicating that there are multiple effectors/paths in IT *(122)*. Gene expressions that lead to apoptosis may be related to ischemic tolerance. Expression of *p53* is correlated with neuronal apoptosis after cerebral ischemia. A recent report showed that ischemia activates *p53* gene expression along with its downstream genes, and preconditioning ischemia markedly reduces this activation *(118)*. Procedures that induce ischemic tolerance enhance Bcl-2 pro-tein expression *(120)*, and antisense treatment against Bcl-2 mRNA blocked tolerance *(122)*. Since Bcl-2 expression is known to be a key step that inhibits apoptosis, its increase may well be related to ischemic tolerance *(122)*. Although many researchers are actively charting the signaling pathways of PC/IT in the nervous system, our knowledge of PC/IT is still too patchy to justify direct extrapolation of laboratory results to the patient. However, PC/IT can be used to identify/discover new targets and pathways of brain protection.

Glial cell proliferation has been implicated in the induction of ischemic tol-erance in cortical neurons *(24,123,124)*. Although this is clearly an important phenomenon caused by PC in focal stroke *(66)*, did not occur in the hippocam-pus during the period when ischemic tolerance was acquired *(125)*.

It must be emphasized that increased expression of stress proteins, particu-larly Hsp70, may not be absolutely required at all time-points of ischemic toler-ance. Cerebral ischemia, even if it is sublethal for neurons, is now recognized to induce an increasingly complex range of gene expression *(126,127)*. Increase in Hsp70 could simply be a coincidental increment of gene expression on cellular stress. Evidence not exclusively supporting the "stress protein" hypothesis has been documented *(128)*. Spreading depression conferred tolerance in cortical neurons to ischemia in the rat, but it failed to induce the expression of Hsp70 mRNA in regions of tolerance *(129)*. Gene expression was also studied in a gerbil ischemia model and was correlated with the timing of anoxic depolarization by DC potential in the hippocampus and acquisition of ischemic tolerance. This experiment demonstrated that the threshold depolarization that was required to induce tolerance correlated well with gene expression of transcription factors such as junB but not with Hsp70 *(127)*. These experiments have therefore shown that the presence of Hsp70 is not mandatory for ischemic tolerance. Nevertheless, it is widely accepted that Hsp70 can make neurons resistant to noxious stimuli such as ischemia when it is expressed in excess. Focal cerebral ischemia in the same transgenic mice revealed that overexpression of Hsp70 protects the brain from ischemic brain injury *(129–131)*.

1.7. In Vitro PC/IT Models: Pro-Survival Signaling

Ischemic tolerance has also been described in cultured neurons and in brain slices. A neuroprotective effect of sublethal hypoxia on hypoxic neuronal damage in cortical cultures correlated with increased fibroblast growth factor *(132)*. Sublethal oxygen–glucose deprivation or Na+/K+ ATPase inhibition in cortical culture also showed a neuroprotective effect *(133)*. Cultured mouse cortical neurons preconditioned by sublethal oxygen–glucose deprivation developed resistance to a longer exposure to oxygen–glucose deprivation. The tolerance was blocked by an NMDA antagonist *(134)*. Other in vitro models include metabolic/chemical PC methods that induce IT *(135–137)*. Neuroprotection by hypoxic preconditioning requires activation of both vascular endothelial growth factor and Akt *(138)*. Recently, a model has been developed to identify the importance of activated caspase 3 and elevated Hsp70 in the mechanism(s) of IT *(139)*, as will be described in **Subheading 4.2.**

An emerging body of evidence suggests that the upregulation of pro-survival elements within preconditioned cells is dependent on activation of signaling pathways typically associated with degeneration. For example, in cardiomyocytes, generation of reactive oxygen species (ROS) within the mitochondria is critical for the expression of protection as blockade of ROS production decreases ischemic tolerance *(140–142)*. In a similar manner, models of brain ischemic preconditioning initiated by stimuli such as mild ischemia or low-level ceramide, amyloid-β peptide, or mitochondrial toxin exposure share ROS generation as a critical component of their mechanism of action. The blockade of ROS generation can block the expression of ischemic preconditioning in neurons as well *(143,144)*.

Energetic dysfunction also probably contributes to preconditioning induced protection. Loss of ATP leads to opening of the K_{ATP} channels that reside in the inner mitochondrial membrane *(144)*. However, during energetic crisis, such as that which occurs during ischemia, channel activation can increase ROS generation *(145,146)*. In neurons, the protective effects of preconditioning can generally be attenuated with K_{ATP} antagonists including glibenclamide *(147)*. Conversely, K_{ATP} activators, such as cromakalim, can mimic the neuroprotective effects of preconditioning stress and are protective against excitotoxic and ischemic cell death *(148)*. In spite of an abundance of evidence suggesting an important contribution of energetic dysfunction and ROS generation in preconditioning-induced protection, little is known about how far these pathways progress before being halted or the mechanism by which they are blocked.

2. Materials and Methods

Following this overview of PC/IT, it is now appropriate to provide some actual data from both in vivo and in vitro studies. This will help us to understand the experimental models, which can be used to evaluate gene and protein changes related to PC-induced neuroprotection. The methods for these studies are given below.

2.1. In Vivo Studies

2.1.1. Animals

Adult male spontaneously hypertensive rats (290–350 g) were utilized in the present studies. Prior to surgical procedures, the rats were anesthetized with sodium pentobarbital (60 mg/kg ip). All the animals were cared for in accordance with the *Guide for the Care and Use of Laboratory Animals* of the US Department of Health and Human Services. The Institutional Animal Care and Use Committee of GlaxoSmithKline approved the procedures and the protocol for using laboratory animals in the present studies.

2.1.2. Focal Ischemic Preconditioning

1. Transient MCAO or sham surgery was carried out in SHRs, 300–350 g in weight, under sodium pentobarbital anesthesia as described previously *(8,24,149–151)*. SHR were utilized because of their increased sensitivity and decreased variability in responding to focal ischemia *(151)*.
2. All animals were allowed free access to food and water prior to and after surgery. Body temperature was maintained at 37°C using a heating pad throughout the surgery procedure and during postsurgery recovery.
3. Briefly, following craniotomy using stereotaxic procedures, the dura was opened over the right MCA.
4. The hooked tip of a platinum–iridium (0.0045-in diameter) mounted on a micromanipulator was placed under the MCA (at the level of the inferior cerebral vein) and used to lift the artery away from the brain surface to occlude blood flow temporarily, as verified previously by monitoring cortical microvascular perfusion *(151)*.

5. A period of 10 min of temporary MCAO was utilized for focal PC based on previous occlusion time response data *(8)* and exploratory studies demonstrating that 10 min of temporary MCAO produced no brain injury but tended to reduce the response to ischemic injury 24 h later.
6. Sham surgery (SS) was conducted in place of PC (i.e., the cranium was removed over the artery) and served as the control group for all subsequent comparisons.

2.1.3. Permanent Focal Ischemia

1. In order to systematically evaluate the effects of PC to induce ischemic tolerance, pMCAO was performed at various times (i.e., 6 and 12 h, and 1, 2, 7, 14, and 21 d; $n = 7$–9 per group) of reperfusion following PC.
2. pMCAO was carried out as described previously *(8,24,150,151)*. The MCA was simultaneously occluded and cut dorsal to the lateral olfactory tract also at the level of the inferior cerebral vein.
3. SS was performed 24 h following PC for comparative purposes ($n = 7$).
4. Following SS, PC, and pMCAO surgeries, the temporalis muscle and skin were closed in two layers.
5. The control groups of animals received SS in place of PC prior to pMCAO and received pMCAO at all the various time following SS (i.e., in a counterbalanced manner to optimize the groups as a comparative controls; $n = 22$).

2.1.4. Neurological Deficits

1. Twenty-four hours following pMCAO, a neurological examination was performed as previously reported *(8,16,150,151)*.
2. Briefly, forelimb scores were zero (no observable deficit), 1 (any contralateral forelimb flexion when suspended by the tail), and 2 (reduced resistance to lateral push toward the paretic, contralateral side.
3. A hindlimb placement test consisted of pulling the contralateral hindlimb away from the rat over the edge of a table. A normal response is an immediate placement of the limb back onto the table, and an abnormal/deficit response is no limb placement/movement.

2.1.5. Brain Injury Analysis

1. Following the neurological evaluation, rats were euthanized with an overdose of sodium pentobarbital, and the forebrain was sliced into seven coronal slices (2 mm thick) that were immediately immersed in 1% triphenyltetrazolium chloride (TTC) as described previously *(8,103)*.
2. Stained tissues were then fixed by infiltration, photographed, and quantitated for ischemic damage using an image analysis system *(8,151)*, and the degree of brain damage was corrected for the contribution made by brain edema/swelling, as described previously *(152,153)*. Briefly, infarct size was expressed as the percent infarcted tissue in reference to the contralateral hemisphere, and infarct volume (mm^3) was calculated from the infarct areas on forebrain sections.

2.1.6. Effects of PC on Potential Early IT

1. To determine whether PC produced an earlier IT-protective effect in this focal ischemia model, a separate experiment was conducted where by SS or PC was performed only 2 h prior to permanent MCAO (n = 7–8 per group).
2. Animals were then evaluated as described above for neurological deficits and brain injury 24 h after permanent MCAO.

2.1.7. Preconditioning and Cell Death

1. In order to examine whether the PC procedure *per se* produced any significant brain injury that could not be detected by gross, TTC histological evaluation, forebrain tissue was also prepared as histological sections (6 μm), stained with H&E, and then evaluated microscopically for injury as described previously *(154)*.
2. In addition, apparent apoptotic or DNA reparative cell changes were evaluated using *in situ* end labeling 1 and 2 d and 2 and 4 wk following PC, as described previously *(155)*.
3. Following fixation, the brains were cryoprotected in 30% sucrose and frozen, and sections were cut at 10 μm on a cryotome. Sections on slides were stored at –70°C.
4. Before staining, sections were dried onto the slides at 37°C for 30 min, and then endogenous peroxidase activity was quenched by 10 min incubation in 1% H_2O_2 in methanol.
5. Following rehydration, sections were incubated for 15 min in 10 μg/mL proteinase K in 50 mM EDTA, 100 mM Tris-HCl, pH, 8.0 buffer.
6. The reaction was terminated by washing sections for 5 min in 100 mM Tris-HCl buffer at pH 8.0 (TB).
7. Following TB washes, sections were incubated for 90 min with a solution containing 400 pmol biotinylated dATP (biotin-14-dATP; Gibco-BRL, Gaithersburg, MD), 0.1 μM $CoCl_2$ (Boehringer Mannheim, Indianapolis, IN), and 25 U of terminal transferase (TdT; Boehringer Mannheim) in TdT buffer (Boehringer Mannheim).
8. The end-labeling reaction was terminated by washing sections twice in TB.
9. Sections were then incubated for 1 h at room temperature in avidin-biotin-horseradish peroxidase complex (1:100; Vector).
10. After three 5-min TB washes, diaminobenzidine (0.02%) and H_2O_2 (0.02%) were used to visualize the catalyzed reaction product.
11. After multiple rinses in TB, sections were counterstained with hematoxylin, dehydrated, cleared, and coverslipped.

2.1.8. Cross-Hemispheric PC

The effects of focal ischemic PC in one hemisphere on susceptibility to focal ischemic injury in the contralateral hemisphere was determined as follows.

1. PC or SS was carried out using the right MCAO as described above, and 24 h later pMCAO was performed on the left MCA.

2. Rats were then sacrificed and their forebrains analyzed (n=3-4 per group) as described in **Subheading 2.1.5.**

2.1.9. Cerebral Blood Flow

Laser-Doppler flometry was used to verify occlusion and reperfusion of the MCA by monitoring local cortical microvascular perfusion in the ischemic cortex receiving blood supply from the artery, as described previously *(8,151)*.

1. A 2–3-mm diameter hole was drilled through the skull above the cortical area receiving blood supply from the artery (i.e., centered at AP = 0 mm, L = 4 mm from bregma with level skull.
2. The probe (1-mm diameter) of a laser-Doppler perfusion monitor (Periflux PF3; Stockholm, Sweden) was then positioned on the surface of the dura, and the local cortical perfusion was monitored before and after MCAO and during MCA reperfusion on d 1 (for 10 min temporary MCAO-PC or SS groups; $n = 9$ per group); before and after pMCAO on d 2 for both groups; and on d 3 (24 h post pMCAO) for both groups.
3. Extreme care was taken to position the perfusion monitor probe in exactly the same cortical location on each day.
4. Animals were anesthetized with pentobarbital and positioned in the stereotaxic unit as described in **Subheading 2.1.2.**, and the cortex was closed with sutures between recordings.
5. The calibrated output of the perfusion monitor was connected to a Beckman R711 polygraph and averaged in 5-min periods for comparison between groups. Basal perfusion was recorded prior to PC or SS on d 1, and all data on d 1 and subsequent days were normalized as a percent of that value.

2.1.10. PC and Protein Synthesis

Cycloheximide, a protein synthesis inhibitor, was utilized to study the role of protein synthesis in the PC induction of IT in two separate studies, as described previously *(156,157)*.

1. In the first study, cycloheximide, used at a dose demonstrated previously to block protein synthesis in vivo for approx 24 h (*157*; 1 mg/kg ip), or vehicle (distilled water) was administered to SHRs as a dose of 2.5 mL/kg, 30 min prior to PC or SS, as described above (i.e., protein synthesis was inhibited primarily for the 24-h period following PC and before pMCAO; $n = 13$–17 per group). Twenty-four hours following either PC or SS, all animals underwent pMCAO as described above. Twenty-four hours following the pMCAO, a neurological examination was performed as described above; animals were then sacrificed, the brain was removed and stained with TTC, and infarct size and volume was quantitated as described above.
2. In the second study, cycloheximide (at the same dose) or vehicle was administered much later, 30 min prior to pMCAO in SHRs that had received PC or SS 24 h earlier (i.e., the completely counterbalanced experimental design was identical

to the first experiment except protein synthesis was primarily inhibited for the 24-h period after pMCAO; $n = 7$–8 per group). Twenty-four hours following the pMCAO, a neurological examination was performed, animals were then sacrificed, and infarct size and infarct volume were quantitated as described in **Subheading 2.1.5.**

2.1.11. IL-1ra mRNA Expression

1. PC was performed as described in **Subheading 2.1.2.** and then forebrains were removed for cortical dissection at 6, 12, 24, or 48 h following PC ($n = 4$ per group), or 24 and 48 h following sham surgery ($n = 4$).
2. The PC frontal–parietal cortex was dissected from the ipsilateral hemisphere. The contralateral cortex was dissected as the nonischemic control from the same rat *(8,151)*.
3. The cortical samples were immediately frozen in liquid nitrogen and stored at –80°C.
4. Total cellular RNA was prepared from cortical samples and processed as previously described *(41,158)* and initially subjected to quantitative reverse-transcription polymerase chain reaction (RT-PCR) analysis.
5. Briefly, the cellular RNA (5 μg) isolated from the cortical samples at the indicated time-points following PC was reverse-transcribed with 200 U of RNase H⁻ SuperScript II reverse transcriptase (Gibco-BRL) for 60 minutes at 37 oC primed with 1 mg of oligo(dT)12-18 (GIBCO BRL) at conditions recommended by the manufacturer.
6. The RT products were extracted with phenol/chloroform, ethanol-precipitated, resuspended in 200 mL of TE (10 m*M* Tris-HCl and 1 m*M* EDTA, pH 7.5), and stored at –20°C.
7. The quantitative PCR was carried out in a similar fashion as that described in detail previously *(158)*.
8. A reference gene (rpL32) previously demonstrated to exhibit constant expression throughout the time-course following MCAO *(113,158)* was used as an internal control for co-amplification with the targeted gene, IL-1ra.
9. PCR primers used for amplification of IL-1ra and rpL32 were synthesized according to published sequences *(8,46)*. The optimal amplification conditions were determined as described previously *(158)*, and the linear portions of the amplification for both IL-1ra and rpl32 were used for PCR reactions in a total 50 μL of reaction mixture (i.e., RT products from 0.1 μg RNA, 28 cycles of amplification, containing 1×10^6 cpm ^{32}P-labeled antisense primer for IL-1ra and 5×10^4 for rpL32, together with 100 ng of each nonradioactive sense and antisense primers.
10. The amplification was carried out using 2.5 U of TaqAmpli polymerase (Perkin-Elmer Cetus) in a thermocycler (Perkin-Elmer Cetus) according to the conditions described previously *(158)*: initial denaturation, 3 min at 94°C; initial annealing, 1 min at 54°C; initial extension, 3 min at 72°C. The subsequent cycles were: denaturation, 15 s at 94°C; annealing, 20 s at 54°C; extension, 1 min at 72°C.

11. The PCR product (10 μL) was electrophoresed through a 6% polyacrylamide gel.
12. The gel was dried and subjected to autoradiography at room temperature.
13. The signal intensity was quantitated using PhosphorImager (Molecular Dynamics, Sunnyvale, CA) analysis, and the relative mRNA levels were determined by calculating the ratio of IL-1ra to rpL32 in each coamplified sample.
14. To confirm the quantitative RT-PCR data, we applied Northern blot analysis using poly(A) RNA isolated from PC and contralateral control cortex of 50 rats 24 h following PC. Poly(A) RNA (10 μL/lane) was electrophoresed through formaldehyde agarose gel and transferred to a GeneScreen Plus membrane (Du Pont-New England Nuclear).
15. For Northern analysis, the cDNA fragments for IL-1ra and rpL32 were gel-purified after PCR amplification (as described above) and uniformly labeled with [gggg-^{32}P]dATP (3000 Ci/mmol; Amersham) using a random-priming DNA labeling kit (Boehringer Mannheim).
16. Hybridization and washing were performed as described in detail in **refs.** *41* and *158*.

2.1.12. IL-1ra Protein Expression

1. Following PC or SS, rats were allowed to recover for 6, 12, 24, and 48 h ($n = 3$–4 per group).
2. Rats were overdosed with sodium pentobarbital and perfused via the aorta with 100 mM phosphate-buffered saline for 5 min followed by 100 mM phosphate-buffered saline containing 2% paraformaldehyde and 0.2% glutaraldehyde (4°C) for 15 min.
3. The brain was then removed and postfixed in 100 mM phosphate buffer containing 2% paraformaldehyde for 6 h as described previously *(8,24)*.
4. Brains were stored at 4°C in 120 mM sodium phosphate buffer containing 0.06% sodium azide.
5. Each brain was sectioned (50 μm) on a vibratome in ice-cold 100 mM Tris-HCl buffer, pH 7.6.
6. Immunohistochemistry was performed on free floating sections. Sections were incubated in 1% H_2O_2/Tris-HCl for 1 h, and rinsed in Tris-HCl and then in Tris-HCl containing 0.1% Triton X-100, and in Tris containing 0.1% Triton X-100 and 0.005% bovine serum albumin (BSA).
7. Sections were incubated for 1 h in 10% horse serum in Tris-HCl containing 0.1% Triton X-100 and 0.005% BSA to block nonspecific immunoreactivity.
8. After two Tris-HCl buffer washes, sections were incubated overnight at 4°C in Tris-HCl buffer containing a sheep-anti rat IL-1ra affinity-purified antibody (1:10,000; National Institute for Biological Standards and Control, Herts, UK).
9. After two Tris buffer washes, sections were incubated in Tris-HCl buffer containing a biotinylated horse polyclonal antibody raised against sheep IgG (1:500 Vector).
10. After two Tris-HCl buffer washes, sections were incubated for 1 h in avidin-biotin-horseradish peroxidase complex (1:1,000; Vector).

11. After three Tris-HCl buffer washes, sections were immersed for 20 min in diaminobenzidine-tetrachloride (DAB, 0.02%; Sigma) made up in Tris-HCl buffer containing 0.15 mg/100 mL glucose oxidase, 40 mg/mL ammonium chloride, and 200 mg/100 ml β-D(+)glucose (Sigma).
12. Appropriate control experiments were utilized in these studies (e.g., optimal antibody dilutions were determined by serial dilutions on control and experimental sections, and IL-1ra [R&D Systems] was added to demonstrate adsorbtion out of the immunoreactivity signal observed in the PC cortex).
13. Incubation of sections in solution were conducted using a shaker bath. Immunohistochemically stained sections were mounted onto coated slides, air-dried overnight, dehydrated, and coverslipped.
14. Analysis was carried out using an Olympus microscope, and representative sections were photographed for illustration *(8)*.

2.1.13. IL-1b mRNA Expression

1. The ipsilateral and/or contralateral cortex was processed for RNA extraction as described previously *(41,159)*.
2. The PCR primers and TaqMan probes for IL-1b and rpL32 were designed using a Software program from Perkin-Elmer according to the published rat IL-1β and rpL32 cDNA sequences as indicated previously *(159)*.
3. The TaqMan probes were labeled with a reporter fluorescent dye, FAN (6-carboxyfluorescein), at the 5' end and a fluorescent dye quencher, TAMRA (6-carboxy-tetramethyl-rhodamine), at the 3' end.
4. The specificity of PCR primers was tested under normal PCR conditions in a thermocycler (model 9700, Perkin-Elmer) prior to TaqMan quantitation.
5. RT and PCR were carried out using a TaqMan Reverse Transcription Reagents and TaqMan Gold RT-PCR kit (Perkin-Elmer Applied Biosystems) according to the manufacturer's specification.
6. A two-step RT-PCR was performed. The RT reaction used 3 µg total RNA in a total volume of 40 µL containing 1X TaqMan ET buffer, 5.5 mM MgCl$_2$, 500 µM of each dNTP, 2.5 µM oligo d(T)16 primers, 0.4 U/µL RNase Inhibitor and MultiScribe Reverse Transcriptase.
7. The RT reaction was carried out at 25°C for 10 min, 48°C for 30 min, and 95°C for 5 min.
8. A thermal stable AmpliTaq Gold DNA Polymerase was used for the second-strand cDNA synthesis and DNA amplification.
9. Real-time PCR was performed with 4 µL of RT products (300 ng total RNA), 1X TaqMan buffer A, 5.5 mM MgCl2, 200 µM dATP/dCTP/dGTP, 400 µM dUTP, 200 nM primers (forward and reverse), 100 nM TaqMan probe, 0.01 U/µL AmpErase, and 0.025 U/mL AmpliTaq Gold DNA Polymerase in a total volume of 50 µL.
10. PCR was performed at 50°C for 2 min (for AmpErase UNG incubation to remove any uracil incorporated into the cDNA), 95°C for 10 min (for AmpliTaq Gold activation) and then run for 40 cycles at 95°C for 15 s, 60°C for 1 min on the ABI PRISM 7700 Detection System.

The principle of the real-time detection is based on the fluorogenic 5' nuclease assay was illustrated previously *(159)*. During the PCR reaction, the AmpliTaq Gold DNA Polymerase cleaves the TaqMan probe at the 5' end and separates the reporter dye from the quencher dye only if the probe hybridizes to the target. This cleavage results in the fluorescent signal generated by the cleaved reporter dye and directly monitored by the ABI PRISM 7700 Detection System. The increase in the fluorescence signal is proportional to the amount of the specific PCR product.

1. To determine the absolute copy number of the target transcript, a cloned plasmid DNA for IL-1β and rpL32 *(159)* was used to produce a standard curve.
2. IL-1β plasmid contains 373 bp of cDNA insert in a cloning vector (2958 bp), and rpL32 contains a 465-bp insert.
3. To determine the copy numbers of DNA template, the molecular weights of the plasmids were calculated according to 6.6×10^5 per kb average value and then converted into the copy numbers based on the Avogadro number, i.e., 1 mol = 6.022×10^{23} molecules.
4. The cloned plasmid DNA was serially (every fivefold) diluted, ranging from 16.4 fg to 64 pg (or 1.65–7.24 log molecules) and from 0.409 pg to 0.32 ng (or 3.04–7.93 log molecules) for IL-1β and rpL32, respectively.
5. Each sample was run in duplicate, and the ΔRn (the ratio for the amount of reporter dye emission to the quenching dye emission) and threshold cycle (C_t) values were averaged from each reaction.
6. Data were analyzed using a Sequence Detector V1.6 program (Perkin-Elmer) *(159)*.

2.1.14. IL-1β Protein Expression

1. Each brain was sectioned (50 μm) on a vibratome in ice-cold 100 m*M* Tris-HCl buffer (pH 7.6).
2. Immunohistochemistry was performed on free-floating sections.
3. Sections were incubated in 1% H_2O_2/Tris-HCl for 1 h, rinsed in Tris-HCl, then in Tris-HCl containing 0.1% Triton X-100, and in Tris-HCl containing 0.1% Triton X-100 and 0.005% BSA.
4. Sections were blocked with 10% goat serum in Tris-HCl containing 0.1% Triton X-100 and 0.005% BSA to block nonspecific immunoreactivity.
5. After two Tris-HCl buffer washes, sections were incubated overnight in a primary rabbit polyclonal antibody raised against recombinant rat IL-1β (1:1000; Endogen, Cambridge, MA) and then in a biotinylated secondary goat antibody raised against rabbit IgG (1:400; Vector).
6. IL-1β immunoreactivity was then revealed using the avidin-biotin-DAB reaction. Specifically, after two Tris-HCl buffer washes, sections were incubated for 1 h in avidin-biotin-horseradish peroxidase complex (1:1000; Vector).
7. After three Tris-HCl buffer washes, sections were immersed for 20 min in (0.02% DAB (Sigma) made up in Tris-HCl buffer containing 0.15 mg/100 mL glucose

oxidase, 40 mg/mL ammonium chloride, and 200 mg/100 mL β-D(+)glucose (Sigma).
8. Immunohistochemically stained sections were mounted onto coated slides, air-dried overnight, dehydrated, and coverslipped *(24,159)*.
9. For IL-1β and glial fibrillary acidic protein (GFAP) double labeling, sections were incubated for 1 h in 100 mM Tris-HCl containing 10% goat serum.
10. After three 100 mM Tris-HCl rinses, sections were incubated overnight in 100 mM Tris-HCl containing either the rabbit polyclonal IL-1β antibody (1:1000; Endogen) and a mouse monoclonal GFAP antibody (1:800; Sigma).
11. After three 100 mM Tris-HCl washes, sections were incubated with Texas Red-conjugated goat antibody raised against mouse IgG (1:50; American Qualex) and fluorescein-conjugated goat antibody raised against rabbit IgG (1:50; American Qualex) to detect the primary antibodies.
12. Immunofluorescent sections were examined using a Olympus microscope equipped with an incident-light fluorescence illuminator and filters suitable for viewing the fluorochromes Texas Red and fluorescein *(159)*.

2.1.15. Early Response Gene Expression

1. In four separate groups of SHRs, SS or PC surgery was followed 24 h later with pMCAO.
2. After 2 h of pMCAO, ipsilateral (ischemic) and contralateral (control) cortex samples (corresponding to the control infarction region) were dissected as described previously *(8,160)* and immediately frozen in liquid nitrogen and stored at –80°C for evaluation of early response gene expression ($n = 4$ per group). The 2-h time-point was selected based on previous data demonstrating optimum early response gene expression in this model *(160)*.
3. Total cellular RNA was extracted, and RNA samples (20 μg/lane) were electrophoresed through formaldehyde–agarose slab gels and transferred to membranes.
4. Northern hybridization and stripping of the cDNA probes for c-fos, zif268, and rpL32 were carried out as previously described *(8,160)*.
5. PhosphorImager was used to quantify the band intensities of the Northern blots, which were analyzed by using computer image analysis, and relative mRNA levels (%) for c-fos and zif268 early response genes were normalized to the rpL32 mRNA signals in each sample as described previously *(8,113)*.

2.1.16. Northern Analysis for Hsp70 and Hsp27

1. The ipsilateral and/or contralateral cortex was dissected and immediately frozen in liquid nitrogen at various times after PC or sham surgery and stored at –80°C.
2. Total RNA was prepared by homogenizing the cortical tissues in an acid guanidinium thiocyanate solution and extracted with phenol and chloroform as previously described *(8,24)*.
3. For Northern analysis, 30 μg/lane total cellular RNA was electrophoresed through formaldehyde–agarose gel and transferred to a GeneScreen Plus membrane (Dupont-New England Nuclear).

4. cDNA fragments for Hsp70, Hsp27, and rpL32 were generated by RT-PCR using gene specific primers synthesized according to the published rat sequences.

5. Forward (5'-TGCTGACCAAGATGAAGGAG-3') and reverse (5'-AACAGAGAGTCGATCTCCAG-3') primers were used for the Hsp70 cDNA (496 bp) amplification in the coding sequence (GenBank L16764; *24,159*); a forward (5'-GCCTCTTCGATCAAGCTTTC-3') and a reverse (5'-GCAAGCTGAAGGCTTCTACT-3') primer was used for Hsp27 cDNA (554 bp) amplification (GenBank M86389; *24,159*).

6. The cDNA probes were uniformly labeled with [P-^{32}P]dATP (3000 Ci/mmol; Amersham) using a random-priming DNA labeling kit (Boehringer Mannheim).

7. Hybridization and washing were performed as described in detail in **Subheading 2.1.11**.

8. Quantitation was carried out using PhosphorImager analysis *(24)*.

2.1.17. Immunohistochemistry for Hsp70

1. Following PC as described above, rats were allowed to recover for 6 and 12 h, and 24 h, 2, 4, 7 d, and 2 or 4 wk.

2. Sham surgery was conducted in place of the PC procedure (i.e., the cranium was removed over the artery); animals were sacrificed at 12 h or 2 d after surgery and served as a control group.

3. Naive rats were also examined as controls. Rats were overdosed with sodium pentobarbital and perfused via the aorta with 100 mM phosphate-buffered saline for 5 min to remove blood and then with 100 mM phosphate-buffered saline containing 2% paraformaldehyde and 0.2% glutaraldehyde (4°C) for 15 min as described previously *(24)*.

4. The brain was then removed and postfixed in 100 mM phosphate buffer containing 2% paraformaldehyde for 6 h. Brains were stored at 4°C in 120 mM sodium phosphate buffer containing 0.06% sodium azide.

5. Each brain was sectioned (50 μm) on a vibratome in ice-cold 100 mM Tris-HCl buffer, pH 7.6.

6. Immunohistochemistry was performed on free floating sections as described previously *(16)*.

7. Sections were incubated in 1% H$_2$O$_2$/Tris-HCl for 1 h and rinsed in Tris-HCl, then in Tris-HCl containing 0.1% Triton X-100, and in Tris-HCl containing 0.1% Triton X-100 and 0.005% BSA.

8. Sections were incubated for 1 h in 10% horse serum in Tris-HCl containing 0.1% Triton X-100 and 0.005% BSA to block nonspecific immunoreactivity.

9. After two Tris-HCl buffer washes, sections were incubated overnight at 4°C in Tris-HCl buffer containing a mouse monoclonal antiserum that recognizes Hsp70 (1:10,000; StressGen Biotechnologies, Victoria, British Columbia, Canada).

10. After two Tris-HCl buffer washes, sections were incubated in Tris-HCl buffer containing a biotinylated horse polyclonal antibody raised against mouse IgG (1:500; Vector).

11. After two Tris-HCl buffer washes, sections were incubated for 1 h in avidin-biotin-horseradish peroxidase complex (1:1000; Vector).

12. After three Tris-HCl buffer washes, sections were immersed for 20 min in 0.02% DAB; Sigma) made up in Tris-HCl buffer containing 0.15 mg/100 mL glucose oxidase, 40 mg/mL ammonium chloride, and 200 mg/100 mL β-D(+)glucose (Sigma) *(24)*.

2.1.18. Immunohistochemistry for Hsp27

1. Hsp27 immunohistochemistry sections were incubated as described for Hsp70 (*see* **Subheading 2.1.17.**), except that sections were blocked with 10% goat serum, and incubated overnight in a primary rabbit polyclonal antibody raised against mouse Hsp25 (1:10,000; StressGen Biotechnologies) and then in a biotinylated secondary goat antibody raised against rabbit IgG (1:400; Vector).
2. Hsp27 immunoreactivity was then revealed using the avidin-biotin-DAB reaction as described in **Subheading 2.1.17.** for Hsp70 immunostaining.
3. Immunohistochemically stained sections were mounted onto coated slides, air-dried overnight, dehydrated, and coverslipped *(24)*.

2.1.19. In Situ *End Labeling*

1. *In situ* end labeling of DNA fragments was done on brain sections as described previously *(154)*.
2. For comparison with PC treatment, forebrains of neonatal rats and adult rats subjected to permanent MCAO for 2 d were studied.
3. Brains were cryoprotected in 30% sucrose, and frozen, and sections were cut at 10 μm on a cryotome.
4. Sections on slides were stored at –70°C. Before staining, sections were dried onto the slides at 37°C for 30 min, and then endogenous peroxidase activity was quenched by 10-min incubation in 1% H_2O_2 in methanol.
5. Following rehydration, sections were incubated for 15 min in 10 μg/mL proteinase K, and 0.05 *M* EDTA in 100 m*M* Tris-HCl, pH 8.0 buffer.
6. After washes, sections were incubated for 90 min with a solution containing 400 pmol biotinylated dATP (biotin-14-dATP; Gibco-BRL), 0.1 μ*M* $CoCl_2$ (Boehringer Mannheim), and 25 U of TdT, (Boehringer Mannheim) in TdT buffer (Boehringer Mannheim). Sections were washed, and biotinylated dATP was revealed using the avidin-biotin-DAB reaction as described above in **Subheading 2.1.17.** for Hsp70 immunostaining *(24)*.

2.1.20. Immunofluorescence for Hsp27 and GFAP

1. For Hsp27 and GFAP double labeling, sections were incubated for 1 h in 100 m*M* Tris-HCl containing 10% goat serum.
2. After three 100 m*M* Tris-HCl rinses, sections were incubated overnight in 100 m*M* Tris-HCl containing the rabbit polyclonal antibody Hsp25 antibody (1:10,000; StressGen Biotechnologies) and a mouse monoclonal GFAP antibody (1:800; Sigma).

3. After three 100 mM Tris washes, sections were incubated with Texas Red-conjugated goat antibody raised against mouse IgG (1:50; American Qualex) and fluorescein-conjugated goat antibody raised against rabbit IgG (1:50; American Qualex) to detect the primary antibodies. Immunofluorescent sections were examined using a Olympus microscope equipped with an incident-light fluorescence illuminator and filters suitable for viewing the fluorochromes Texas Red and fluorescein *(24)*.

2.1.21. Caspase 3 Immunohistochemistry

1. Animals were transcardially perfused with 0.9% saline to remove blood components followed by 4% paraformaldehyde in 100 mM phosphate buffer to fix the tissue.
2. The brains were then removed, postfixed in 4% paraformaldehyde at 4°C, cryoprotected in 30% sucrose, then frozen for 2 min in chilled isopentane and cut into 12 μm coronal sections using a cryostat.
3. Sections were rinsed in 100 mM PBS, pH 7.4, and incubation in 0.2% Triton X-100 for 30 min.
4. Endogenous peroxidases were blocked by incubating in 3% hydrogen peroxide for 20 min.
5. Sections were incubation in PBS containing 1% BSA and 10% normal horse serum for 1 h at room temperature.
6. Incubation with anti-active caspase 3 (1:500) was performed overnight at 4°C.
7. Sections were then washed and incubated with an anti-rabbit biotinylated secondary antibody (1:1000; Santa Cruz) followed by ABC complex (Vector) with DAB as the chromagen.
8. Negative controls were included in each experiment in which the primary antibody was omitted.
9. Sections were then dehydrated, cleared, and mounted in DPX for microscopic analysis.
10. Monoclonal HSP 70 antibody (SP-810), Hsp70, Hsp25, and caspase 3 rabbit polyclonal antibodies were purchased from StressGen.
11. An antibody to the activated/cleaved form of caspase 3 was purchased from Promega (Madison, WI) *(139)*.

2.1.22. In Vivo Statistical Analysis

All data are presented as mean ± SEM. Comparisons between multiple groups and or time periods were made using an analysis of variance with follow-up analyses using the least significant difference or dunnett tests. Similar analyses were carried out for RNA expression data. However, the relative mRNA levels in the ipsilateral PC cortex at different time-points were compared with the ipsilateral sham-operated cortex. Also, the levels in the ipsilateral PC cortex were compared with those in the contralateral control cortex at the same time-point. Nonparametric data (i.e., neurological deficits/scores) were analyzed by the χ^2 test. Differences were considered significant at $p < 0.05$.

2.2. In Vitro Studies

2.2.1. Chemicals and Reagents

1. All cysteine protease inhibitors were purchased from Enzyme Systems (Livermore, CA).
2. All media and media supplements were from Gibco-BRL.
3. Western blotting gels, transfer apparatus, and standards were from Bio-Rad (Hercules, CA).
4. Bcl_{xl}, constitutive inhibitor of apoptosis (cIAP)1, and cIAP2 antibodies were purchased from Trevigen.
5. Calbindin antibody and all secondary antibodies were from Santa Cruz Biotechnology.
6. Polyvinylidene difluoride (PVDF) membranes and enhanced chemiluminescence kit for immunoblotting were from Amersham.
7. All other chemicals were from Sigma.
8. An antibody to the activated/cleaved form of caspase 3 was purchased from Promega.

2.2.2. Tissue Culture

1. Cortical cultures were prepared from embryonic d-16 Sprague-Dawley rats as previously described *(139,160)*.
2. Briefly, cortices were dissociated, and the resultant cell suspension was adjusted to 250,000 cells/well (24-well treated tissue culture plates) or 600,000 cells/well (6-well tissue culture plates containing five 12-mm poly-L-lysine–treated coverslips/well) in growth medium.
3. This media was composed of a volume-to-volume mixture of 80% Dulbecco's modified Eagle medium (MEM), 10% Ham's F12 nutrients, 10% bovine calf serum (heat-inactivated, iron-supplemented; Hyclone, Logan, UT) with 25 mM HEPES, 24 U/mL penicillin, 24 µg/mL streptomycin, and 2 mM L-glutamine.
4. Cultures were maintained at 37°C, 5% CO_2, and medium was partially replaced every 2–3 d.
5. Glial cell proliferation was inhibited after 2 wk in culture with 1–2 µM cytosine arabinoside, after which the cultures were maintained in growth medium containing 2% serum and without F12 nutrients.
6. All experiments were conducted 1 wk following mitotic inhibition (25–29 d in vitro), when excitotoxicity is fully expressed in this system.

2.2.3. Preconditioning Cultures With Chemical Ischemia

1. As described previously *(139,160)*, cortical cultures were exposed to 3 mM KCN prepared in sterile, glucose-free balanced salt solution (150 mM NaCl, 2.8 mM KCl, 1 mM $CaCl_2$, and 10 mM HEPES; pH 7.2) for 90 min at 37°C, 5% CO_2.

2. Cyanide treatment was terminated by rinsing cells (200:1) and replacing the wash solution with maintenance medium (2% serum, no F12).

3. Preliminary experiments revealed that this KCN concentration and exposure time is the most extreme ischemic conditions that result in no detectable neuronal toxicity (*139*).

4. To test the efficacy of various agents at altering the neuroprotective effects of preconditioning, compounds were added either 1 h prior to KCN treatment, during the 90-min KCN treatment, and/or in the 24-h period after KCN treatment while the cells were in maintenance media.

5. These were referred to as *preincubation, coincubation,* and *postincubation* conditions respectively.

6. Glibenclamide, an ATP-dependent K$^+$ channel antagonist (1 μM; Research Biochemicals) was added during and after exposure, whereas the free radical spin trap *N*-tert-butyl-α-(2-sufophenyl)-nitrone (PBN; 500 μM; Aldrich, Milwaukee, WI), a free radical spin trap, was only present during KCN treatment (coexposure). BOC-aspartate-fluoromethylketone (BAF; 10 μM), Ac-tyrosine-valine-alanine-aspartate-AFC (YVAD; 20 μM), Ac-aspartate-glutamate-valine-asparate-AFC (DEVD; 20 μM) and Z-phenlyalanine-alanine-fluoromethylketone (zFa; 20 μM) and Ac-VDVAD (100 μM) were added pre-, co-, and postincubation to ensure maximal peptide absorption (*139*).

2.2.4. Excitotoxicity and Cell Viability Assays

1. Twenty-four hours after the preconditioning stimulus, cells were treated with the glutamate receptor agonist NMDA (*139*).

2. Immediately prior to agonist treatment, cells were rinsed (200:1) in MEM with Earle's salt (0.01% BSA and 25 mM HEPES, without phenol red).

3. Cells were then exposed for 60 min to 100 μM NMDA in the presence of 10 μM glycine at 37°C, 5% CO$_2$ unless otherwise noted.

4. Treatment was halted by serial dilution (200:1) in MEM.

5. MEM was then added at 1 mL per well, and the cultures were returned to the incubator.

6. Neuronal viability was determined 18–20 h later by measuring lactate dehydrogenase (LDH) release with an in vitro toxicology assay kit (Sigma).

7. Samples of medium (40 μL) were assayed spectrophotometrically (490:630 nm) according to the manufacturer's protocol, to obtain a measure of cytoplasmic LDH released from dead and dying neurons (*160*).

8. LDH results were confirmed qualitatively by visual inspection of the cells and, in several instances, quantitatively by cell counts (*139*).

2.2.5. Immunoblot Analysis of Protein

1. At various times after KCN treatment, cultures were harvested to assess the extent protein induction.

2. Cells were placed on ice, washed twice with ice-cold PBS, and scraped from the dish using a rubber policeman in 500 mL of TNEB (50 mM Tris-HCl, pH 7.8, 2 mM EDTA, 100 mM NaCl and 1% NP-40).

3. Of this, 200 mL was saved for protein determination, and the remaining 300 mL was resuspended in an equal volume of Laemmli buffer, heated to 95°C for 5 min, and stored at –20°C.
4. Protein concentrations were determined spectrophotometrically by using a microprotein assay kit (Bio-Rad).
5. Positive controls for Hsp70 induction were generated by placing neurons for 80 min in an incubator that had been preheated to 42°C.
6. Cells were then returned to 37°C and harvested 6 or 24 h later.
7. We also evaluated several promising target proteins from several families of proteins (e.g., calbindins, the Bcl-2 family, the IAPs, and the heat shock proteins) from in vivo PC animals for comparison with in vitro results, which will also be described below in **Subheading 3.2.5.**
8. Control and PC cortex tissues was harvested at various time points following PC (i.e., same 10 min MCAO models) and processed as described previously *(139)*.
9. Equal concentrations of proteins were run by sodium dodecyl sulfate polyacrylamide gel electrophoresis (SDS-PAGE).
10. Gels were transferred to PVDF membranes and blocked for 1 h in 5% blocking grade nonfat milk dissolved in PBS with 0.02% Tween (PBST).
11. Membranes were then washed for a total of 35 min in PBST and exposed overnight at 4°C to primary antibodies diluted 1:1000 in PBST.
12. Membranes were washed again as above, and proteins were visualized using horseradish peroxidase (HRP)-labeled anti-rabbit or anti-mouse secondary antibodies in combination with electrochemiluminescence (ECL) *(139)*.

2.2.6. Overexpression of Hsc70 by Addition of Exogenous Protein

1. To increase the expression of constitutively expressed heat shock protein 70 (Hsc 70) in cultures during the preconditioning period, cells were incubated in the presence of high levels of purified, biotinylated Hsp70 protein (StressGen) prior to KCN exposure.
2. Cultures were placed in MEM with Earle's salt (0.01% BSA and 25 mM HEPES, without phenol red) in the presence or absence of 75 µg/mL Hsc70 for 48 h prior to exposure to KCN.
3. Approximately 40% of cells incorporated substantial amounts of heat shock proteins after 3 h of incubation in 20 µg/mL purified protein.
4. Moreover, incubation with comparable protein concentration results in reduced apoptotic cell death when cells are immediately placed in the broad-spectrum kinase inhibitor staurosporine or exposed to axotomy-induced cell death *(161)*.
5. Control experiments were performed in the presence or absence of the permeabilizing agent Triton X-100 (0.3%) to ensure that the Hsc that was recognized in our immunohistochemical experiments was taken up by cells (data not shown).
6. Over the 72-h period between the initiation of Hsp overexpression and NMDA exposure, we observed a marked decrease in the expression of this protein in our cultures as well as in supernatant of cells continuously exposed to the protein which was attributable to protein degradation (data not shown).

7. Furthermore, the medium, from cultures that had been previously incubated in Hsc70 was harvested prior to NMDA exposure to ensure that all exogenous Hsc had been removed from these cells.

8. We observed no detectible soluble Hsc in this fraction, probably because of the extensive washing associated with the preconditioning treatment.

2.2.7. In Vitro Statistical Analysis

1. Pooled data were normalized as the ratio of LDH in the NMDA-treated cultures to the vehicle-treated cells ($LDH_{NMDA}/LDH_{Vehicle}$). Data expressed in this manner are referred to as *relative toxicity*.

2. The data were pooled using this method *(139)*. NMDA treatment in cells that were not preconditioned resulted in a nearly 5-fold increase in LDH release compared with control values, whereas KCN treatment provided significant neuroprotection, resulting in only a 2.5-fold increase in LDH release.

3. Cell count experiments were also performed to support the LDH release experiments.

4. LDH measurements were confirmed by counting the number of live neurons assessed by trypan blue staining *(139)*.

5. Neurons were counted 18–20 h after NMDA exposure, and the results confirmed the presence of excitotoxic tolerance in KCN-treated cells (data not shown).

3. Results

3.1. In Vivo Results

3.1.1. Ischemic Tolerance to Brain Injury

Brain injury in this model of focal stroke involves development of a consistent infarction restricted primarily to the cortex. This is typical for the SHR and has been characterized many times. PC produced significant tolerance to subsequent focal ischemia as reflected by reduced cortical infarct size when performed at certain times prior to pMCAO. Percent hemispheric infarct produced by 24 h of pMCAO was significantly ($p < 0.01$) reduced if PC was performed 1 d (58.4%), 2 d (58.1%). or 7 d (59.4%) previously (data not shown). However, percent hemispheric infarct was not affected if PC was performed 2 h, 6 h, 12 h, 14 d, or 21 d prior to pMCAO. Note that there was no neuroprotective effect of PC on brain injury or neurological function when pMCAO was initiated this soon after PC. PC alone produced no tissue injury that could be detected by TTC staining (data not shown; $n = 7$), and PC performed 24 h prior to SS (i.e., the PC–SS group) produced absolutely no infarction. Identical results were observed for infarct volume, which was significantly reduced if PC was performed 1 d (57.5%), 2 d (56.4%) or 7 d (56.2%), but not 6 h, 12 h, 14 d or 21 d prior to pMCAO (**Fig. 1A**). The profile of infarct areas throughout cortical sections (Fig. 1, Right side) illustrate that cortical infarct protection

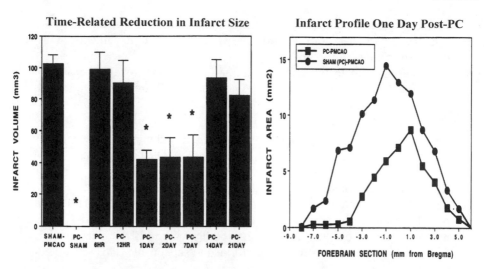

Fig. 1. Ischemic preconditioning (PC) reduces absolute infarct volume produced by permanent middle cerebral artery occlusion (pMCAO). **(Left)** Infarct volume is significantly reduced when pMCAO is performed at 1, 2 and 7 d post PC. SHAM-pMCAO group ($n = 22$: sham surgery conducted at various times prior to pMCAO and grouped together as the appropriate control condition. PC-SHAM group ($n = 7$: PC conducted 1 d prior to sham surgery demonstrating that PC does not damage the brain. All other groups had pMCAO performed at the times indicated post PC ($n = 7$–9 per group). *, $p < 0.01$ compared with SHAM-pMCAO; ANOVA with Dunnett follow-up test. Similar protection from neurological deficits were demonstrated as described in the text. **(Right)** Profile of infarct area over forebrain slices at various distances from the skull landmark Bregma corresponding to SHAM-pMCAO and PC1-DAY (pre-pMCAO) groups. Data demonstrate the significant neuroprotection/brain tolerance exhibited throughout the forebrain owing to PC.

was significant throughout the forebrain (i.e., the spatial profile of percent hemispheric infarct illustrates that protection extended throughout all cortical sections when pMCAO is performed 1 d following).

3.1.2. No Contralateral Effects of PC

The effect of PC on induction of ischemic tolerance was restricted to the area made ischemic by PC, and no significant neuroprotection was observed in the contralateral cortex (data not shown). PC (compared with SS) of the right MCA did not significantly modify the degree of infarction 24 h after left PMCAO (i.e., did not affect the degree of brain injury to the left hemisphere owing to pMCAO when performed 24 h later).

3.1.3. Ischemic Tolerance to Neurological Deficits

PC, when performed at certain times prior to pMCAO, produced significant reductions in neurological deficits that parallel the reductions in infarct size. The forelimb deficit grade quantitates the contralateral hemiparalysis and hemiparesis that can be a consequence of focal ischemia owing to the ipsilateral cortical infarction. Forelimb deficit was significantly reduced if PC was performed 2 d (31.0%), 7 d (31.0%), 14 d (64%), or 21 d (31.0%), but not 6 h, 12 h or 1 d prior to pMCAO. PC alone produced no forelimb deficit ($n = 7$), and PC performed 24 h prior to SS (i.e., the PC-SS group) produced no significant forelimb deficits. Hindlimb deficit grade quantitates a deficit in proprioception associated with loss of ipsilateral cortical function. Hindlimb deficit to 24 h of pMCAO was significantly reduced if PC was performed 1 d (52%), 2 d (100%), 7 d (100%), or 14 d (62%) previously. However, hindlimb deficit was not affected if PC was performed 6 h, 12 h, or 21 d prior to pMCAO. PC alone produced no hindlimb deficit (data not shown; $n = 7$), and PC performed 24 h prior to SS (i.e., the PC–SS group) produced absolutely no hindlimb deficit.

3.1.4. PC Does Not Induce Cell Death

No significant injury was identified owing to PC. On H&E-stained histologic sections, only a localized injury was identified at the surgical MCA site. Also, no difference in the incidence of end-labeled neurons was observed between the ipsilateral and contralateral cortices (i.e., rarely, *in situ* end-labeled neurons and/or glial cells were observed similarly in both PC and contralateral control cortex). Also, a few *in situ* end-labeled neurons could sometimes be identified right at the surgical site 1–2 d post PC. Similar results were also observed in sham surgery rats sacrificed 1 d post PC. No end-labeled or abnormal cells were detected at 2 wk or 4 wk post PC. Therefore, no significant tissue injury or cell death was identified that could be attributed to the PC procedure (data not shown; $n = 3–4$ rats per time-point).

3.1.5. Cerebral Blood Flow Unaffected by PC

On d 1, SS produced no change from 100% basal cortical flow, and PC produced an immediate decrease in flow to below 30% of basal flow, which recovered to 100% immediately upon reperfusion and persisted for at least 60 min of continuous monitoring, as described previously in this model (data not shown). On d 2, a similar basal flow between groups prior to pMCAO was observed, and a similar decrease in blood flow was observed for both PC and SS groups after pMCAO. This lack of any significant difference between groups was observed for at least 60 min of post-pMCAO monitoring. Although blood flow was reduced on d 3 in both groups, no significant difference in cortical perfusion was observed between SS or PC groups at this time (i.e., 24 h post pMCAO; data not shown).

3.1.6. Preconditioning Requires Protein Synthesis

In the first study, administration of cycloheximide 30 min prior to PC blocked its protective effects, as demonstrated by infarct size, infarct volume, and neurological deficits (**Fig. 2A**). Vehicle-treated PC animals demonstrate a reduced ($p < 0.05$) percent hemispheric infarct compared with vehicle-treated SS animals, whereas cycloheximide-treated PC animals did not exhibit a reduced hemispheric infarct compared with cycloheximide-treated SS animals (**Fig. 2A**, top left). Animals that underwent sham surgery all had significantly greater ($p < 0.05$) and similar-sized infarcts compared with PC vehicle-treated animals. These relationships were also evident for infarct volume (**Fig. 2A**, top right). PC animals treated with cycloheximide did not exhibit a reduced infarct volume (i.e., they did not exhibit IT/were not protected). In addition, cyclohex-imide also blocked the protective effect of preconditioning on neurological outcome. Both forelimb and hindlimb deficits (**Fig. 2A**, bottom left and bottom right, respectively) were significantly reduced ($p < 0.05$) in vehicle-treated PC animals compared to vehicle-treated SS rats. No protective effects of PC were observed in animals treated with cycloheximide. Therefore, blocking protein synthesis early following PC prior to pMCAO abolished PC-induced IT.

In the second study the results were completely different. The administration of cycloheximide 30 min prior to pMCAO did not block PC-induced IT-protective effects (**Fig. 2B**). Both vehicle- and cycloheximide-treated PC animals demonstrate a reduced ($p < 0.05$) percent hemispheric infarct compared with SS animals (**Fig. 2B**, top left). Animals that underwent sham surgery all had significantly greater ($p < 0.05$) and similar sized infarcts compared to PC-treated animals. These relationships were also evident for infarct volume (**Fig. 2B**, top right). In addition, cycloheximide also did not block the protective effect of PC on neurological outcome when administered later prior to pMCAO. Both forelimb and hindlimb deficits (**Fig. 2B**, bottom left and bottom right, respectively) were significantly reduced ($p < 0.05$) in both vehicle- and cycloheximide-treated PC animals compared with SS rats. Therefore, blocking protein synthesis much later after PC (i.e., essentially after pMCAO) did not block PC-induced IT *(8)*.

3.1.7. Preconditioning Induces IL-1ra mRNA Expression

Quantitative analysis of IL-1ra mRNA ($n = 4$) was performed, and data were normalized to the internal standard. Sham-operated samples were taken at 24 and 48 h. Only very low levels of IL-1ra mRNA were detected in the sham-operated animals or in the contralateral (control) cortex, as well as in the early time-points of the ipsilateral (PC) cortical samples. The level of IL-1ra mRNA was markedly increased in the PC cortex at 24 h ($p < 0.01$) and 48 h ($p < 0.01$) following PC. The expression of IL-1ra in the ipsilateral PC and contralateral control cortex 24 h after PC was confirmed by Northern analysis (i.e., was in close agreement with the data generated by quantitative RT/PCR (data not shown) *(8)*.

3.1.8. Preconditioning Induces IL-1ra Protein Expression

IL-1ra protein was not observed in the PC cortex at 6 and 12 h post-PC or in the contralateral control cortex at any time following. However, at 24 and 48 h post-PC, significant, specific IL-1ra labeling was observed in cells scattered throughout the PC cortex (data not shown). The significant IL-1ra protein expression at 24 and 48 h appeared to be restricted within cells having histologically identified neuronal morphology in the PC cortex and was not detected at all or was only lightly present in a few cells immediately adjacent to the surgical wound in SS rats ($n = 3$) *(8)*.

3.1.9. Preconditioning Induces IL-1β mRNA Expression

To quantify the levels of IL-1β mRNA expression after PC accurately, we applied a real-time PCR method and further refine the technique using cloned plasmid DNA as a standard. Based on the known molecular weight of the cloned DNA, the concentration of the template applied in each reaction can be calculated, and the copy number of the template can be determined based on the Avogadro number. Therefore, the correlation between the C_t value of PCR and the copy numbers of the templates can determined. A series of diluted DNA templates can thus serve as a standard for TaqMan amplification. Standards for IL-1β and rpL32 used the cloned plasmid DNA as a template, of which IL-1β was in a dilution series of 1.65–7.24 log molecule (or 16.4 fg–64 pg) copies and rpL32 at 3.04–7.93 log molecules (or 0.409 pg–0.32 ng). The serially diluted DNA samples produced different amplification plots (difference between the Rn+ value and the Rn– value vs the cycle numbers) for both IL-1β and rpL32. The amplification (as indicated by the C_t value) was in a linear relationship with the initial template. All our tested samples were located within this linear amplification range *(159)*.

To evaluate IL-1β mRNA expression after PC, RNA samples were reversibly transcribed and used for real-time PCR analysis. The specificity of PCR primers was evaluated using a standard protocol prior to the TaqMan amplification. The reversibly transcribed cDNA samples were then amplified using the ABI Prism 7700 Detection System, and their C_t values were determined. The copy numbers of the templates in each sample was thus determined by the C_t value according to the DNA standard.

On the other hand, since the inaccuracy of RNA concentration measurements may result in false information reflecting the level of RNA expression,

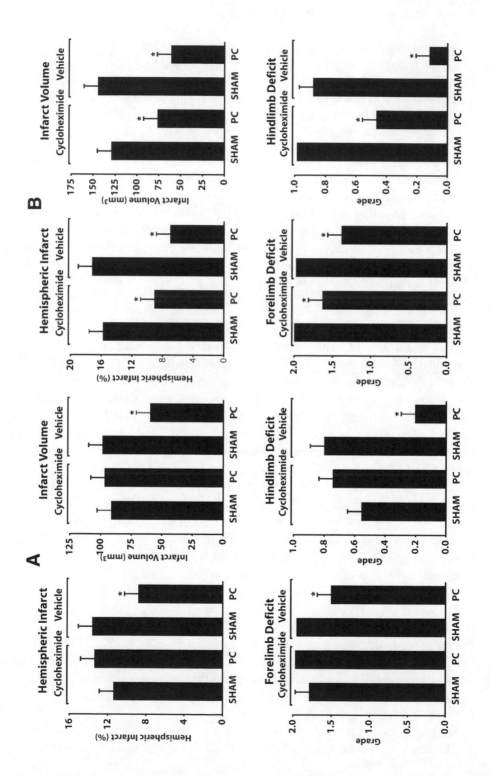

Fig. 2. (A) Blocking protein synthesis following ischemic preconditioning (PC) eliminates preconditioning-induced ischemic tolerance effects of reduced brain injury and neurological deficits. Completely counterbalanced study conducted to evaluate effects of protein synthesis inhibition using cycloheximide treatment. Cycloheximide administered 30 min prior to PC (as described in **Subheading 2.**, Methods) demonstrated that new protein(s) synthesized during the period following PC was essential to PC-induced IT-neuroprotective effects on later permanent MCAO. Top left, the significant neuroprotective effect of reduced percent hemispheric infarct in PC compared with SHAM surgery performed 1 d prior to pMCAO in animals treated with vehicle was not observed in animals treated with cycloheximide. Cycloheximide treatment did not alter the degree of ischemic injury owing to pMCAO (i.e., in SHAM surgery animals), but it blocked the ability of PC to reduce injury (i.e., blocked PC-induced IT when administered pre-PC). Top right, Similar results for infarct volume. Bottom left, the significant neuroprotective effect of reduced forelimb deficit in PC compared with SHAM surgery performed 1 d prior to pMCAO in animals treated with vehicle was not observed in animals treated with cycloheximide. Cycloheximide treatment did not alter the degree of forelimb deficit owing to pMCAO (i.e., in SHAM surgery animals), but it blocked the ability of PC to reduce deficits (i.e., blocked PC-induced IT when administered pre-PC). Bottom right, Similar results for hindlimb deficit ($n = 13–17$ per group). *, $p < 0.05$ different from all other groups; ANOVA with least significant difference follow-up test. (B) Blocking protein synthesis following pMCAO (but not during the period after PC preceding pMCAO) does not affect preconditioning-induced ischemic tolerance effects of reduced brain injury and neurological deficits. Completely counterbalanced study conducted to evaluate effects of protein synthesis inhibition using cycloheximide treatment. Cycloheximide administered later, 30 min prior to permanent MCAO (as described in **Subheading 2.**, Methods), demonstrated that new protein synthesized later following pMCAO in PC animals was not important for PC-induced IT. Top left, A significant neuroprotective effect of reduced percent hemispheric infarct in PC compared with SHAM surgery performed 1 d prior to pMCAO was observed in animals treated with cycloheximide or vehicle. Cycloheximide treatment did not alter the degree of ischemic injury owing to pMCAO (i.e., in SHAM surgery animals) and did not affect the ability of PC to reduce deficits (i.e., did not block PC-induced IT when administered much later after PC, 30 min before pMCAO). Top right, Similar results for infarct volume. Bottom left, A significant neuroprotective effect of reduced forelimb deficit in PC compared to SHAM surgery performed 1 d prior to pMCAO was observed in animals treated with cycloheximide or vehicle. Cycloheximide treatment did not alter the degree of forelimb deficit owing to MCAO (i.e., in SHAM surgery animals) and did not affect the ability of PC to reduce deficits (i.e., did not block PC-induced IT when administered much later after PC, 30 min before permanent MCAO). Bottom right, Similar results for hindlimb deficit ($n = 7–8$ per group). *, $p < 0.05$ different from both sham groups; ANOVA with least significant difference follow-up test.

especially the use of a very sensitive PCR method for RNA detection, the differences among the RNA samples are often corrected with a housekeeping gene that is consistently expressed in particular conditions. In the present study, rpL32 was selected as the internal control housekeeping gene as it was known to be consistently expressed in cortical samples after PC and focal stroke. The levels of rpL32 mRNA expression in the cortical samples were 1.01–1.53 × 10^5 copies of transcript per ng total RNA (or μg tissue). After normalizing to the rpL32 transcript, the quantitative data ($n = 4$) for IL-1β mRNA expression in cortical samples after PC or sham-operated animals indicated low levels of IL-1β mRNA (one to four copies of mRNA per ng RNA or μg tissue) were observed in the contralateral cortex. While the levels of IL-1β mRNA expression were elevated in the ipsilateral cortex after PC and even in sham-operated samples, significant induction in IL-1β mRNA was observed only at 6 h after PC (87 copies of mRNA per μg tissue, or fivefold increase compared with sham, $p < 0.001$).

3.1.9.1. COMPARISON OF IL-1B MRNA EXPRESSION IN ISCHEMIC PC WITH ISCHEMIC STROKE

To evaluate whether the levels of IL-1β mRNA expression in PC are different from those in ischemic cortex and thus correlate with its neuroprotective or neurotoxic role, we directly compared its peak expression in both models, i.e., 6 h after PC or 12 h after permanent MCAO *(41)* using a TaqMan PCR approach. The results indicate a significantly higher level of IL-1β mRNA expression after focal stroke (546 copies per μg ischemic tissue) compared with 87 copies after PC ($p < 0.01$, $n = 4$) *(159)*.

3.1.10. Preconditioning Induces IL-1β Protein Expression

To define further the cellular components and the upregulation of IL-1β peptide after preconditioning, immunohistochemical techniques were applied using a mouse anti-rat IL-1β antibody to examine IL-1β expression in sham or PC brain tissues at 6, 12, and 24 h and 2 and 14 d ($n = 3$). IL-1β immunoreactivity was detectable in a few glia along the surface of the cortex at 6 h after PC. IL-1β-immunoreactive glia were evident throughout most of the ipsilateral PC cortex including the cingulate cortex at 24 and 48 h after PC (**Fig. 3**). IL-1β-immunoreactive glia in the ipsilateral cortex were observed significantly over the next week and were still evident at 2 and 28 d after PC. The immunoreactivity appeared to be in the cell body and large processes of astroglia (**Fig. 3**). A very weak immunoreactive signal of IL-1β was detected in sham-operated cortex. Control sections stained with IL-1β-preabsorbed antisera, or exclusion of the primary antibody, did not demonstrate detectable staining for IL-1β (data not shown). Double immunofluorescence for IL-1β and GFAP

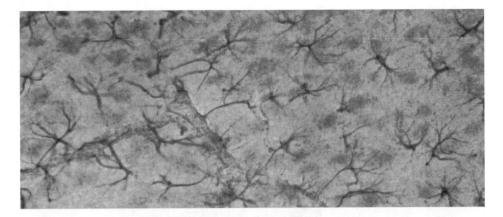

Fig. 3. Ischemic preconditioning increases the expression of IL-1β. Real-time PCR was used to quantify the significant induction of IL-1β mRNA post PC. Increases in the PC cortex at 6 h (87 ± 9 copies of the mRNA per microgram of brain tissue compared with 16 ± 5 copies in sham-operated samples, $p < 0.001$, $n = 4$) and 8 h (46+4 copies, $p < 0.01$, n = 4) after PC by means of real-time TaqMan PCR. The peak expression of IL-1β mRNA after PC was significantly ($p < 0.01$) lower than that after permanent occlusion of the middle cerebral artery, i.e., 87 ± 9 and 546 ± 92 copies of RNA per microgram tissue at peak levels for PC and focal stroke, respectively *(159)*. Immunohistochemistry studies revealed a parallel induction of IL-1β in the PC cortex. The maximal expression of IL-1β protein was observed during the first week post PC and paralleled the duration of ischemic tolerance *(159)*. However, a photomicrograph of IL-1β-labeled glial cells in the PC cortex 2 wk later is presented here. Significant labeling was also observed at 4 wk post-PC. Glial cell labeling was verified using glial fibrillary acidic protein as described in the text.

clearly demonstrated that IL-1β was expressed in GFAP-positive astrocytes (data not shown) *(159)*.

3.1.11. Ischemic Tolerance to Early Response Gene Expression

PC 24 h prior to pMCAO produced a significant reduction in the expression of early response genes 2 h following pMCAO. We previously reported that c-fos mRNA expression is increased in the ischemic cortex compared with the control cortex 2 h following MCAO; however, previous PC treatment significantly reduced this response (**Fig. 4A**). Similar results were also observed for zif268, the expression of which was significantly reduced owing to previous PC at 2 h following MCAO (**Fig. 4B**). Northern blots were similar to those presented previously from this laboratory *(8,160)* (data not shown). Basal levels of c-fos and zif268 mRNA in the control cortex were not altered by PC compared with SS and were similar to basal levels described previously in the control cortex or following sham surgery *(8,160)*.

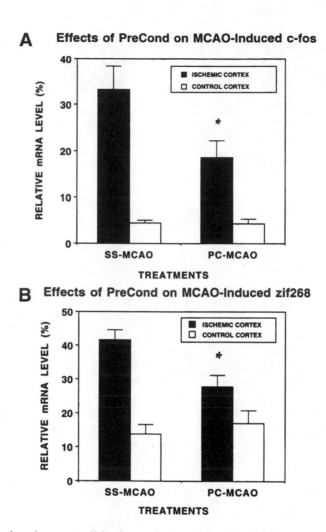

Fig. 4. Ischemic preconditioning reduces permanent middle cerebral artery occlusion (pMCAO)-induced increases in early response gene expression. A 1-d previously applied short (10-min) period of MCAO (ischemic preconditioning [PC])
significantly reduced the ischemic cortex early response gene expression produced
by 2 h of permanent MCAO (PC-MCAO) compared with sham PC surgery (SS)
followed by 2 h of MCAO (SS-MCAO). **(A)** Results for c-fos mRNA (n = 4 per
group). **(B)** Results for zif268 mRNA (n = 4 per group). Northern blots were quantitated by PhosphorImager analysis and normalized to ribosomal protein L32 mRNA
bands for samples loaded in each lane as described in the text and previously *(37)*.
The relative mRNA level for each probe is displayed graphically with a sum of 100%
for all cortical samples. *, $p < 0.05$ compared with SS-MCAO ischemic cortex;
ANOVA with Dunnett follow-up test.

3.1.12. Expression of Hsp70 and Hsp27 mRNA After PC

The quantitative data for Hsp70 and Hsp27 mRNA ($n = 6$), after normalizing autoradiographs of Northern blots to a housekeeping gene rpL32, are summarized graphically in **Fig. 5**. Sham-operated samples were taken at 6 ($n = 3$) and 24 h ($n = 3$) post surgery. Expression levels at both time points were similar, and the data were pooled. Practically no Hsp70 mRNA was detected under basal/sham conditions. Hsp27 mRNA was constitutively expressed in the contralateral cortex (basal levels). The expression of Hsp70 mRNA was similar to basal levels in the ipsilateral cortex of sham-operated animals. A significant induction of Hsp70 mRNA was observed at 6 h (26-fold increase over control, $p < 0.001$) and diminished to basal levels at 24 h in the ipsilateral control cortex after PC (**Fig. 5**). Hsp27 mRNA was induced, even in sham-operated ipsilateral or 6 h PC cortex (i.e., nonsignificant trend to increase over the contralateral control cortex). The expression of Hsp27 was significantly elevated later after PC (i.e., **Fig. 5B**; 12.7-fold increase at 24 h and 16.7-fold at 5 d post PC compared with contralateral cortex; $p < 0.05$) *(24)*.

3.1.13. Hsp70 Immunoreactivity After PC

Immunoreactive Hsp70 was observed at 6 h after PC in only a few glial cells immediately adjacent to the MCAO site (**Fig. 6A**). After 12 h, numerous Hsp70-immunoreactive neurons were evident in mid cortical layers and adjacent to the occlusion site. At this time, immunoreactive Hsp70 was localized to neuronal cell bodies, and a few Hsp70 immunoreactive glia were also evident adjacent to the occlusion site. After 1 and 2 d (**Fig. 6A**, middle), many immunoreactive cortical neurons were detected in the MCA territory. These neurons often had long apical dendrites filled with immunoreactive Hsp70. At 4 d, Hsp70-immunoreactive neurons were still detectable (**Fig. 6**, bottom) in the ischemic cortex. By 7 d (and at 2 and 4 wk) after PC, Hsp70 immunoreactivity was not detected in neurons in the cortex (data not shown) *(24)*.

3.1.14. In Situ End-Labeling After PC

Eleven brains were examined using *in situ* end-labeling following 10 min of MCA occlusion, with 2 d ($n = 4$), 2 wk ($n = 4$), and 4 wk ($n = 3$) of recovery. In addition, as positive controls for technique, sections were included from a neonatal rat brain ($n = 1$) and from an adult rat brain, 2 d after permanent MCA occlusion ($n = 1$). The neonatal brain had numerous *in situ* end-labeled (DAB-positive) nuclei throughout the cortex. The brain with 2-d permanent MCA occlusion had large numbers of *in situ* end-labeled nuclei in the periphery of the infarcted area (data not shown). Two days after PC, there were only a few *in situ* end-labeled nuclei immediately adjacent to the site of MCA occlusion in the cortex. At high magnification, the DAB reaction product appeared to be

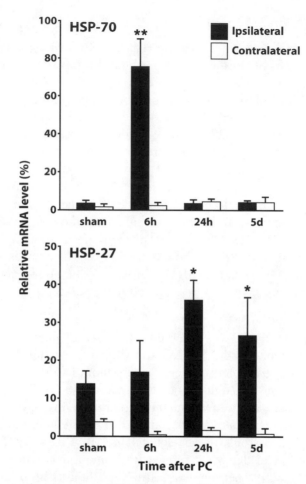

Fig. 5. Temporal expression of Hsp70 and Hsp27 mRNA in cortical samples after 10-min MCAO (PC) in rats. Total cellular RNA (40 µg/lane) was resolved by electrophoresis, transferred to a nylon membrane, and hybridized to the indicated cDNA probe. Presented here are the resulting quantitated Northern blot data ($n = 6$ per time-point/condition) for Hsp70 and Hsp-27 mRNA expression after PC. The data were generated using PhosphorImager analysis and displayed graphically (mean + standard errors) after normalizing with rpL32 mRNA signals. **, $p < 0.001$ compared with ipsilateral sham-operated animals at different timepoints and compared with contralateral control cortex at the same time-points. *, $p < 0.05$ compared with contralateral control at the same time-points.

localized over the nuclei of neurons. *In situ* end-labeled cells were not seen in the ischemic area were neurons that were Hsp70-positive. No obvious *in situ* end-labeled cells were detected in the ischemic area at 2 wk and 4 wk after 10-min MCA occlusion (data not shown) *(24)*.

Hsp 70   Hsp 27

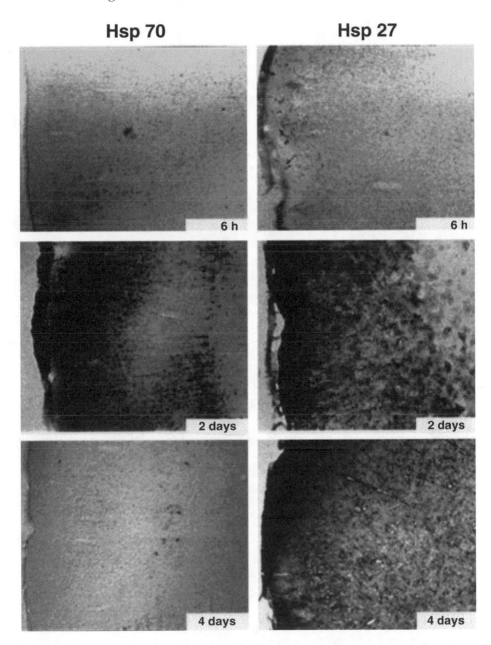

Fig. 6. Immunoreactive heat shock protein expression increases following PC. (**A**) Hsp70 expression in PC cortex at 6 h, 2 d and 4 d. (**B**) Hsp70 expression in PC cortex at 6 h, 2 d and 4 d. Hsp70 was expressed in neurons for the first week post-PC. Hsp27 was expressed primarily in astrocytes for 4 wk post-PC. Text describes results in detail.

3.1.15. Hsp27 Immunoreactivity After PC

Hsp27 immunoreactivity was first observed in the cortex at 6 h after PC in a few astrocytes immediately adjacent to the MCAO site on the cortical surface (**Fig. 6B**, top). After 12 h, 1 d, and 2 d (**Fig. 6B**, middle), Hsp27 immunoreactivity was detected in occasional neurons among numerous astrocytes. There was a progressive increase in the number of Hsp27 immunoreactive astrocytes to 4 d (**Fig. 6B**, bottom) after PC. Hsp27-immunoreactive astrocytes were detected in the ischemic zone but also in nonischemic ipsilateral cortex and frequently in astrocytes of the contralateral cortex. Hsp27-immunoreactive astroglia were detected in the ipsilateral and contralateral cortex at 1, 2, and 4 wk after PC. Hsp27 immunoreactivity was localized mainly to the cell bodies, the large processes, and also the fine processes of astroglia. Although Hsp27 immunoreactivity was seen mostly in astroglia, it was also seen in a few neurons *(24)*.

3.1.16. Distribution of Hsp70 and Hsp27 After PC

At 2 d post-PC, Hsp70 was detected in neurons in the region of the cortex corresponding to the ischemic area, whereas Hsp27 was detected in glia not only in the ischemic area of the cortex but also in nonischemic areas of the ipsilateral cortex *(24)*.

3.1.17. Double Immunofluorescence for Hsp27 and GFAP After PC

Four days after PC, Hsp27-immunopositive cells in the cortex (**Fig. 6B**, bottom) and were also GFAP-immunopositive (**Fig. 7**). Hsp27 immunoreactivity was localized in the numerous branching processes of the GFAP-positive cells. The degree of astrocyte reaction/gliosis at 2 wk after PC was remarkable (**Fig. 8**), with Hsp27/GFAP-positive cells extending throughout the whole ipsilateral cortex extending well beyond the MCA perfusion territories *(24)*.

The time-related and cell-type–associated expression of Hsp70, Hsp27, and IL-1β (i.e., as related to the IT interval post PC) is depicted in **Fig. 9**. Although induced and related to IT, none of the proteins absolutely correlated in time with IT. However, Hsp70 expression does correlate much more than the others do.

3.1.18. Ischemic Preconditioning Activates Caspase 3

Given that events such as ROS generation and energetic failure appear to contribute to PC, we sought to determine the extent of the cell death pathway activation in the in vivo model of ischemic preconditioning. Animals were treated to the preconditioning stimulus in the absence of a subsequent pMCAO and sacrificed at various times. Again, the preconditioning treatment alone did not result in any damage as assessed 24 h later (infarct size was $0 \pm 0\%$; $n = 4$). This result is consistent with our prior observations: no significant brain injury

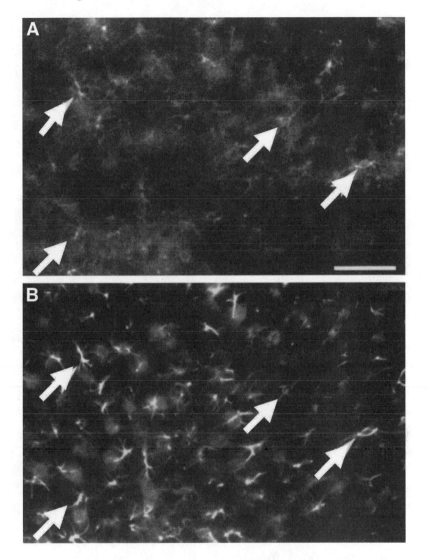

Fig. 7. Double immunofluorescence for Hsp27 and glial fibrillary acidic protein (GFAP). At 4 d after PC, Hsp27 was localized to the cell bodies, main processes, and fine bushy processes of astrocytes. **(B)** GFAP, a marker of astrocytes, was localized to the cell bodies and main processes of the Hsp27-immunopositive cells. Arrows indicate only four of several Double-labeled astrocytes observed in the identical fields of **(A)** and **(B)**. Scale bar 50 μm in A (also applies to B).

was observed, as shown by gross histological evaluation and *in situ* end-labeled neurons after PC. Twenty-micron coronal sections were taken from preconditioned animals and stained with an antibody that recognized the proteolytically

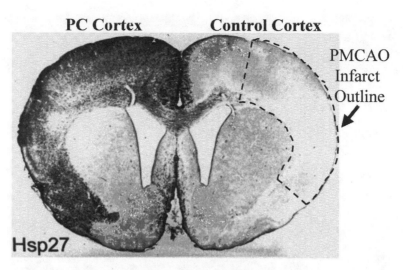

Fig. 8. Full frontal section illustrating the remarkable degree of Hsp27-immunore-active glial cells (and the degree of gliosis) present in the ischemic preconditioning (PC) cortex (compared with the control cortex) 2 wk after PC. The reactive brain area (clearly seen as the darkened area in the PC cortex and beyond) extends far beyond the MCA perfusion territory (i.e., beyond the area made ischemic by the PC stimulus). To provide some perspective for these studies, the area that typically becomes infarcted (i.e., at this forebrain level) following a permant middle cerebral artery occlusion (pMCAO) in this SHR stroke model is outlined by a dotted line and labeled in the control hemisphere.

cleaved/activated p20 subunit of caspase 3. Following sham PC, or at times when there was no preconditioning neuroprotection (6 h and 14 d), we did not observe appreciable caspase cleavage in the parietal cortex either contra- or ipsilateral to the PC (**Fig. 10**). Surprisingly, many cells expressed caspase 3 cleavage products in animals sacrificed at 1 and 7 d after preconditioning, times at which maximal neuroprotection is observed post-PC if animals are given subsequent pMCAO (**Fig. 10**).

3.2. In Vitro Results

3.2.1. Development of an In Vitro Model for Ischemic Preconditioning

Brief chemical ischemia induces neuroprotection against subsequent excitotoxicity in dissociated cortical neurons. To assess the role of caspase activation in preconditioning as well as to understand the biochemical events that contribute to this process, we developed an in vitro model of precondition-ing using a combination of energetic and oxidative stressors. In preliminary experiments, we observed that 90-min exposure to the mitochondrial toxin KCN

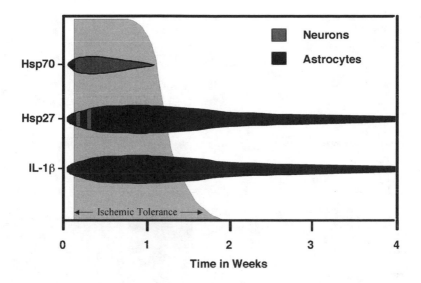

Fig. 9. Protein expression and ischemic tolerance following focal ischemic preconditioning. Schematic representation depicting the time-course and cellular immunoreactive labeling for Hsp70, Hsp 27, and interleukin-1β (IL-1β) following PC and in
reference to the period of brain protection from focal stroke that follows (i.e., period
of significant IT).

(3 mM) in glucose-free media was the maximal time we could treat cultures
without inducing any appreciable toxicity as measured by LDH release. These
conditions were then used as a preconditioning stimulus against a subsequent
excitotoxic dose of NMDA (100 mM for 60 min), which was delivered 24 h
later. There was near total neuronal loss in cultures that were not preconditioned but were then treated with NMDA. LDH release from dead and dying
neurons was assessed 24 h following the addition of NMDA. KCN-precondi-
tioned cells had significantly lower LDH values when subsequently NMDA-
treated compared with the naive cultures (**Fig. 11**). Importantly, as in the in
vivo model of preconditioning, there was little background cell death caused by
KCN in the absence of NMDA treatment, as the LDH values in this group were
virtually identical to the control non-NMDA-treated cultures (data not shown).

To assess the ability of 3 mM KCN to inhibit the toxicity produced by an even
greater secondary insult, additional experiments ($n = 5$) were performed in sister
cultures exposed to 200 μM NMDA for 18–24 h. Surprisingly, the KCN pretreatment was still able to provide a significant level of neuroprotection even in
cultures subjected to this extreme agonist exposure. Control cultures not exposed
to KCN or NMDA had a confluent bed of glia with many well-dispersed, phase-
bright neurons. When cultures were treated with KCN alone, without subsequent

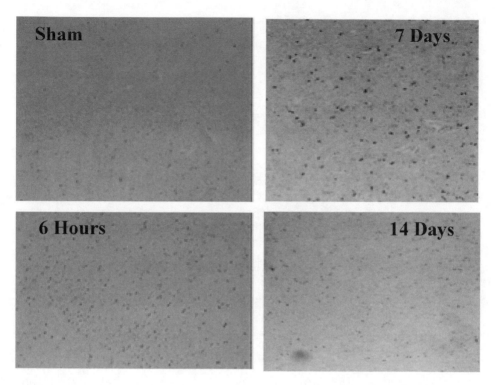

Fig. 10. Activated caspase 3 is increased in the PC cortex post PC. Animals were treated to a nontoxic preconditioning treatment with a 10-min MCA in the absence of a subsequent pMCAO and sacrificed at various times. Coronal sections were taken from rostral-caudal 4.4 mm from the vertical zone plane within the parietal cortex (i.e., it was previously shown that in this distal SHR pMCAO model, brain infarction is restricted to the cortex and observed primarily in the parietal lobe). Photomicrographs were taken at approx 0.5 mm in the dorsal-ventral plane on the side ventral to the horizontal plane and 6 mm in the medial-lateral coordinates. Sections were stained with an antibody to the activated form of caspase 3. No staining was observed in the cortex contralateral to PC at any of the time-points examined (data not shown). Similarly, at times when there was less preconditioning neuroprotection (6 h and 14 d), we did not observe any caspase cleavage in the parietal cortex ipsilateral to PC. We did observe remarkable caspase 3 cleavage (as shown by brown chromogenic reaction product) in animals sacrificed 1 d and 7 d (as shown above) after PC (i.e.) a time during which neuroprotection is observed. This was not observed at 6 h or at 14 d (i.e., when IT is not observed) as shown in **Subheading 3.1.18.** Sham animals were sacrificed 7 d after surgical procedure without PC and had no labeling in the cortex either ipsi- or contralateral to the skull opening.

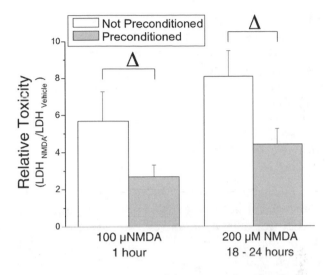

Fig. 11. Development of an *in vitro* model to assess the role of caspases in ischemic preconditioning. To ensure that the preconditioning stimulus alone was not toxic to cells, cortical cultures were treated for 90 min with 3 mM KCN in a glucose-free balanced salt solution (chemical ischemia) or with control solution, and lactate dehydrogenase (LDH) release was assessed 24 h later. Although no LDH release was observed in cortical neurons 18–20 h after KCN treatment, N-methyl-D-aspartate (NMDA)-treated cells subject to chemical ischemia 24 h earlier are significantly protected from excitotoxic cell death. Data represent LDH values taken 18–20 h following NMDA exposure. The pooled data from over 25 experiments (i.e., with results expressed by normalizing the ratio of LDH levels in the NMDA-treated cultures over those in the vehicle-treated cultures [LDH$_{NMDA}$/LDH$_{Vehicle}$], a measure of percent toxicity) exhibited significant neuronal tolerance/protection under these conditions. In this figure, cortical cultures were preconditioned with 3 mM KCN (PC) and then 24 h later were exposed to either 100 μM NMDA for 1 h or 200 μM NMDA for 18–20 h to determine whether preconditioning could also protect cells from very strong excitotoxic stressors/stimulation. Data were normalized as just described, and results demonstrate the neuroprotective effects of the PC in neurons subject to both types of exposure. Data represent the mean ± SEM. Statistical significance was assessed using paired *t*-tests between bracketed groups ($n = 5$; Δ, = $p < 0.05$).

NMDA exposure, neurons remained morphologically intact, and there were no obvious pathological changes throughout the culture. Cultures that were not preconditioned, but were then exposed to NMDA, underwent massive neuronal cell death, with large clusters of cellular debris evident following the excitotoxic insult and exhibited increased LDH (i.e., relative toxicity, as described in **Subheading 2.2.7.**) (**Fig. 11**). In contrast, NMDA-treated cultures that had been preconditioned exhibited much less cell death, with many intact, phase-bright neurons remaining that had significantly less relative toxicity (**Fig. 11**). The observed neuroprotection is consistent for the LDH values, compared with viable cell count experiments, and demonstrate that approx 50% of the cells are saved following NMDA exposure in preconditioned cultures (139). This was similar to the approx 50% reduction of infarct size exhibited in the in vivo PC-induced IT studies described in **Subheading 3.3.3.**

3.2.2. Ischemic Preconditioning In Vitro Is Time- and Protein Synthesis-Dependent

Ischemic preconditioning affords a limited window of efficacy and is protein synthesis-dependent in vivo *(8)*. Thus, we next assessed the time frame during which KCN treatment could provide neuroprotection in vitro. Cultures were treated for 60 min to 100 m*M* NMDA either 24, 48, or 72 h following the 90-min KCN treatment. For comparison across conditions, data are expressed as % toxicity, with values for NMDA exposure set as 100% cell death. Although approx 50% of the neurons were protected from NMDA toxicity with a 24-h interval between the KCN pretreatment and the excitotoxic exposure, there was no appreciable neuroprotection when agonist exposure occurred either 48 or 72 h following the KCN treatment. We also observed that preconditioning at the 24-h time point could be blocked when cells were treated with the preconditioning stimulus in the presence of 1 mg/mL of the protein synthesis inhibitor cyclohexamide. These data suggest that our in vitro model of preconditioning is comparable to other models including our in vivo system *(8,139)* in that it has a limited window of efficacy, which requires new protein synthesis.

3.2.3. Blocking Caspases During Preconditioning Increases Vulnerability to Subsequent Excitotoxicity

The purpose of developing an in vitro model of preconditioning that was comparable to our in vivo system was to determine the mechanism by which caspases could be activated in the absence of cell death in the preconditioning period. To investigate this process, we first sought to determine whether caspase activation was also important to the expression of excitotoxic tolerance. As with our previous experiments, cells were exposed for 90 min to 3 m*M* KCN in glucose-free media. Although this stimulus is nontoxic to the cells, we observed appreciable caspase 3 cleavage by both immunoblotting and immunohistochemistry. Immunoblots taken from cells harvested 4 h after the onset of chemical ischemia were probed with an antibody that recognizes both the pro-/uncleaved form of caspase 3 (p35) as well as its proteolytically cleaved product (p20), which is biologically active. These immunoblots showed not only that there is appreciable caspase 3 activation in cells that have only received the nontoxic preconditioning stimulus, but also that a substantial percentage of the protease remains intact. We also observed appreciable caspase 3 activation 6 h following the KCN exposure (data not shown) *(139)*.

To determine both localization and the extent of caspase 3 activation within the in vitro system, we performed immunohistochemistry on cultures 4 h following the initiation of chemical hypoxia. Using an antibody that only recognized the activated form of caspase 3, we observed that a substantial number of neurons within our cultures were immunopositive for caspase 3 activation after the onset of chemical ischemia. These observations are notable not only because there is no toxicity observed in the cells that have only received the preconditioning stimulus, but also because the caspase 3-immunoreactive cells comprise only a fraction of the total neurons within the cultures. Given that only 50% of the neurons within the cultures are spared by prior preconditioning, this may suggest that caspase 3 activation is critical for cell fate determination following preconditioning. Moreover, in spite of the caspase cleavage observed within these cells, immunoreactive neurons remained morphologically intact and had no indication of pyknosis or gross loss of cellular cytoarchitecture. Based on these experiments, as well as experiments in which cultures were stained for caspase 3 and GFAP, we determined that there was no appreciable caspase activation in the underlying glial bed *(139)*.

3.2.4. Ischemic Preconditioning is Blocked by Caspase Inhibitors, ROS Scavengers, and K_{ATP} Blockers

To determine the importance of limited caspase 3 proteolysis, K_{ATP} opening, and ROS generation for the expression of ischemic tolerance, we used a K_{ATP} blocker (glibenclamide), a free radical spin trap (*N*-tert-butyl-α-phenylnitrone [PBN]), a pan-caspase inhibitor (BOC-aspartate fluoromethylketone [BAF]), or a more specific caspase 3, 6, 7, 8, 10 inhibitor, Asp-Glu-Val-Asp-Ala (DEVD), during preconditioning. When these compounds are applied during preconditioning, there is an appreciable enhancement in the amount of toxicity observed when cells are subsequently exposed to NMDA. Cultures exposed to KCN pretreatment in the presence or absence of the broad-spectrum caspase inhibitor BAF demonstrated that there was no effect of prior treatment with the caspase inhibitor on NMDA toxicity (data not shown) *(139)*.

We also tested a number of more specific caspase inhibitors including zVDVADfmk (50 μ*M*), a caspase 2 inhibitor. Caspase 2 may perform the unique role of providing neuroprotection when activated in vivo. Caspase 2, like caspases 3 and 7 prefers the substrate recognition site DEVD, but it also requires a P5 amino acid in the peptide for efficient cleavage. The requirement for a fifth amino acid in substrates for caspase 2 means that inhibitors of the DEVD type, although they inhibit caspases 3 and 7, would have little effect on caspase 2 activity. As we observed no effect of zVDVADfmk that would suggest it either enhanced or abrogated the preconditioning (data not shown), it seems that caspase 2 activation does not contribute to neuroprotection induced by tolerance. Similarly, the caspase 1 and 4 specific inhibitor YVAD fmk (20 μM) provided only a 3 ± 9% block (n = 6; NS) of protection, which was not statistically significant, and the fmk control peptide (zFAfmk; 20 μM) slightly increased toxicity by 7 ± 4% (n = 3; NS). The caspase inhibitor that provided the most dramatic blockade of preconditioning was 10 μM DEVD (i.e., inhibitor of caspase 3; data not shown) *(139)*.

3.2.5. Bcl$_{Xl}$ and Hsp70 Are Induced at Times When Preconditioning Is Observed In Vivo

Caspase activation is normally associated with cell death. However, multiple cell signaling pathways may provide neuroprotection against cell death following caspase cleavage. These factors can be divided into four groups: the calbindins, the Bcl-2 family, the inhibitors of apoptosis (IAPs), and the heat shock proteins *(162,163)*. To determine which, if any, of these factors contributed to blocking cell death subsequent to the caspase cleavage in our model of preconditioning, we evaluated the effects of several promising target proteins from each of these families on proteins harvested at various time-points following preconditioning only (10-min MCAO) in our in vivo model of tolerance. Neither IAP 1,2 nor calbindin was consistently increased in the parietal cortex ipsilateral to the transient MCAO (PC). Similarly, no alterations in their expression were observed that would be consistent with a protective role in PC (data not shown). Although the mitochondrial heat shock protein Hsp25 was increased at 7 d, the increase was even more robust at 14 d, a time at which we have observed no appreciable preconditioning-induced protection (i.e., and consistent with our previous in vivo data). However, both Bcl$_{xl}$ and Hsp70 were consistently increased in the ipsilateral parietal cortex at times when protection is present if pMCAO is performed. These data suggest that both elevated Bcl$_{xl}$ and Hsp70 correlate well with the temporal profile of protection in vivo *(131)*.

3.2.6. Blocking Subtoxic Activation of Death-Associated Pathways Also Blocks Protection

We next assessed the temporal profile of the upregulation of Bcl$_{xl}$ and Hsp70 following preconditioning in vitro and sought to determine whether agents that blocked preconditioning also blocked induction of these proteins. We first established that these proteins were also upregulated in vitro by harvesting cells at various times after exposure to chemical ischemia. We observed a substantial increase in Bcl$_{xl}$ expression at both 24 and 72 h following KCN exposure.

Neuroprotection was only observed when cells were treated with NMDA at 24 h, but not 72 h, after KCN exposure, suggesting that Bcl_{xl} expression does not match the temporal profile of a protein that would be a critical mediator of ischemic tolerance in our in vitro system. We next assessed the effects of various inhibitors on the observed increase in Bcl_{xl} expression at the 24-h time point. Although both BAF and glibenclamide blocked the increase in Bcl_{xl} expression, PBN had no effect. Taken together, these data suggest that although the temporal profile of Bcl_{xl} is consistent with the observed neuroprotection in vivo, it is not consistent with our in vitro system. Further, given that at least one agent that has been shown to block preconditioning does not block Bcl_{xl} expression, the increase in this antiapoptotic protein may not be relevant to the observed neuroprotection *(139)*.

We next assessed the role of Hsp70 in our *in vitro* system. For these experiments, positive controls were generated by treating separate cultures to a 42°C heat shock for 90 min to assess the relative amount of Hsp induction by KCN. We observed that Hsp70 was only increased at a time when we observed protection (24 h) and that this induction was blocked to a large extent by each of the conditions that block preconditioning protection (data not shown) *(139)*. These observations support a critical role for cell stress caused by ROS production, energetic stress, and caspase activation in leading to induction of Hsp70. Moreover, the increased expression of Hsp70 appears to be one, if not the, critical component of IT under these conditions.

A final series of experiments were designed to test the ability of heat shock protein over expression during preconditioning to block neuroprotection. We hypothesized that activated caspases were held in check by pre-existing constitutively expressed heat shock proteins (Hscs) within the cell and that depletion of these chaperones was a critical signal to induce enhanced expression of Hsp70 (i.e., the induction of this protein). This would suggest that if sufficient Hsc70 were present during preconditioning, enhanced expression of Hsp70 would not be induced. To test this theory, we exposed cultures to high levels of biotinylated Hsc70 (75 µg/mL) protein 48 h prior to the initiation of KCN treatment. These experiments demonstrated that at the doses and time-points we used, intracellular Hsc70 expression is markedly enhanced. Immunoblots taken from intracellular homogenates of cultures exposed for 48 h to purified Hsc70 demonstrated a marked enhancement of Hsc70 expression immediately prior to KCN treatment (**Fig. 12A**).

Immunohistochemical detection of Hsc70 also revealed that many of the more intensely Hsc70-positive cells were indeed neurons. To assess the ability of limited Hsc70 over expression to block preconditioning, half of the Hsc overexposed cells were treated to KCN followed by either glycine or NMDA

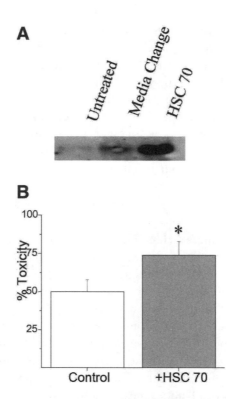

Fig. 12. Heat shock protein 70 overexpression during preconditioning can block ischemic tolerance. Although those agents that block in vitro PC-induced IT also block increased Hsp70 expression, we needed to directly test the hypothesis that caspase activation triggers new protein synthesis by binding to, and depleting, the free pool of heat shock proteins, reducing IT. Therefore, we overexpressed HSC 70 during preconditioning to determine whether it could block the expression of tolerance. (**A**) Cultures were exposed to 75 µg/mL biotinolyated purified Hsc70 or control media washes and harvested 48 h later. Immunoblots for intracellular HSC 70 expression reveal large increases in protein expression at the time when cultures were exposed to preconditioning chemical ischemia (with only small effects caused by media changes alone). (**B**) Toxicity experiments were performed as described in text and represented as percent toxicity. Cultures that were previously exposed to Hsc70 purified protein had increased cell death upon subsequent exposure to NMDA. Data represent the mean ± SEM of three independent experiments performed at least in duplicate. Asterisk denotes a significant difference vs matched naïve (not preconditioned) control with Hsc70 (*, $p <$ 0.05 paired t-test). Immunohistochemical staining of cells exposed to control washes or biotinylated Hsc70 indicated increased intraneuronal expression of the molecular chaperone immediately prior to cells being subject to KCN. Hsc70-pretreated cells that were preconditioned and then subsequently exposed to 200 µM NMDA for 24 h were severely morphologically compromised following excitotoxin exposure, consistent with the percent toxicity data in (**B**) (data not shown).

treatment. The other half of the Hsc overexpressed cells were treated to control washes similar to KCN and then NMDA or glycine. In support of this hypothesis, we observed a significant decrease in the amount of neuroprotection afforded by preconditioning in cells that had been pretreated with Hsc extract ($n = 3$; $p < 0.05$) (**Fig. 12B**).

Direct visualization of cells indicated that Hsc overexpression did not alter cell viability on its own or in the presence of KCN. Furthernore, the enhanced expression of Hsc70 during preconditioning did not alter susceptibility to NMDA exposure in cells that had not been preconditioned. However, there was a marked visible increase in susceptibility to NMDA in preconditioned cells that had received Hsc70 protein. Indeed, these cultures looked quite similar to cultures that received Hsc and NMDA but were not preconditioned (data not shown) and further substantiate the quantified results described above *(139)*.

4. Discussion

In Vivo PC/IT Model. It was initially demonstrated that brief periods of global ischemia having no untoward consequences *per se* resulted in significantly less hippocampal cell death after later longer periods of global ischemia *(7,95)*, or in significantly smaller infarctions after later pMCAO *(74)*. The protective effect of short bouts of global ischemia preceding transient focal ischemia has also been demonstrated *(6)*. A 20-min period of transient MCAO (which produced focal injury) reduced neuronal necrosis following subsequent global ischemia in the rat *(71)*. It has been demonstrated that previous transient focal ischemia reduces subsequent transient focal ischemic injury *(26)*. In this study, three 10-min periods of transient ischemia separated by 45 min of reperfusion (which by itself produced brain injury) reduced the degree of infarction owing to 100 min of transient MCAO if applied 2, 3, and 5 d (but not 1 or 7 d) before this more severe transient focal ischemia *(26)*. In the present studies, we demonstration brain tolerance to permanent focal stroke owing to previous PC with transient focal ischemia. Although it has been demonstrated previously that reperfusion accelerates the tissue response to ischemia *(154)*, the present technique of PC (that was selected from a pilot study of ischemic time–injury response relationships) produced no significant brain injury, as shown by gross histology and by evaluating histologic sections and *in situ* end-labeled neurons after PC. In this respect, our PC paradigm differs from many previous focal models. In addition, unlike IT in the heart, this series of studies indicates that as for cerebral global ischemia, no early (i.e., neuroprotection within 2 h post-PC) protection can be demonstrated for PC on pMCAO injury. Finally, this was the first demonstration of IT in SHRs. Again, additional studies revealed that there are strain differences in PC-induced IT *(101)*, as have been identified for strain differences in ischemic sensitivity *(104,151)*.

The degree of tissue protection in this model is remarkable, exceeding 50% reduction in infarct size from 1 to 7 d post-PC. In this distal pMCAO model, brain infarction is restricted to the cortex. The brain protection associated with PC-induced IT was distributed across the entire forebrain cortical infarction, with a larger reduction observed in the more posterior forebrain cortex (*see* **Figs. 1B** and **2B**). A reduction of neurological deficits owing to IT is also demonstrated for the first time, which is in accord with the degree of tissue protection. Perhaps other factors, such as increased nervous system plasticity following injury *(164)*, are also involved in protection from neurological deficits in PC-induced IT. Growth factors have been demonstrated to preserve function and increase recovery from injury *(165)*, and their expression and involvement in IT needs to be determined. It is important to note that the neurobehavioral measures tend to be more variable than brain injury measurements but do reflect the significant IT effects of PC.

No significant difference was observed in cortical blood flow that could explain the PC-induced neuroprotection. Lack of regional cerebral blood flow changes in brain areas induced for tolerance has also been reported in other models *(6,26)*. Furthermore, no significant contribution of plasma glucose, blood gases, or blood pressure changes could be associated with brain PC and the induction of IT *(6,26)*. In addition, the present study demonstrates that IT is not associated with systemic release of neuroprotective factor(s) since the protection was localized to the ipsilateral preconditioned cortex and was not observed in the contralateral hemisphere.

The molecular mechanisms associated with PC have not yet been elucidated. However, significant changes in gene transcription/translation have been documented following focal stroke that consist of well-defined sequential expression of genes with diverse functions that may bear upon tissue remodeling and resolution of the ischemic brain *(29,30)*. The present data clearly demonstrate that newly synthesized proteins are critical to PC-induced brain tolerance in vivo and in vitro. It has also been demonstrated that the PC-induced IT in the heart is associated with new protein synthesis that occurs within 60 min after PC *(166)* and is important for the later, delayed induction (i.e., "second window") of protection **(167)**. Understanding specific changes in gene expression and the identification of newly translated proteins can be critical to understand the mechanism of tissue protection in this model of IT.

As discussed earlier, ischemic injury has been shown to induce the expression of heat shock proteins. Heat shock proteins (Hsps) are believed to contribute to cellular repair processes by refolding denatured proteins and acting as molecular chaperones in normal processes such as protein translocation and folding *(168)*. The highly inducible member of the family of Hsps, Hsp70, has been cited in association with tolerance to ischemic injury in the brain

(24,26,27,72,78,95,99). Although much of the available data indicates the strong relationship between Hsps and IT, other data also suggest that Hsp is not required for IT (i.e., Hsp expression is not always closely correlated with neuroprotection induced by PC) *(24,26,61)*. We have recently demonstrated increased Hsp70 and Hsp27 expression following PC and currently are trying to understand the relationship of this protein expression to IT (data not shown).

The presence of low levels of IL-1ra in the normal brain and the marked upregulation of IL-1ra mRNA and protein after ischemic injury *(158)* suggest that IL-1ra can serve as a defense system to attenuate inflammatory reactions elicited by brain injury. Neuroprotection produced by IL-1ra has now been demonstrated in many animal models of brain injury *(48–51)*, and the temporal induction profile of IL-1ra following MCAO *(158)* virtually parallels that of IL-1β *(40)*, suggesting that IL-1ra may counteract IL-1β effects after ischemic stroke. Therefore, increased IL-1ra expression induced by PC could interfere with the development of injury associated with ischemia-induced IL-1β production. The present study is the first demonstration that PC can induce expression of this neuroprotective protein and that this increased expression is apparently in neurons, albeit based at this time only on cell morphology. The increased expression of both IL-1ra message and protein correlates in time with PC-induced neuroprotection. Recent data suggest that blocking IL-1 in the striatum using IL-1ra reduces focal stroke cortical injury (i.e., stimulation of IL-1 receptors in the striatum contribute to cortical ischemic injury and blocking those receptors contributes to IL-1ra's neuroprotective effects via a polysynaptic pathway or release of a specific substance[s]) *(169)/* Although we did not study the striatum in the present study, it is interesting to speculate that cortical PC-induced IL-1ra–expressing neurons project to the striatum to mediate the cortical protection observed in the present study. The IL-1ra message and expression data these really only circumstantially linked to IT, and this data is primarily "hypothesis generating." To test such a hypothesis, future studies should evaluate whether blocking IL-1ra can reduce IT under these conditions. These data, together with previous reports on biological effects of IL-1 in brain, suggest a dual role for IL-1β in ischemic injury and ischemic tolerance.

The mRNA encoded by the *c-fos* gene, and its protein product, Fos, provide an index of cell activation *(170)*. Together with Fos, Jun, Zif264, and other related proteins, these transcription-regulating factors can couple diverse stimuli to widespread expression of other genes *(171)*. For example, the dimerization of Fos and Jun forms a functional transcription factor complex (e.g., activator protein 1 [AP-1]) that binds to regulatory DNA sequences located in the upstream regions of target genes and regulates gene transcription *(172)*. The zif268 sequence encodes a 'zinc finger' protein that acts at another class of transcription regulatory sites *(173)*. The increased expression of c-fos has been demonstrated following ischemia *(174–186)*, and Fos- and Jun-like immunore-activity has been identified in neurons and astrocytes in ischemic-tolerant brain tissue *(177–179)*. Early increased expression of zif268 has been demonstrated following ischemia *(113,174,180)*. The present study is the first demonstration of altered ischemia-induced early response gene expression in a model of PC. We do not have evidence of cellular or topographical localization of early re-sponse genes following focal stroke in the present study. Poststroke protein product expression for early response genes has been demonstrated previously by others (*see* reviews in **refs. 29** and **30**), listed in **Subheading 1.4.** How-ever, the present data demonstrating reduced early response gene message ex-pression following focal stroke indicate that even this very early cellular response to ischemia was reduced by previous PC, thus demonstrating the in-creased brain tolerance to insult early following injury. One can expect that this reduced response to more severe ischemia produced by PC can alter down-stream gene expression effects that might contribute to a reduced degree of ultimate tissue injury. Others have also shown that even brief periods of MCAO-induced ischemia, which cause only mild cortical damage, increase c-fos and c-jun mRNA exclusively in the ipsilateral cortex, with a later increase in the DNA binding activity of AP-1 *(178)*. Recently, we have demonstrated that PC produces a small but significant early increase in c-fos and zif268 (unpublished observations) that might be responsible for the PC-induced IL-1ra response and the attenuated early response gene response following pMCAO in the previously PC cortex. However, the present data do not delin-eate the relationship of early response genes to IL-1ra, and changes in early gene expression might even be correlated with increased Hsp expression.

Brain resistance to ischemic injury is not limited to ischemic PC but has also been observed following many other tissue stresses including heat shock treat-ment *(181)* and chemical metabolic stress *(186)*. In addition, IT can be pro-duced by spreading depression (a common ischemia-related phenomenon). Indeed, spreading depression can induce ischemic tolerance *(183–185)*, and this appears to activate glial cells *(185)*, suggesting that increased support by these cells may be involved in subsequent neuroprotection. Available data sug-gest that threshold depolarizations required to induce tolerance are comparable to those that induce transcription factor mRNAs (e.g., c-fos), whereas that inducing Hsp72 approaches closer to the threshold for neuronal injury *(184)*. The coordinated expression of protective antioxidant enzymes *(186)* and nerve growth factors *(17,174,187,188)* may also be involved in IT. In addition, PC might alter tissue metabolism in a manner that sustains IT tissue by providing an increased penumbra and a reduced ischemic core zone, without altering blood flow. Finally, PC might attenuate later postischemic leukocyte adhesion and emigration *(189)*.

In general, the present in vivo model of PC-induced IT results in substantial and prolonged brain protection. The hallmarks of this model of focal ischemia are: (1) a remarkable, delayed yet prolonged neuroprotection against permanent focal stroke, (2) a PC-induced IT that is unrelated to changes in cerebral blood flow, (3) it elimination of the possibility of blood-borne factors in the phenomenon, (4) a neuroprotection/IT that is associated with changes in gene expression and is dependent on newly synthesized protein(s), and (5) a PC-induced resistance/tolerance to focal stroke brain injury that is associated with the increased prestroke expression of the neuroprotective protein IL-1ra and the reduced post-stroke expression of early response genes. It has been suggested that transient ischemic attacks might provide the preconditioning necessary to protect the brain from later, more severe insults *(190)*, and recent data suggests that this indeed might be the case *(191)*. Certainly utilizing controlled TIAs as a potential strategy bears the danger and risk of significant brain injury and is an unrealistic approach to therapy. However, it can be expected that as the mechanism(s) responsible for PC become more fully understood, we will increase our ability to identify novel targets for the post treatment protection from focal stroke brain injury, and/or we will be more likely to discover a safe pharmacological preconditioning agent (i.e., to produce a chemical preconditioning) that can protect the brain in high-risk individuals or prior to invasive cerebral surgical procedures.

Significant new information has emerged regarding in vivo PC-induced IT and Hsp's. In these studies, PC transiently upregulated Hsp70 mRNA and protein but produced a more prolonged upregulation of Hsp27 and IL-1β. Hsp27 mRNA expression was particularly sensitive to brain stimulation (i.e., nonsignificant trend to an increase due to surgery alone). These data were consistent with protein expression data. An early elevated neuronal Hsp70 immunoreactivity was observed in the ischemic area and a later, remarkable Hsp27 astrologic immunoreactivity within and outside the ischemic area was observed following PC. Such Hsp70 expression has been used as an indicator of cellular stress *(192)*. However, expression of Hsp70 does not always indicate whether neurons will die or survive following cerebral ischemia *(193)*. Interestingly, injury that induces IT also induces expression of mRNA for immediate early genes and transcription factors, whereas more severe injury that approaches the threshold for ischemic neuronal injury is required to induce mRNA for hsp70 strongly *(89)*. Although Hsp70 may facilitate cellular recovery after injury and contribute to IT, it does not appear to be responsible alone for IT. Hsp27 immunoreactivity was remarkably elevated in astrocytes of the ischemic cortex both within and outside the MCA territory. Although Hsp70 immunore-activity had returned to control levels by 7 d, Hsp27 and IL-1β immunoreactiv-ity was detectable, even quite remarkably at 4 weeks after PC, and both were localized in astrocytes, as verified by GFAP double labeling.

Finding Hsp70 expression within the ischemic area suggests that its induction is directly caused by consequences of the PC ischemic stimulus. For example, protein denaturation caused by pH changes or free radical generation would ini-tiate the stress response. As damaged protein accumulates in the cells, the cog-nate chaperone Hsc70 is recruited to refold or renature such damaged protein. As Hsc70 is recruited, it releases the heat shock transcription factor (HSF1), allow-ing HSF1 to trimerize, and bind to the heat shock element in the promoter region of Hsp70 genes *(194)*. Transcription of the inducible Hsp70 is initiated. Although it has been shown that increased expression of Hsp70 in the brain can provide neuroprotection, and several studies have shown that Hsp70 is expressed in PC-induced. IT can occur without increased Hsp70 expression. In the present study, little or no neuronal Hsp70 was detected 7 d after PC when significant IT, i.e., reduced cortical infarct size and reduced functional deficits, is observed *(8)*. This suggests that many factors may be involved in IT. We have previously indicated that altered immediate-early gene expression and/or increased neuronal expres-sion of the neuroprotective IL-1ra may also contribute to IT *(8)*.

Although PC induced the transient expression of Hsp70 in neurons, it pro-duced a much more remarkable and longer lasting astrogliosis, H&E as indi-cated by the increased astrocytic expression of Hsp27. It must be emphasized that little or no tissue/cell injury could be identified by standard histological techniques, e.g., hematoxyllin and eosin (data not shown), in the cortex owing to PC. However, using immunohistochemistry for Hsp27, a remarkable activa-tion of astrocytes was observed. Expression of Hsp27 both within and outside the ipsilateral MCA territory suggests that this protein is regulated differently following PC compared with Hsp70. Cortical spreading depression caused by the brief ischemia used for PC may activate astrocytes and induce Hsp27 out-side the ipsilateral MCA territory. Glutamate toxicity caused by glutamate release in injured neurons may be propagated throughout the cortex, and astro-cytes take up the glutamate to buffer its toxicity. Such glutamate uptake by astrocytes may induce expression of Hsp27 and contribute to neuroprotection. With use of the glutamate analog kainic acid, an antagonist of the NMDA receptor, to induce epileptic seizures, increased Hsp27 expression in astrocytes has been shown in the hippocampus and cortex *(82)*. Spreading depression induced by the cortical application of potassium chloride induces Hsp27 in astrocytes not only around the application site but also throughout the ipsilat-eral cortex *(84)*. The induction of cortical spreading depression and Hsp27 expression by potassium chloride can be suppressed by MK-801, a blocker of the NMDA receptor *(84,195)*. Therefore, it is possible that PC-/brief ischemia-induced spreading depression could be responsible for the development of IT. Furthermore, as indicated earlier, cortical spreading depression can protect the brain from injury, i.e., can produce IT to subsequent focal cerebral ischemia. Interestingly, spreading depression-induced early response gene expression can also be blocked by MK-801 *(96,195)*.

It is likely that the PC-induced astrogliosis contributes, at least in part, to the resulting IT/neuroprotection. Astrocytes certainly can protect neurons from injury *(196)*, and early astrocytic death in focal stroke may contribute to neuronal cell loss/infarction *(197)*. Glia dysfunction has been suggested to contribute to neuronal death in the penumbra area *(24,123,124,198,199)*; thus increasing or activating astrocytic function(s) might be expected to provide increased neuroprotective function. Glial protective action(s) have been demonstrated previously in vitro *(196)*. Astroglial activation may enhance astroglial function to provide tolerance to injury through such mechanisms as improved blood–brain barrier function, increased capacity for glutamate uptake and glutamine synthetase (i.e., glutamate inactivation) activity, enhanced maintenance of water and ionic homeostasis *(197,200–202)*, and increased expression of protective and reparative growth factors to enhance neuronal survival *(197,200–202)*. Similarly, in reactive astrogliosis, neuroprotection may be increased via the increased antioxidant activity of astrocytes including increased expression of antioxidant enzymes, increased reducing equivalent biosynthesis and recycling of vitamin C *(203–205)*. It has already been shown that PC can induce both IT and increased activity of the antioxidant enzyme SOD *(206)*. Spreading depression that produces IT also increases astroglial activation/gliosis *(207–209)*. In gerbil global cerebral ischemia, astrogliosis was also associated with ischemic tolerance of vulnerable CA1 hippocampal neurons *(123)*. Although Hsps can protect the brain and heart, there is evidence that they are not the only molecules/mechanisms responsible for brain tolerance *(210,211)*. They may play significant roles at different times. For example, early Hsp70 may protect neurons from stress, and later Hsp27 may protect astrocytes from stress (i.e., thus allowing increased support functions) in ischemia following PC. In addition, reduced microglial activation/phagocytic transformation was observed following severe global ischemia in animals previously exposed to PC. Microglia were not evaluated in the present study, but they may play a role in IT (i.e., by exhibiting a reduced activation) and should be evaluated in the future. The present data do suggest that longer term neuron–astrocyte interactions are modified by PC and that increased astrocytic support may play a critical role in increased neuronal survival in IT.

The activation of glia and the induction of Hsp27 also may be mediated through the known expression of several inflammatory cytokines *(29,30)*. More recently, the expression of TNF-α and IL-1β has been demonstrated soon after PC and suggested to contribute to the development of IT *(1,4,159,212)*. The long-term astrogliosis and Hsp27 expression is in line with the more chronic brain inflammatory response that may be associated with the remodeling and plasticity/recovery of the brain that occurs post stroke *(29,30)*. As discussed earlier, the pleotrophic cytokine TNF-α is increased and is apparently involved in PC/IT mechanisms *(212)*. For IL-1β, there also appears to be a paradox between the protective *(213)* vs the destructive *(43)* influences in IT and stroke injury. However, data are available that helps us understand its role(s) under different conditions. IL-1β can protect neurons from in vitro excitotoxicity, and this neuroprotection can be reversed by IL-1ra or nerve growth factor antibody neutralization *(214)*. However, high concentrations of IL-1β in vitro are neurotoxic, which can be reversed by IL-1ra *(213)*, suggesting a dual role for IL-1 in neuronal survival. Thus involvement of cytokines such as IL-1β in the protective effects induced by PC may be complex. Along the same lines, caspase 3 increases post PC when the brain is protected. The in vitro assay of neuronal tolerance allowed determination of the mechanism of neuronal protection under these conditions.

The induction of brain tolerance appears to involve many ongoing, simultaneous mechanisms. The present data clearly demonstrate the degree to which the brain can react to even minor, noninjurious stimulation. For us to identify the factor(s) most important in IT, we must understand changes in gene expression that occur following PC and during brain tolerance *(215)*. Then we will be able to develop tools such as antisense oligonucleotides, viral transfection vectors, appropriate transgenic animals, and/or selective blocking antibodies that can interfere with these differentially expressed genes and gene products to suppress programmed cell death pathways or enhance protective mechanisms.

In general, we have demonstrated that after a PC stimulus (i.e., 10-min MCAO) that induces a significant brain tolerance/neuroprotection, there is increased Hsp70 expression in neurons in the PC area, and Hsp27 expression in activated astrocytes within and extending significantly outside the originally ischemic field in the PC cortex. IL-1β is expressed in a manner similar to Hsp27. IT might be mediated, in part, by PC owing to its induction of neuronal Hsp70 and its activation of astrocytes with increased Hsp27; thus IT of neurons may be associated with and dependent on astrocyte-neuronal interactions. Hsps and other protective proteins may participate in the induction of IT. The expression of caspase 3 during IT is interesting. For example, the association of neuronal IL-1ra expression known to protect the brain from injury, the upregulation of IL-1β and TNF-α (factors known to exacerbate injury) protective growth, and the altered expression of transcription factors and inhibitors of proteolytic enzyme activity that occurs with PC-induced IT may contribute to brain tolerance. At this time, it is difficult to understand the prolonged upregulation of Hsp27 and IL-1β and their roles in PC-induced IT, although these are now open to testable hypotheses. In the future, the identification of the minimal brain PC stimuli that can produce IT (i.e., that required to upregulate protective gene transcription and translation) will need to be identified in order to elucidate the mechanism(s) of induced brain tolerance. This should be facilitated by differential genomic expression technology but might also require proteomics as well *(216)*.

4.2. In Vitro Model and Proposed IT Mechanisms

Given that traditional cell death activators such as ROS and energetic dysfunction can contribute to subsequent neuroprotection, the purpose of this work was to test the hypothesis that other cellular elements normally associated with neuronal cell death may be involved in neuroprotection in ischemic preconditioning. In the current study, we observed appreciable caspase 3 cleavage during the protective period in ischemic preconditioning in vivo in the absence of any subsequent cell death. To determine whether activation of caspases and ROS were important for the expression of tolerance, we developed a model of ischemic preconditioning in dissociated neuronal cultures. In this system, cells were exposed to cyanide in glucose-free media, resulting in a subtoxic chemical ischemic insult that provided protection against subsequent exposure to NMDA.

Chemical inhibition of mitochondrial function has been used by a number of groups to condition neurons against subsequent ischemia *(1,217)*. Cyanide is a potent mitochondrial toxin that prevents the oxidation of cytochrome *c* oxidase, thereby obstructing the electron transport chain and oxidative phosphorylation. It has been estimated that 90% of the cellular oxygen consumption occurs by the conversion of O_2 to H_2O by cytochrome oxidase. Inhibition of this reaction provides an effective means of inducing a controlled ischemic insult in dissociated cells. Cyanide treatment of neurons can be used both as a model of hypoxia and to characterize the relationship between oxygen starvation and excitotoxicity.

To validate this in vitro model and to understand the signaling pathways that contribute to preconditioning, we evaluated several hallmark features of ischemic tolerance including a limited time window of efficacy, a critical role for new protein synthesis, and a requisite role of K_{ATP} channels and ROS production. Based on our current observations, as well as those we have previously reported, we believe that these factors are also critical to both our in vivo and our in vitro preconditioning model. For instance, the acquisition of tolerance following ischemic conditioning is influenced by the interval between conditioning stimulus and subsequent injury. In the current study, we determined that a 24-h delay following chemical hypoxia provided maximal neuroprotection against NMDA receptor-mediated excitotoxicity, which is consistent with other reports *(218)*. Similarly, in our in vivo system, dramatic neuroprotection is observed when animals are given brief MCAO 24 h to 7 d prior to pMCAO. Although a rapid preconditioning phenomenon has been described in cardiomyocytes (starting within minutes of preconditioning and lasting less than 3 h), a comparable phenomena has not been reported in brain nor observed in these studies. The expression of a delayed "second window" of tolerance beginning approx 24 h after ischemic preconditioning has also been observed in cardiomyocytes and seems to parallel the events described in this work more closely.

A critical role for opening of K_{ATP} channels has been established in a variety of neuronal and non-neuronal models of preconditioning (reviewed in **ref. *219***). Phosphotransfer reactions regulate the behavior of K_{ATP} channels, which determines membrane excitability in response to metabolic stress (reviewed in **ref. *220***). Energetic dysfunction induced by exposure to 2 mM KCN in glucose free conditions is sufficient to open K_{ATP} channels within the cell membrane *(221)*. Opening of the K_{ATP} channel within the inner mitochondrial membrane decreases the mitochondrial membrane potential established by the proton pump and thereby accelerates electron transfer, resulting in a net oxidation of the mitochondrial matrix *(222)*. Oxidizing agents such as *p*-diethylaminoethylbenzoate or free radicals can also open K_{ATP} channels and contribute to pre-conditioning *(223)*. Moreover, ROS generation is itself then increased by opening of K_{ATP} channels *(224,225)*, further supporting a role for production of ROS in the expression of ischemic tolerance. Indeed, a critical role for ROS is also supported by our data in which we observed a significant decrease in the amount of preconditioning-induced protection when a free radical scavenger was present during the preconditioning period.

The role of K_{ATP} channels in the activation of mitochondrially initiated signals associated with apoptosis (such as Smac/Diablo, cytochrome *c*, apoptosis-inducing factor [AIF], EndoG, and HtrA2) remains unclear. Moreover, the contribution of potassium influx into the mitochondria in initiating cell death is not well understood. Activation of a cytoplasmic potassium conductance and the subsequent loss of this cation from the cell has been associated with neuronal apoptosis *(226–228)*. Mitochondrial potassium and proton gradients result in "active" cytochrome c release from the mitochondria *(227)*. These events can be blocked by the antiapoptotic protein Bcl$_{xl}$. Moreover, it has recently been shown that neuroprotective doses of the K_{ATP} opener diazoxide may increase the release of cytochrome c from the mitochondria in neurons, although this observation remains controversial *(226–229)*.

Mitochondria-orchestrated signaling has been shown to be critical in determining cell fate following exposure to a variety of stressors. Proapoptotic proteins within the mitochondria can be redistributed into the cytosol in response to injurious stimuli. The best characterized of these events is the mitochondrial release of cytochrome c. Once in the cytosol, cytochrome c associates with cytosolic apoptosis protease-activating factor-1 (Apaf-1) *(230,231)*. Association of these two proteins in the presence of dATP or ATP results in initiation of the formation of a multiprotein complex called the apoptosome *(232)*. This complex recruits the zymogen form of caspase 9, and promotes its cleavage. Once activated, caspase 9 can then directly cleave and activate procaspase 3.

Activation of caspases has historically been viewed as downstream of the "commitment to die" decision. However, emerging evidence now suggests that multiple cell signaling pathways may provide neuroprotection against both apoptosome assembly and activation of cell death elicited by cleaved caspases. These factors fall into four groups: calbindin, the Bcl-2 family, IAPs, and members of the heat shock protein family *(232,233)*. The calcium binding protein calbindin 28k has been shown to block apoptosis induced by a number of different triggers *(233,234)*, and purified calbindin 28k can bind to and inhibit the action of activated caspase 3 *(233)*. Although we observed increased calbindin expression 24 h after preconditioning with 10-min MCAO, this elevation was not specific to the occluded side of the cortex.

The Bcl-2 family has been shown to play a complex and varied role in apoptotic cell death, with members of the family both promoting and inhibiting caspase activation. Bcl_{xL} and Bcl-2 are thought to act, at least in part, by inhibiting proapoptotic members of the Bcl-2 family through heterodimerization. Another key function of Bcl-2 family members is regulation of release of proapoptotic factors, in particular cytochrome c, from the mitochondria into the cytosol *(235)*. It has been suggested that tolerance is conferred on cells by upregulation of Bcl-2 and Bcl_{xL} *(123)*. Although the upregulation of Bcl_{xl} did, to limited extent, parallel the time-course of neuroprotection observed in both our models of ischemic preconditioning, several points suggest that this protein does not solely contribute to the expression of tolerance. For instance, the induction of Bcl_{xl} was not entirely dramatic, particularly in vivo. Moreover, the free radical spin trap PBN, which blocks preconditioning, did not block the induction of Bcl_{xl}. These data do not rule out a contribution of this antiapoptotic protein to the expression of ischemic tolerance but do suggest that other factors probably play a larger role in this neuroprotective phenomenon.

The IAPs are another group of proteins that are capable of blocking activated caspases from inducing death and thus were investigated for their role in preconditioning. IAPs are a family of cytosolic proteins containing one or more characteristic baculoviral IAP repeat (BIR) domains. The BIR domains in XIAP, cIAP-1, and cIAP-2, bind to the p10 amino terminal of activated caspases 3, 7, and 9 and functionally sequester them *(236,237)*. Apoptosis induced by serum or potassium withdrawal can be delayed but not blocked by XIAP, IAP1, or IAP2 overexpression. Our observation that there was not a significant amount of induction of IAP1 or -2 in our in vivo model of preconditioning is consistent with recent studies suggesting that hypoxic insults more severe than our preconditioning stimuli are required to induce IAP2.

The final group of proteins we evaluated for their contribution to ischemic tolerance was the heat shock family. The Hsp family is comprised of highly conserved, abundantly expressed proteins with diverse functions including the assembly and sequestering of multiprotein complexes, transportation of nascent polypeptide chains across cell membranes, and regulation of protein folding. We first chose to evaluate the contribution to PC of the small Hsps. These proteins are inducible by cell stress, are phosphorylation-sensitive, and are capable of oligomerization. We were particularly interested in Hsp27, based on the in vivo expression of this protein and the fact that it has been shown to protect against apoptotic cell death triggered by a variety of stimuli including hyperthermia, oxidative stress, staurosporine, Fas ligand activation, and cytotoxic drugs (238,239). Hsp27 has been shown to contribute to cytoskeletal rearrangement following preconditioning, to decrease the production of ROS by increasing intracellular glutathione levels, and to block the activation of cell death downstream of cytochrome c release (240,241). We saw appreciable upregulation of Hsp27 following PC. However, the most dramatic increase in Hsp27 expression was 14 d after preconditioning, a time at which there is appreciably less ischemic tolerance. Although these this data confirm that Hsp27 can be induced by low-level insults, they also suggest that Hsp27 is not correlated with neuroprotection afforded by preconditioning and corroborate the previous in vivo immunohistochemistry data.

We next evaluated the role of the most abundant inducible and constitutively expressed heat shock proteins—the Hsp70 family (referred to as Hsp70 and Hsc70, respectively). Emerging evidence suggests that in addition to important functions in protein refolding and transport following cell stress, these highly homologous Hsps are also capable of binding and sequestering the activated caspases Apaf and AIF, making them particularly appealing targets for a role in ischemic tolerance. Hsc70 is the most abundant Hsp found in cells. Hsp70 is the major inducible Hsp found in all cells (242). Not surprisingly, injuries that induce protein denaturation (including heat shock, heavy metal exposure, hypoglycemia, low pH, and a variety of disease states) increase Hsp70 expression (243). Furthermore, ROS production and ischemic injury also lead to the induction of Hsp70. Moreover, induction of Hsp70 has been commonly linked with the expression of tolerance to ischemic brain injury (244). Overexpression of Hsp70 protects the brain against ischemia induced by MCAO or kainite-induced seizures *(244,245)*. Indeed, Hsp70 gene transfer can provide neuroprotection even when delivered several hours after the onset of ischemia. We have previously reported an increase in Hsp70 expression in our in vivo model of preconditioning and now present evidence that the upregulation of Hsp70 correlates well with the expression of tolerance. When in vivo tolerance is most fully expressed, from 1 to 7 d following PC, an appreciable early enhancement of Hsp70 expression is observed. Similarly, in our *in vitro* model of preconditioning, prior preconditioning induces appreciable neuroprotection when administered 24 h before excitotoxin exposure, a time at which there is maximal induction of Hsp70. Moreover, agents that block preconditioning were also effective at ameliorating the increase in Hsp70 expression, suggesting a critical role for activation of K_{ATP} channels, ROS, and caspases in the pathway leading to upregulation of this neuroprotective protein.

Taken together, our observations suggested that caspases and ROS are activated during preconditioning but are somehow held in check so as not to elicit cell death. Moreover, if activation of these processes is blocked, then there is no protection afforded by preconditioning. As new protein synthesis is required for preconditioning, we developed a model by which subthreshold caspase activation could upregulate neuroprotective pathways and not cause cell death on its own (**Fig. 13**). This model suggests that the initial energetic stress put upon cells during preconditioning generates ROS as well as opening of mitochondrial K_{ATP} channels. These events appear to lead to limited cytochrome c redistribution and activation of caspase 3. Once activated, caspases are held in check by sequestration using pre-existing proteins such as Hsc70. We hypothesized that binding to activated caspases depletes the free pool of Hsc70, resulting in the initiation of a positive feedback cycle and leading to increased production of Hsps. This allows for the expression of Hsps at a time when the second stressor is put upon the cells. Thus, preconditioning-associated protein synthesis is able to block a normally lethal exposure to excitotoxic insults. This model would predict that if either Hsp70 or Bcl_{xl} is critical to the expression of tolerance, blocking the events that trigger preconditioning (ROS generation, K_{ATP} opening, or caspase activation) should also block induction of these proteins. This indeed appears to be the case for Hsp70 expression.

The nature of the cell death pathway induced by the second insult following the preconditioning stimulus is important when considering potential neuroprotective mechanisms of Hsp70. Pairing KCN with NMDA results in a vast potentiation of cell death *(246,247)*, which strongly suggests that there is convergence in the cell death pathways induced by mitochondrial inhibition and NMDA. Caspase activation has been shown to contribute to both excitotoxicity and ischemic injury. The relative contribution of apoptotic pro-

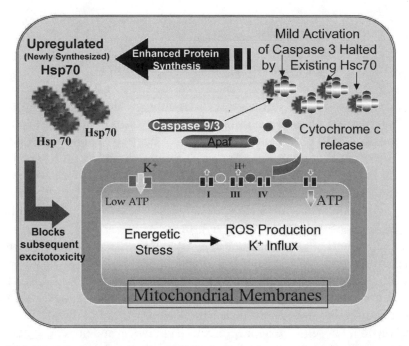

Fig. 13. Model of ischemic preconditioning based on available in vivo and in vitro data. Based on our observations that preconditioning elicits caspase cleavage and reactive oxygen species (ROS) generation, which are required for expression of protection, and that this protection requires new protein synthesis, we propose the following pathway. We hypothesize that the initial energetic stress put upon cells generates ROS and activation of mitochondrial K+ channels. The culmination of these events probably leads to limited cytochrome c redistribution and activation of caspase 3. These activated caspases are probably held in check by pre-existing proteins such as Hsc70. When these proteins are depleted, available data suggest that this results in the activation of a positive feedback cycle leading to increased production of Hsps. This upregulation is then able to block subsequent (i.e., later) caspase activation following a normally lethal exposure to excitotoxic insults. This model would predict that if either Hsp70 or Bcl$_{xl}$ is critical to the expression of tolerance, then blocking the events that trigger preconditioning (ROS generation, K$_{ATP}$ opening, or caspase activation) should block induction of these proteins, which is consistent with these studies and the available literature.

cesses to NMDA-induced toxicity is still not well understood, although it has been suggested that moderate activation of these glutamate receptors leads to reversible energetic failure and mitochondrial dysfunction followed by apoptosis, which can be attenuated with caspase inhibitors *(248)*. Moreover, caspase 3 inhibitors are capable of providing neuroprotection even when delivered several hours after an ischemic insult *(249)*.

Thus, in this in vitro model, preconditioning-induced protection occurs downstream of caspase activation, as in a variety of neuronal models of protection. It is, however, important to note that the second insult need not be apoptotic in nature to be blocked by Hsp70 overexpression. Given the diverse mechanisms by which Hsp70 can sequester, refold, and reprocess stressed proteins, caspase activation itself may not be the only critical component for this group of molecular chaperones to block subsequent injury associated with the process of brain tolerance.

There is great interest in further understanding the transcriptional process by which Hsp70 is induced by ischemic injury. As mentioned earlier, Hsps transcriptionally activate the induction of Hsp70. Hsp90 normally sequesters the Hsfs, but they are released when these chaperones bind to denatured proteins. Following their release, they are phosphorylated by unidentified kinases and then bind to the promoter region of the Hsp70 gene and induce its production. Gaining further insight into these transcriptional processes as well as understanding mechanisms that can induce Hsp expression in brain are of intense clinical interest. As the gene expression and signaling pathways responsible for ischemic tolerance become more fully understood, novel targets for the post-treatment protection from focal stroke brain injury might be identified, thus leading to the development of novel, safer pharmacological preconditioning agents that can protect the brain in high-risk individuals or before invasive cerebral surgical procedures. It will be of great interest to determine whether blockade of cell death pathways, particularly caspases, will also block cardiac preconditioning, given the important clinical ramifications of this phenomena for invasive thoracic surgery. Taken together, our data suggest that intervention against the neurological events (which result in subtoxic activation of traditional cell death pathways that can result in later protection) may actually increase vulnerability to the subsequent stressors. A clear understanding of the mechanisms and paths involved in these processes is certainly warranted.

4.3. Conclusions

The present line of research describes the use of in vivo and in vitro models of brain protection tolerance to elucidate important endogenous mechanisms of brain (and potentially other end organs for that matter). Together these model systems provide opportunities to test key hypotheses and build a significant knowledge base regarding brain tolerance. Future work should concentrate on identifying differential gene expression profiles (including microarrays) and verifying protein expression changes (including proteomics) and their cellular locations in order to help increase understanding of cellular/molecular processes of this endogenous neuroprotection process that can provide opportunities for end-organ injury in the future. Remarkable and long-term effects on expression systems can be observed that are not yet easy to understand (e.g., Hsp27 and IL-1β). However, the brain can be reformatted in many ways following subthreshold stimuli associated with injury, and, beyond tolerance, one should also determine the relevance of increased glial protein expression, proliferation, and activation to support systems associated with brain plasticity and recovery of function.

Acknowledgments

The author would like to thank his long-time collaborators in these studies. Although we are no longer working directly together at this time, it is especially important for me to acknowledge Drs. R. William Currie, Julie A. Ellison, James J. Velier, Giora Z. Feuerstein, BethAnn McLaughlin, Patricia Spera, and Xinkang Wang. Without their continued friendship and intellectual stimulation, this chapter would not have been possible. Others that need to be thanked include Drs. Jeffrey J. Legos, Joseph A. Erhardt, Ray F. White, Karen A. Hartnett, and Elias Aisenman. Special thanks go to Dr. BethAnn McLaughlin. It is her continued interest in collaborating on the in vitro and in vivo mechanistic studies that I expect will provide a new understanding of endogenous brain tolerance.

References

1. Dirnagl, U., Simon, R. P., and Hallenbeck, J. M. (2003) Ischemic tolerance and endogenous neuroprotection. *Trends Neurosci.* **26,** 248–254.
2. Chen, J., Graham, S. H., Zhu, R. L., and Simon, R. P. (1996) Stress proteins and tolerance to focal cerebral ischemia. *J. Cereb. Blood Flow Metab.* **16,** 566–577.
3. Dawson., V. l. Dawson, T. E. (2000) Neuronal ischaemic preconditioning. *Trends Pharmacol. Sci.* 21, 423–424.
4. Kirino, T. (2002) Ischemic tolerance. *J. Cereb. Blood Flow Metab.* 22, 1283–1296.
5. Schaller, B. and Graf, R. (2002) Cerebral ischemic preconditioning: an experimental phenomenon or a clinical important entity of stroke prevention? *J. Neurol.* 249, 1503–1511.
6. Matsushima, K. and Hakim A. M. (1995) Transient forebrain ischemia protects against subsequent focal cerebral ischemia without changing cerebral perfusion. *Stroke* **26,** 1047–1052.
7. Kitagawa, K., Matsumoto, M., Tagaya, M., et al. (1990) "Ischemic tolerance" phenomenon found in the brain. *Brain Res.* **528,** 21–24.
8. Barone, F. C., White, R. F., Spera, P. A., Currie, R. W., Wang, X. K., and Feuerstein, G. Z. (1998) Ischemic preconditioning and brain tolerance: temporal histologic and functional outcomes, protein synthesis requirement, and IL-1ra and early gene expression. *Stroke* **29,** 1937–1951.
9. Stagliano, N. E., Perez-Prinz, M. A., Moskowitz, M. A., Huang, P. L. (1999) Focal ischemic preconditioning induces rapid tolerance to middle cerebral artery occlusion in mice. *J. Cereb. Blood Flow Metab.* 19, 757–761.

10. Yellon, D. M., Baxter, G. F., Garcia-Dorado, D., Heusch, G., Sumeray, M. S. (1998) Ischaemic preconditioning: present position and future directions. *Cardiovasc Res.* **37,** 21–23.

11. Lawson, C. S. and Downey, J. M. (1993) Ischemic preconditioning: state of the art myocardial protection. Cardiovasc. Res. 27:542-550.

12. Alkhulaifi, A. M., Pugsley, W. B., Yellon, D. M. (1993) The influence of the time period between preconditioning ischemia and prolonged ischemia on myocardial iprotection. *Cardioscience* **4,** 163–169.

13. Yellon, D. M. and Baxter, G. F. (1995) A "second window of protection" or delayed preconditioning phenomenon: Future horizons for myocardial protection? J. Mol. Cellul. Cardiol. 27:1023-1034.

14. Prass, K., Wiegand, F., Schumann, P., et al. (2000) Hyperbaric oxygenation induced tolerance against focal cerebral ischemia in mice is strain dependent. *Brain Res.* 871, 146–150.

15. Zimmermann, C., Ginnis, I., Furuya, K., et al. (2001) Lipopolysaccharide-induced ischemic tolerance is associated with increased levels of ceramide in brain and in plasma. *Brain Res.* 895, 59–65.

16. Blondea, N., Widmann, C., Lazdunsk, M., Heurteaux, C. (2002) Polyunsaturated fatty acids induce ischemic and epileptic tolerance. *Neuroscience* 109, 231–241.

17. Wiegand, F., Liao, W., Busch, C., et al. (1999) Respiratory chain inhibition induces tolerance to focal cerebral ischemia. *J. Cereb. Blood Flow Metab.* 19, 1229–1237.

18. Chopp, M., Chen, H., Ho, K. L., et al. (1989) Transient hyperthermia protects against subsequent forebrain ischemic cell damage in the rat. *Neurology* 39, 1396–1398.

19. Nishio, S., Yunoki, M., Chen, Z. F., Anzivino, M. J, Lee, K. S. (2000) Ischemic tolerance in the rat neocortex following hypothermic preconditioning. *J. Neurosurg.* 93, 845–851.

20. Kobayashi, S., Harris, V. A., and Welsh, F. A. (1995) Spreading depression induces tolerance of cortical neurons to ischemia in rat brain. *J. Cereb. Blood Flow Metab.* **15,** 721–727.

21. Galea, E., Glickstein, S. B., Feinstein, D. L., Golanov, E. V., Reis, D. J. (1998) Cerebellar stimulation reduces inducible nitric oxide synthase expression and protects brain from ischemia. *Am. J. Physiol.* 274, H2035–H2045.

22. Kato, H., Liu, Y., Araki, T., Kogure, K. (1991) Temporal profile of the effects of pretreatment with brief cerebral ischemia on the neuronal damage following secondary ischemic insult in the gerbil: cumulative damage and protective effects. *Brain Res.* **553,** 238–242.

23. Tomida, S., Nowak, T.S., Vass, K., Lohr, J. M., Klatzo, I. (1987) Experimental model for repetitive ischemic attacks in the gerbil: The cumulative effect of repeated ischemic insults. *J. Cereb. Blood Flow Metab.* **7,** 773–782.

24. Currie, R. W., Ellison, J. A., White, R. F., et al. (2000) Benign focal ischemic preconditioning induces neuronal hsp70 and prolonged astrogliosis with expression of Hsp27. *Brain Res.* **863,** 169–181.

25. Liu, Y., Kato, H., Nakata, N., Kogure, K. (1992) Protection of rat hippocampus against ischemic neuronal damage by pretreatment with sublethal ischemia. *Brain Res.* **586,** 121–124.

26. Chen, J., Graham, S. H., Zhu, R. L., Simon, R. P. (1996) Stress proteins and tolerance to focal cerebral ischemia. *J. Cereb. Blood Flow Metab.* **16,** 566–577.

27. Kitagawa, K., Matsumoto, M., Kuwabara, K., et al. (1991) 'Ischemic tolerance' phenomenon detected in various brain regions. *Brain Res.* **561,** 203–211.

28. Kogure, K. and Kato, H. (1993) Altered gene expression in cerebral ischemia. *Stroke* **24,** 2121–2127.

29. Barone, F. C. (1998) Emerging therapeutic targets in focal stroke and brain trauma: Cytokines and the brain inflammatory response to injury. *Emerg. Ther. Targ.* 2, 1–23.

30. Barone, F. C. and Feuerstein, G. Z. (1999) Inflammatory mediators and stroke: New opportunities for novel therapeutics (Review). *J. Cereb. Blood Flow Metab.* **15,** 819–834.

31. Read, S. J., Parsons, A. A., Harrison, D. C., et al. (2001) Stroke genomics: approaches to identify, validate and understand adaptive gene expression changes in ischemic stroke (Review). *J. Cereb. Blood Flow Metab.* **21,** 755–778.

32. Trendelenburg, G., Prass, K., Priller, J., et al. (2002) Serial analysis of gene expression identifies metallothionein-II as major neuroprotective gene in mouse focal cerebral ischemia. *J. Neurosci.* **22,** 5879–5888.

33. Lu, A., Tang, Y., Ran, R., et al. (2003) Genomics of the periinfarction cortex after focal ischemia. *J. Cereb. Blood Flow Metab.* **23,** 786–810.

34. Weih, M., Kallenberg, K., Bergk, A., et al. (1999) Attenuated stroke severity after prodromal TIA: a role for ischemic tolerance in the brain? *Stroke* 30, 1851–1854.

35. Moncayo, J., de Freitas, G. R., Bogousslavsky, J., Altieri, M., van Melle, G. (2000) Do transient ischemic attacks have a neuroprotective effect? *Neurology* 54, 2089–2094.

36. Moncayo, J., de Freitas, G.R., Bogousslavsky, J., Altieri, M., van Melle, G. (1995) Essential role of adenosine, adenosine A1 receptors, and ATP-sensitive K^+ channels in cerebral ischemic preconditioning. *Proc. Natl. Acad. Sci. USA* 92, 4666–4670.

37. Bordet, R., Deplanque, D., Maboudou, P., et al. (2000) Increase in endogenous brain superoxide dismutase as a potential mechanism of lipopolysaccharide-induced brain ischemic tolerance. *J. Cereb. Blood Flow Metab.* 20, 1190–1196.

38. Ohtsuki, T., Ruetzler, C.A., Tasaki, K., and Hallenbeck, J. M. (1996) Interleukin-1 mediates induction of tolerance to global ischemia in gerbil hippocampus CA1 neurons. *J. Cereb. Blood Flow Metab.* **16,** 1137–1142.

39. Buttini, M., Sauter, A., Boddeke, H. W. G. M. (1994) Induction of interleukin-1β mRNA after cerebral ischemia in the rat. *Mol. Brain Res.* **23,** 126–134.

40. Liu, T., McDonnell, P. C., Young, P. R., et al. (1993) Interleukin-1β mRNA expression in ischemic rat cortex. *Stroke* **24,** 1746–1751.

41. Wang, X., Yue, T. L., Barone, F. C., et al. (1994) Concomitant cortical expression of TNF-α and IL-1 β mRNAs follows early response gene expression in transient focal ischemia. *Mol. Chem. Neuropathol.* **23,** 103–114.
42. Saito, K., Suyama, K., Nishida, K., Sei, Y., Basile, A. S. (1996) Early increases in TNF-α, IL-6 and IL-1β levels following transient cerebral ischemia in gerbil brain. *Neurosci. Lett.* **206,** 149–152.
43. Yamasaki, Y., Matsuura, N., Shozuhara, H., Onodera, H., Itoyama, Y., Kogure, K. (1992b) Interleukin-1 as a pathogenetic mediator of ischemic brain damage in rats. *Stroke* **26,** 676–680.
44. Hara, H., Friedlander, R. M., Gagliardini, V., et al. (1997) Inhibition of interleukin 1beta converting enzyme family proteases reduces ischemic and excitotoxic neuronal damage. *Proc. Natl. Acad. Sci. USA* **94,** 2007–2012.
45. Hara, H., Fink, K., Endres, M., Friedlander, R.M., Gagliardini, V., Yuan, J., Moskowitz, M.A. (1997) Attenuation of transient focal cerebral ischemic injury in transgenic mice expressing a mutant ICE inhibitory protein. J. Cereb Blood Flow Metab. 17: 370-375.
46. Eisenberg, S. P., Brewer, M. T., Verderber, E., et al. (1991) Interleukin 1 receptor antagonist is a member of the interleukin 1 gene family: evolution of a cytokine control mechanism. *Proc. Natl. Acad. Sci. USA* **88,** 5232–5236.
47. Dinarello, C. A. and Thompson, R. C. (1991) Blocking IL-1: interleukin 1 receptor antagonist *in vivo* and *in vitro*. *Immunol. Today* **12,** 404–410.
48. Relton, J. K. and Rothwell, N. J. (1992) Interleukin 1 receptor antagonist inhibits ischemic and exciotoxic neuronal damage in the rat. *Brain Res. Bull.* **29,** 243–246.
49. Rothwell, N. J. and Relton, J. K. (1993) Involvement of interleukin-1 and lipocortin-1 in ischemic brain damage. *Cerebrovasc. Brain Metab. Rev.* **5,** 178–198.
50. Toulmond, S. and Rothwell, N. J. (1995) Interleukin-1 receptor antagonist inhibits neuronal damage caused by fluid percussion injury in the rat. *Brain Res.* **671,** 261–266.
51. Loddick, S. A. and Rothwell, N. J. (1996) Neuroprotective effects of human recombinant interleukin-1 receptor antagonist in focal cerebral ischemia in the rat. *J. Cereb. Blood Flow Metab.* **16,** 932–940.
52. Wong, G. H. and Goeddel, D. V. (1988) Induction of manganous superoxide dismutase by tumor necrosis factor: possible protective mechanism. *Science* 242, 941–944.
53. Goodman, Y. and Mattson, M. P. (1996) Ceramide protects hippocampal neurons against excitotoxic and oxidative insults, and amyloid β-peptide toxicity. *J. Neurochem.* 66, 869–872.
54. Nawashiro H., Tasaki K., Ruetzler C. A., Hallenbeck J. M. (1997) TNF-α pretreatment induces protective effects against focal cerebral ischemia in mice. J. Cereb. Blood Flow Metab. 17:483–490.
55. Liu, J., Ginis, I., Spatz, M., Hallenbeck, J.M. (2000) Hypoxic preconditioning protects cultured neurons against hypoxic stress via TNF-α and ceramide. Am. *J. Physiol. Cell Physiol.* 278, C144–C153.
56. Ginis, I., Jaiswal, R., Klimanis, D., Liu, J., Greenspon, J., Hallenbeck, J. M. (2002) TNF-α-induced tolerance to ischemic injury involves differential control

of NF-B transactivation: the role of NF-B association with p300 adaptor. *J. Cereb. Blood Flow Metab.* 22, 142–152.

57. Cadenas, A., Morro M.A., Leza, J.C., et al. (2002) Upregulation of TACE/ ADAM17 after ischemic preconditioning is involved in rain tolerance. *J. Cereb. Blood Flow Metab.* **22**, 1297–1302.

58. Dienel, G. A., Kiessling, M., Jacewicz, M., and Pulsinelli, W. A. (1986) Synthesis of heat shock proteins in rat brain cortex after transient ischemia. *J. Cereb. Blood Flow Metab.* **6**, 505–510.

59. Gonzalez, M. F., Lowenstein, D., Fernyak, S., Hisanaga, K., Simon, R., and Sharp, F. R. (1991) Induction of heat shock protein 72-like immunoreactivity in the hippocampal formation following transient global ischemia. *Brain Res. Bull.* **26**, 241–250.

60. Sharp, F. R., Lowenstein, D., Simon, R., and Hisanaga, K. (1991) Heat shock protein hsp72 induction in cortical and striatal astrocytes and neurons following infarction. *J. Cereb. Blood Flow Metab.* **11**, 621–627.

61. Simon, R. P., Cho, H., Gwinn, R., and Lowenstein, D. H. (1991) The temporal profile of 72-kDa heat-shock protein expression following global ischemia. *J. Neurosci.* **11**, 881–889.

62. Vass, K., Welch, W. J., and Nowak, T. S., Jr. (1988) Localization of 70-kDa stress protein induction in gerbil brain after ischemia. *Acta. Neuropathol. (Berl.)* **77**, 128–135.

63. Lindquist, S. (1988) The heat-shock proteins. *Annu. Rev. Genet.* **22**, 631–677.

64. Becker, J. and Craig, E. A. (1994) Heat-shock proteins as molecular chaperones. *Eur. J. Biochem.* **219**, 11–23.

65. Hartl, F. U. (1996) Molecular chaperones in cellular protein folding. *Nature* **381**, 571–580.

66. Schatz, G. and Dobberstein, B. (1996) Common principles of protein translocation across membranes. *Science* **271**, 1519–1526.

67. Currie, R. W. and White, F. P. (1981) Trauma-induced protein in rat tissues: a physiological role for a "heat shock" protein? *Science* **214**, 72–73.

68. Currie, R. W., Karmazyn, M., Kloc, M., and Mailer, K. (1988) Heat-shock response is associated with enhanced post-ischemic ventricular recovery. *Cir. Res.* **63**, 543–549.

69. Marber, M. S., Mestri, R., Chi, S. H., Sayen, M. R., Yellon, D. M., and Dillmann, W. H. (1995) Overexpression of the rat inducible 70-kD heat stress protein in a transgenic mouse increases the resistance of the heart to ischemic injury. *J. Clin. Invest.* **95**, 1446–1456.

70. Plumier, J. C. L., Ross, B.M ., Currie, R. W., et al. (1995) Transgenic mice expressing the human heat shock protein 70 have improved post-ischemic myocardial recovery. *J. Clin. Invest.* **95**, 1854–1860.

71. Glazier, S. S., O'Rourke, D. M., Graham, D. I., and Welsh, F. A. (1994) Induction of ischemic tolerance following brief focal ischemia in rat brain. *J. Cereb. Blood Flow Metab.* **14**, 545–553.

72. Nishi, S., Taki, W., Uemura, Y., et al. (1993) Ischemic tolerance due to the induction of HSP70 in a rat ischemic recirculation model. *Brain Res.* **615**, 281–288.

73. Plumier, J. C. L., Krueger, A. M., Currie, R. W., Kontoyiannis, D., Kollias, G., and Pagoulatos, G. N. (1997) Transgenic mice expressing the human inducible Hsp70 have hippocampal neurons resistant to injury. *'Cell Stress Chaperones* **2,** 162–167.

74. Simon, R. P., Niiro, M., and Gwinn, R. (1993) Prior ischemic stress protects against experimental stroke. *Neurosci. Lett.* **163,** 135–137.

75. Kitagawa, K., Matsumoto, M., Kuwabara, K., et al. (1991) Hyperthermia-induced neuronal protection against ischemic injury. *J. Cereb. Blood Flow Metab.* **11,** 449–452.

76. Kitagawa, K., Matsumoto, M., Kuwabara, K., et al. (1991) ,'Ischemic tolerance' phenomena detected in various brain regions. *Brain Res.* **561,** 203–211.

77. Liu, Y., Kato, H., Nakata, N., and Kogure, K. (1992) Protection of rat hippocampus against ischemic neuronal damage by pretreatment with sublethal ischemia. *Brain Res.* **586,** 121–124.

78. Liu, Y., Kato, H., Nakata, N., and Kogure, K. (1993) Temporal profile of heat shock protein 70 synthesis in ischemic tolerance induced by preconditioning ischemia in rat hippocampus. *Neuroscience* **56,** 921–927.

79. Nowak, T. S., Osborne, O. C., Suga, S. (1993) Stress protein and proto-oncogene expression as indicators of neuronal pathophysiology after ischemia. *Prog. Brain Res.* **96,** 195–208.

80. Mehlen, P., Préville, X., Chareyron, P., Briolay, J., Klementz, R., and Arrigo, A.P. (1995) Constitutive expression of human Hsp27, *Drosophila* Hsp27, or human alpha B-crystallin confer resistance to TNF- and oxidative stress-induced cytotoxicity in stably transfected murine L929 fibroblasts. *J. Immunol.* **154,** 363–374.

81. Landry, J., Chrétien, P., Lambert, H., Hickey, E., and Weber, L.A. (1989) Heat shock resistance conferred by expression of the human HSP27 gene in rodent cells. *J. Cell. Biol.* **109,** 7–15.

82. Kalwy, S. A., Akbar, M. T., Coffin, R. S., deBelleroche, J., Latchman, D. S. (2003) Heat shock protein 27 delivered via herpes simplex virus vector can protect neurons in the hippocampus against kainic-acid-induced cell loss. *Brain Res.* **111,** 91–103.

83. Plumier, J. C. L., Armstrong, J. N., Landry, J., Babity, J. M., Robertson, H. A., and Currie, R. W. (1996) Expression of the 27-kDa heat shock protein (Hsp27) following kainic acid-induced status epilepticus in the rat. *Neuroscience* **75,** 849–856.

84. Plumier, J. C. L., Armstrong, J. N., Wood, N. I., et al. (1997) Differential expression of c-fos, hsp70 and hsp27 after photothrombotic injury in the rat brain. *Mol. Brain Res.* **45,** 239–246.

85. Plumier, J. C. L., David, J. C., Robertson, H. A., and Currie, R. W. (1997) Cortical application of potassium chloride induces the low molecular weight heat shock protein (Hsp27) in astrocytes. *J. Cereb. Blood Flow Metab.* **17,** 781–790.

86. Kato, H., Kogure, K., Liu, X-H., Araki, T., Kato, K., and Itoyama, Y. (1995) Immunohistochemical localization of the low molecular weight stress protein Hsp27 following focal cerebral ischemia in the rat. *Brain Res.* **679,** 1–7.

87. Kato, H., Liu, Y., Kogure, K., and Kato, K. (1994) Induction of 27-kDa heat shock protein following cerebral ischemia in a rat model of ischemic tolerance. *Brain Res.* **634,** 235–244.

88. Abe, H. and Nowak, T. S., Jr. (2000) Postischemic temperature as a modulator of the stress response in brain: dissociation of heat shock protein 72 induction from ischemic tolerance after bilateral carotid artery occlusion in the gerbil. *Neurosci. Lett.* **295,** 54–58.

89. Abe, H. and Nowak, T. S., Jr. (1996) Gene expression and induced ischemic tolerance following brief insults. *Acta Neurobiol. Exp. (Warsz.)* **56,** 3–8.

90. Sommer, C., Gass, P., Kiessling, M. (1995) Selective c-JUN expression in CA1 neurons of the gerbil hippocampus during and after acquisition of an ischemia-tolerant state. *Brain Pathol.* **5,** 135–144.

91. Belayev, L., Ginsberg, M.D., Alonso, O.F., Singer, J.T., Zhao, W., Busto, R. (1996) Bilateral ischemic tolerance of rat hippocampus induced by prior unilateral transient focal ischemia: relationship to c-fos mRNA expression. *Neuroreport* **8,** 55–59.

92. Massa, S. M., Swanson, R. A., Sharp, F. R. (1996) The stress gene response in brain. *Cerebrovasc. Brain Metab. Rev.* **8,** 95–158.

93. Morimoto, R. I. and Santoro, M. G. (1998) Stress-inducible responses and heat shock proteins: new pharmacologic targets for cytoprotection. *Nat. Biotechnol.* **16,** 833–838.

94. Paschen, W. and Doutheil, J. (1999) Disturbances of the functioning of endoplasmic reticulum: a key mechanism underlying neuronal cell injury. *J. Cereb. Blood Flow Metab.* **19,** 1–18.

95. Kirino, T., Tsujita, Y., and Tamura, A. (1991) Induced tolerance to ischemia in gerbil hippocampal neurons. *J. Cereb. Blood Flow Metab.* **11,** 299–307.

96. Kato, H., Liu, Y., Araki, T., Kogure, K. (1992) MK-801, but not anisomycin, inhibits the induction of tolerance to ischemia in the gerbil hippocampus. *Neurosci. Lett.* **139,** 118–121

97. Bond, A., Lodge, D., Hicks, C. A., Ward, M. A., O'Neill, M. J. (1999) NMDA receptor antagonism, but not AMPA receptor antagonism, attenuates induced ischemic tolerance in the gerbil hippcampus. *Eur. J. Pharmacol.* **380,** 91–99.

98. Lobhert D. and Choi, D. W. (1996) Preincubation with protein synthesis inhibitors protects cortical neurons against oxygen-glucose deprivation-induced death. *Neuroscience* **72,** 335–341.

99. Yagita, Y., Kitagawa, K., Ohtsuki, T., Tanaka, S., Hori, M., Matsumoto, M. (2001) Induction of the HSP110/105 family in the rat hippocampus in cerebral ischemia and ischemic tolerance. *J. Cereb. Blood Flow Metab.* **21,** 811–819.

100. Mullins, P. G., Reid, D. G., Hockings, P. D., et al. (2001) Ischemic preconditioning in the rat brain: a longitudinal magnetic resonance imaging (MRI) study. *NMR Biomed.* **14,** 204–209.

101. Purcell, J. E., Lenhard, S. C., White, R. F., Schaeffer, T., Barone, F. C., and Chandra, S. (2003) Strain-dependent response to cerebral ischemic preconditioning: Differences between spontaneously hypertensive and stroke-prone spontaneously hypertensive rats. *Neurosci. Lett.* **339,** 151–155.

102. Masada,T., Hua, Y., Xi, G., Ennis, S. R., Keep, R. F. (2001) Attenuation of ischemic brain edema and cerebrovascular injury after ischemic preconditioning in the rat. *J. Cereb. Blood Flow Metab.* **21,** 22–33.

103. Barone, F. C., Maguire, S., Strittmatter, R., et al. (2001) Longitudinal MRI measures brain injury and its resolution: reduced neurological recovery post-stroke and decreased brain tolerance following ischemic preconditioning in stroke-prone rats. *J. Cereb. Blood Flow Metab.* **21(suppl. 1),** S230.
104. Maguire, S., Stritmatter, R., Chandra, Barone, F. C. (2003) Stroke Prone Rats Exhibit Prolonged Neurological Deficits Indicating Disruption of Post-Stroke Brain Recovery. *Neurosci. Lett.* **339,** 151–159.
105. Volpe, B. R., Pulsinelli, W. A., Tribuna, J., Davis, H. P. (1984) Behavioral performance of rats following transient forebrain ischemia. *Stroke* **15,** 558–562.
106. Kiyota, Y., Miyamoto, M., and Nagaoka, A. (1991) Relationship between brain damage and memory impairment in rats exposed to transient forebrain ischemia. *Brain Res.* **538,** 295–302.
107. Ohno, M. and Watanabe, S. (1996) Ischemic tolerance to memory impairment associated with hippocampal neuronal damage after transient cerebral ischemia in rats. *Brain Res. Bull.* **40,** 229–236.
108. Sommer, C., Goss, P., Kiessling, M. (1995) Selective c-Jun expression in CA1 neurons og the gerbil hyppocampus during and after acquisition of an ischemic tolerant state. *Brain Pathol.* **58,** 135–144.
109. Beleyev, B., Ginsberg, M. D., Alonso, O.C., Sungan, J. T., Zhao, W., Busto, R. (1996) Bilateral ischemic tolerance of rat hippocampus induced by prior unilateral focal ischemia: Relations to c-fos expression. *Neuroreport* **8,** 55–59.
110. Ikeda, J., Nakajima, T., Osborne, O. C., Mies, G., Nowak, T. S. (1994) Coexpression of *c-fos* and hsp70 mRNAs in gerbil brain after ischemia: induction threshold, distribution and time course evaluated by in situ hybredization. *Mol. Brain Res.* **26,** 249–258.
111. Kindy, M. S., Carney, J. P., Dempsey, R. J., Carney, J. M. (1991) Ischemic induction of protooncogene expression in gerbil brain. *J. Mol. Neurosci.* **2,** 217–228.
112. Nowak, T. S., Ikeda, J., Nakajima, T. (1990) 70 kilodalton heat shock protein and c-fos gene expression following transient ischemia. *Stroke* **21(suppl. III),** 107–111.
113. Wang, X. K., Yue, T. L., Young, P. R., Barone, F. C., Feuerstein, G. Z. (1995) Expression of interleukin 6, c-fos and zif268 mRNA in rat ischemic cortex. *J. Cereb. Blood Flow Metab.* **15,** 166–171.
114. Shamloo, M. and Wieloch, T. (1999) Changes in protein tyrosine phosphorylation in the rat brain after cerebral ischemia in a model of ischemic tolerance. *J. Cereb. Blood Flow Metab.* **19,** 173–183.
115. Shamloo, M., Kamme, F., Wieloch, T. (2000) Subcellular distribution and autophosphorylation of calcium/calmodulin-dependent protein kinase II-α in rat hippocampus in a model of ischemic tolerance. *Neuroscience* **96,** 665–674.
116. Shamloo, M., Rytter, A., Wieloch, T. (1999) Activation of the extracellular signal-regulated protein kinase cascade in the hippocampal CA1 region in a rat model of global cerebral ischemic preconditioning. *Neuroscience* **93,** 81–88.

117. Yano, S., Morioka, M., Fukunaga, K., et al. (2001) Activation of Akt/protein kinase B contributes to induction of ischemic tolerance in the CA1 subfield of gerbil hippocampus. *J. Cereb. Blood Flow Metab.* **21,** 351–360.

118. Tomasevic, G., Shamloo, M., Israeli, D., Wieloch, T. (1999) Activation of p53 and its target genes p21(WAF1/Cip1) and PAG608/Wig-1 in ischemic preconditioning. *Brain Res. Mol.* **70,** 304–313.

119. Walton, M. R. and Dragynow, M. R. (2000) Is CREB a key to neuronal survival? *Trends Neurosci.* 23, 48–53.

120. Shimazaki, K., Ishida, A., Kawai, N. (1994) Increase in bcl-2 oncoprotein and the tolerance to ischemia-induced neuronal death in the gerbil hippocampus. *Neurosci. Res.* **20,** 95–99.

121. Chen, D., Minami, M., and Henshall D. C. (2003) Upregulation of mitochondrial base-excision repair capability within rat brain after brief ischemia. *J. Cereb. Blood Flow Metab.* 23, 88–98.

122. Shimizu, S., Nagayama, T., Jin, K. L., et al. (2001) bcl-2 Antisense treatment prevents induction of tolerance to focal ischemia in the rat brain. *J. Cereb. Blood Flow Metab.* **21,** 233–243.

123. Kato, H., Kogure, K., Araki, T., and Itoyama, Y. (1994) Astroglial and microglial reactions in the gerbil hippocampus with induced ischemic tolerance. *Brain Res.* **664,** 69–76.

124. Liu, J., Bartels, M., Lu, A., Sharp, F.R. (2001) Microglia/macrophages proliferate in striatum and neocortex but not in hippocampus after brief global ischemia that produces ischemic tolerance in gerbil brain. *J. Cereb. Blood Flow Metab.* **21,** 361–373.

125. Wrang, M. L., Moller, F., Alsbo, C. W., Diemer, N. H. (2001) Changes in gene expression following induction of ischemic tolerance in rat brain; detection and verification. *J. Neurosci Res.* **65,** 54–58.

126. MacManus, J. P. and Linnik, M. D. (1997) Gene expression induced by cerebral ischemia: an apoptotic perspective. *J. Cereb. Blood Flow Metab.* **17,** 815–832.

127. Abe, H. and Nowak, T. S. Jr. (1996) The stress response and its role in cellular defense mechanisms after ischemia. *Adv. Neurol.* **71,** 451–466.

128. Graham, S. H. (2002) Expression of the proto-oncogene bcl-2 is increased in the rat brain following kainate-induced seizures. *Restor. Neurol. Neurosci.* 9, 243–250.

129. Plumier, J. C., Krueger, A. M., Currie, R. W., Kontoyiannis, D., Kollias, G., Pagoulatos, G. N. (1997) Transgenic mice expressing the human inducible Hsp70 have hippocampal neurons resistant to ischemic injury. *Cell Stress Chaperones* **2,** 162–167.

130. Rajdev, S., Hara, K., Kokubo, Y., et al. (2000) Mice overexpressing rat heat shock protein 70 are protected against cerebral infarction. *Ann. Neurol.* **47,** 782–791.

131. Yenari, M. A., Fink, S. L., Sun, G. H., et al. (1998) Gene therapy with HSP72 is neuroprotective in rat models of stroke and epilepsy. *Ann. Neurol.* **44,** 58-584,150:591.

132. Sakaki ,T., Yamada, K., Otsuki, H., Yuguchi, T., Kohmura, E., Hayakawa, T. (1995) Brief exposure to hypoxia induces bFGF mRNA and protein and protects rat cortical neurons from prolonged hypoxic stress. *Neurosci. Res.* **23**, 289–229.

133. Bruer, U., Weih, M. K., Isaev, N. K., et al. (1997) Induction of tolerance in rat cortical neurons: hypoxic preconditioning. *FEBS Lett.* **414**, 117–121.

134. Grabb, M. C. and Choi, D. W (1999) Ischemic tolerance in murine cortical cell culture: critical role for NMDA receptors. *J. Neurosci.* **19**, 1657–1662.

135. Ravati, A., Ahlemeyer, B., Becker, A., Krieglstein, J. (2000) Preconditioning-induced neuroprotection is mediated by reactive oxygen species. *Brain Res.* **866**, 23–32.

136. Riepe, M. W., Esclaire, F., Kasischke, K., et al. (1997) Increased hypoxic tolerance by chemical inhibition of oxidative phosphorylation: "chemical preconditioning." *J. Cereb. Blood Flow Metab.* **17**, 257–264.

137. Shou, Y., Gunasekar, P. G., Borowitz, J. L., and Isom, G. E. (2000). Cyanide-induced apoptosis involves oxidative-stress-activated NF-kappaB in cortical neurons. *Toxicol. Appl. Pharmacol.* **164**, 196–205.

138. Wick, A., Wick, W., Waltenberger, J. (2002) Neuroprotection by hypoxic preconditioning requires sequential activation of vascular endothelial growth factor receptor and Akt. *J. Neurosci.* **22**, 6401–6407.

139. McLaughlin, B., Hartnett, K. A., Erhardt, J.A., et al. (2003) Critical role of sublethal caspase activation during ischemic preconditioning. *Proc. Natl. Acad. Sci. USA* **100**, 715–720.

140. Das, D. K., Engelman, R. M., and Maulik, N. (1999). Oxygen free radical signaling in ischemic preconditioning. *Ann. NY Acad. Sci.* **874**, 49–65.

141. Baines, C. P., Goto, M., Downey, J. M. (1997) Oxygen radicals released during ischemic preconditioning contribute to cardioprotection in the rabbit myocardium. *J. Mol. Cell. Cardiol.* **29**, 207–216.

142. Vanden Hoek, T. L., Becker, L. B., Shao, Z., Li, C., Schumaker, P. T. (1998) Reactive oxygen species released from mitochondria during brief hypoxia induce preconditioning in cardiomyocytes. *J. Biol. Chem.* **273**, 18092–18098.

143. Riepe, M. W., Esclaire, F., Kasischke, K., et al. (1997) Increased hypoxic tolerance by chemical inhibition of oxidative phosphorylation: "chemical preconditioning." *J. Cereb. Blood Flow Metab.* **17**, 257–264.

144. Cohen, M. V., Baines, C. P., Downey, J. M. (2000) Ischemic preconditioning: from adenosine receptor of KATP channel. *Annu. Rev. Physiol.* **62**, 79–109.

145. Kaltschmidt, B., Uherek, M., Wellmann, H., Volk, B., Katschmidt, C. (1999) Inhibition of NF-κB potentiates amyloid β-mediated neuronal apoptosis. *Proc. Natl. Acad. Sci. USA* **96**, 9409–9414.

146. Forbes, R. A., Steenbergen, C., and Murphy, E. (2001). Diazoxide-induced cardioprotection requires signaling through a redox-sensitive mechanism. [see comments]. *Circ. Res.* **88**, 802–809.

147. Heurteaux, C., Lauritzin, I., Widman, C., Lazdunski, M. (1995) Essential role of adenosine, adenosine A1 receptors, and ATP-sensitive K+ channels in cerebral ischemic preconditioning. *Proc. Natl. Acad Sci. USA* **92**:4666-4670.

148. Abele, A.E. and Miller, R. J. (1990) Potassium channel activators abolish excitotoxicity in cultured hippocampal pyramidal neurons. *Neurosci. Lett.* **115**, 195–200.

149. Barone, F. C., Hillegass, L. M., Tzimas, M. N., et al. (1995) Time-related changes in myeloperoxidase activity and leukotriene B4 receptor binding reflect leukocyte influx in cerebral focal stroke. *Mol. Chem. Neuropathol.* **24,** 13–30.

150. Barone, F. C., Clark, R. K., Price, W. J., et al. (1993) Neuron specific enolase increases in cerebral and systemic circulation following focal ischemia. *Brain Res.* **623,** 77–82.

151. Barone, F. C., Price, W. J., White, R. F., Willette, R. N., Feuerstein, G. Z. (1992) Genetic hypertension and increased susceptibility to cerebral ischemia. *Neurosci. Biobehav. Rev.* **16,** 219–233.

152. Lin, T. N., He, Y. Y., Wu, G., Khan, M., Hsu, C. Y. (1993) Effect of brain edema on infarct volume in a focal cerebral ischemia model in rats. *Stroke* **24,** 117–121.

153. Swanson, R. A., Morton, M. T., Tsao-Wu, G., Salvalos, R. H., Davidson, C., Sharp, F. R. (1990) A semiautomated method for measuring brain infarct volume. *J. Cereb. Blood Flow Metab.* **10,** 290–293.

154. Clark, R. K., Lee, E. V., White, R. F., Jonak, Z. L., Feuerstein, G. Z., Barone, F. C. (1994) Reperfusion following focal stroke hastens inflammation and resolution of ischemic injured tissue. *Brain Res. Bull.* **35,** 387–391.

155. Davis, E. C., Popper, P., Gorski, R. A. (1996) The role of apoptosis in sexual differentiation of the rat sexually dimorphic nucleus of the preoptic area. *Brain Res.* **734,** 10–18.

156. Lobner, D. and Choi, D. W. (1996) Preincubation with protein synthesis inhibitors protects cortical neurons against oxygen-glucose deprivation-induced death. *Neuroscience* **72,** 335–341.

157. Du, C., Hu, R., Csernansky, C., Hsu, C., Choi, D. (1996) Very delayed infarction after mild focal cerebral ischemia: a role for apoptosis? *J. Cereb. Blood Flow Metab.* **16,** 15–201.

158. Wang, X. K., Barone, F. C., Aiyar, N. V., and Feuerstein, G. Z. (1997) Interleukin-1 receptor and receptor antagonist gene expression after focal stroke. *Stroke* **28,** 155–162.

159. Wang, X. K., Li, X., Currie, R. W., Willette, R. N., Barone, F. C., and Feuerstein, G. Z. (2000) Application of real-time polymerase chain reaction to quantitate the induced expression of interleukin1β mRNA in ischemic brain tolerance. *J. Neurosci. Res.* **59,** 238–246.

160. Harnett, K. A., Stout, A. K., Rajdev, S., Rosenberg, P. A., Reynolds, I. J., Aizenman, E. (1997) NMDA receptor-mediated neurotoxicity: a paradoxical requirement for extracellular Mg^{2+}/Ca^{2+}free solutions in rat cortical neurons in vitro. *J. Neurochem.* **68,** 1836–1845.

161. Houenou, L. J., Li, L., Lei, M., Kent, C. R., and Tytell, M. (1996) Exogenous heat shock cognate protein Hsc 70 prevents axotomy-induced death of spinal sensory neurons. *Cell Stress Chaperones* **1,** 161–166.

162. Davis, E. C., Popper, P., and Gorski, R. A. (1996) The role of apoptosis in sexual differentiation of the rat sexually dimorphic nucleus of the preoptic area. *Brain Res.* **734,** 10–18.

163. Adrain, C. and Martin, S. J. (2001) The mitochondrial apoptosome: a killer unleashed by the cytochrome seas. *Trends Biochem. Sci.* **26,** 390–397.

164. Stroemer, R. P., Kent, T. A., Hulsebosch, C. E. (1995) Neocortical neuronal sprouting, synaptogenesis. and behavioral recovery after neocortical infarction in rats. *Stroke* **26,** 2135–2144.

165. Kawamata, T., Dietrich, W.D., Schallert, T., et al. (1997) Intracisternal basic fibroblast growth factor enhances functional recovery and up-regulates the expression of a molecular marker of neuronal sprouting following focal cerebral infarction. *Proc. Natl. Acad. Sci. USA* **94,** 8179–8184.

166. Das, D. K., Engelman, R. M., Kimura, Y. (1993) Molecular adaptation of cellular defences following preconditioning of the heart by repeated ischemia. *Cardiovasc. Res.* **27,** 578–584.

167. Sun, J. Z., Tang, X. L., Knowlton, A. A., Park, S. W., Qui, Y., Bolli, R. (1995) Late preconditioning against myocardial stunning. An endogenous protective mechanism that confers resistance to postischemic dysfunction 24 h after brief ischemia in conscious pigs. *J. Clin. Invest.* **95,** 388–403.

168. Hartl, F. U. (1996) Molecular chaperones in cellular protein folding. *Nature* **381,** 571–580.

169. Stroemer, R. P. and Rothwell, N. J. (1997) Cortical protection by localized striatal injections of IL-1ra following cerebral ischemia in the rat. *J. Cereb. Blood Flow Metab.* **17,** 597–604.

170. Sagar, S. M., Sharp, F. R., Curran, T. (1988) Expression of c-fos protein in the brain; metabolic mapping at the cellular level. *Science* **240,** 1328–1331.

171. Morgan, J. I. and Curran, T. (1991) Stimulus-transcription coupling in the nervous system: involvement in the inducible protooncogenes fos and jun. *Annu. Rev. Neurosci.* **14,** 421–451.

172. Curran, T. and Franza, B. R. (1988) Fos and Jun: the AP-1 connection. *Cell* **55,** 395–397.

173. Christy, B. and Nathans, D. (1989) DNA binding site of the growth factor-inducible protein Zif268. *Proc. Natl. Acad. Sci. USA* **86,** 8737–8741.

174. Hsu, C. Y., An, G., Liu, J. S., Xue, J. J., He, Y. Y., and Lin, T. N. (1989) Expression of immediate early gene and growth factor mRNAs in a focal cerebral ischemia model in the rat. *Stroke* **24,** I-78–I-81.

175. Onodera, H., Kogure, K., Ono, Y., Igarashi, K., Kiyota, Y., Nagaoka, A. (1989) Proto-oncogene c-fos is transiently induced in the rat cerebral cortex after forebrain ischemia. *Neurosci. Lett.* **98,** 101–104.

176. Uemura, Y., Kowall, N. W., Moskowitz, M. A. (1991) Focal ischemia in rats causes time-dependent expression of c-fos protein immunoreactivity in widespread regions of ipsilateral cortex. *Brain Res.* **552,** 99–105.

177. Kato, H., Kogure, K., Araki, T., Itoyama, Y. (1995) Induction of Jun-like immunoreactivity in astrocytes in gerbil hippocampus with ischemic tolerance. *Neurosci. Lett.* **189,** 13–15.

178. Nowak, T. S., Osborne, O. C., Suga, S. (1993) Stress protein and proto-oncogene expression as indicators of neuronal pathophysiology after ischemia. *Prog. Brain Res.* **96,** 195–208.

179. An, G., Lin, T. N., Liu, J. S., Xue, J. J., He, Y. Y., Hsu, C. Y. (1993) Expression of c-fos and c-jun family genes after focal cerebral ischemia. *Ann. Neurol.* **33,** 457–464.

180. Woodburn, V. L., Hayward, N. J., Poat, J. A., Woodruff, G. N., Hughes, J. (1993) The effect of dizocipine and enadoline on immediate early gene expression in the gerbil global ischemia model. *Neuropharmacology* **32,** 1047–1059

181. Kitagawa, K., Matsumoto, M., Kuwabara, K., et al. (1991) Hyperthermia-induced neuronal protection against ischemic injury. *J. Cereb. Blood Flow Metab.* **11,** 449–452.

182. Riepe, M. W., Esclaire, F., Kasischke, K., et al. (1997) Increased hypoxic tolerance by chemical inhibition of oxidative phosphorylation; "chemical preconditioning." *J. Cereb. Blood Flow Metab.* **17,** 257–264.

183. Matsushima, K., Hogan, M. J., Hakim, A. M. (1996) Cortical spreading depression protects against subsequent focal cerebral ischemia in rats. *J. Cereb. Blood Flow Metab.* **16,** 221–226.

184. Kobayashi, S., Harris, V. A., Welsh, F. A. (1995) Spreading depression induces cortical neurons to ischemia in rat brain. *J. Cereb. Blood Flow Metab.* **15,** 721–727.

185. Kraig, R. P., Dong, L. M., Thisted, R., Jaeger, C. B. (1991) Spreading depression increases immunohistochemical staining of glial fibrillary acidic protein. *J. Neurosci.* **11,** 2187–2198.

186. Toyoda, T., Kassell, N. F., Lee, K. S. (1997) Induction of ischemic tolerance and antioxidant activity by brief focal ischemia. *Neuroreport* **8,** 847–851.

187. Takeda, A., Onodera, H., Sugimoto, A., Kogure, K., Obinata, M., Shibahara, S. (1993) Coordinated expression of messenger RNAs for nerve growth factor, brain-derived neurotrophic factor and neurotrophin-3 in the rat hippocampus following transient forebrain ischemia. *Neuroscience* **55,** 23–31.

188. Chen, J. and Simon, R. (1997) Ischemic tolerance in the brain. *Neurology* **48,** 306–311.

189. Akimitsu, T., Gute, D. C., Korthuis, R. J. (1996) Ischemic preconditioning attenuates postischemic leukocyte adhesion and emigration. *Am. J. Physiol.* **40,** H2052–H2059.

190. Hakim, A. M. (1994) Could transient ischemic attack have a cerebroprotective role? *Stroke* **25,** 715–716.

191. Alteri, M., Melle, G. V., Bogousslavsky, J. (1998) Do transient ischemic attacks protect from severe subsequent stroke? *Stroke* **29,** 320.

192. Tomioka, C., Nishioka, K., and Kogure, K. (1993) A comparison of induced heat-shock protein in neurons destined to survive and those destined to die after transient ischemia in rats. *Brain Res.* **612,** 216–220.

193. Armstrong, J. N., Plumier, J. C. L., Robertson, H. A., and Currie, R. W. (1996) The inducible 70,000 molecular weight heat shock protein is expressed in the degenerating dentate hilus and piriform cortex after systemic administration of kainic acid in the rat. *Neuroscience* **74,** 685–693.

194. Morimoto, R. I. (1993) Cells in stress: transcriptional activation of heat shock genes. *Science* **259,** 1409–1410.

195. Herrera, D. G. and Cuello, A. C. (1992) MK-801 affects the potassium-induced increase of glial fibrillary acidic protein immunoreactivity in rat brain. *Brain Res.* **598,** 286–293.

196. Vibulsreth, S., Hefti, F., Ginsberg, M. D., Dietrich, W. D., and Busto, R. (1987) Astrocytes protect cultured neurons from degeneration induced by anoxia. *Brain Res.* **422,** 303–311.

197. Shigeno, T., Mima, T., Takakura, K., et al. (1991) Amelioration of delayed neuronal death in the hippocampus by nerve growth factor. *J. Neurosci.* **11,** 2914–2919.

198. Liu, D., Smith, D. L., Barone, F. C., et al. (1999) Astrocytic demise precedes neuronal death in focal ischemic rat brain. *Mol. Brain Res.* **68,** 29–41.

199. Largo, C., Cuevas, P., and Herreras, O. (1996) Is glia dysfunction the initial cause of neuronal death in ischemic penumbra? *Neurol. Res.* 18, 445–448.

200. Nakata, N., Kato, H. and Kogure, K. (1993) Protective effects of basic fibroblast growth factor against hippocampal neuronal damage following cerebral ischemia in the gerbil. *Brain Res.* **605,** 458–464.

201. Nicholls, D. and Attwell D. (1990) The release and uptake of excitatory amino acids. *Trends Pharmacol. Sci.* **11,** 462–468.

202. Walz, W. (1989) Role of glial cells in the regulation of the brain ion microenvironment. *Prog. Neurobiol.* **33,** 309–333.

203. Desagher, S., Glowinski, J., and Premont, J. (1996) Astrocytes protect neurons from hydrogen peroxide toxicity. *J. Neurosci.* **16,** 2553–2562.

204. Raps, S. P., Lai, J. C. K., Hertz, L., and Cooper, A. J. L. (1989) Glutathione is present in high concentrations in cultured astrocytes but not in cultured neurons. *Brain Res.* **493,** 398–401.

205. Wilson, J.X. (1997) Antioxidant defense of the brain: a role for astrocytes. *Can. J. Physiol. Pharmacol.* **75,** 1149–1163.

206. Toyoda, T., Kassell, N. F., and Lee, K.S . (1997) Induction of ischemic tolerance and antioxidant activity by brief focal ischemia. *Neuroreport* **8,** 847–851.

207. Bonthius, D. J. and Steward, O. (1993) Induction of cortical spreading depression with potassium chloride upregulates levels of messenger RNA for glial fibrillary acidic protein in cortex and hippocampus: inhibition by MK-801. *Brain Res.* **618,** 83–94.

208. Bonthius, D. J., Lothman, E. W., and Steward, O. (1995) The role of extracellular ionic changes in upregulating the mRNA for glial fibrillary acidic protein following spreading depression. *Brain Res.* **674,** 314-328.

209. Kraig, R. P., Dong, L. M., Thisted, R., and Jaeger, C. B. (1991) Spreading depression increases immunohistochemical staining of glial fibrillary acidic protein. *J. Neurosci.* 11, 218772198.

210. Massa, S. M., Swanson, R. A., and Sharp, F. R. (1996) The stress gene response in brain. *Cerebrovasc. Brain Met. Rev.* 8, 95–158.

211. Sharp, F. R., Massa, S.M. and Swanson, R.A. (1999) Heat shock protein protection. *Trends Neurosci.* 22, 976–99.

212. Wang, X.K., Li, X., Erhardt, J. A., Barone, F.C., Feuerstein, G. Z. (2000) Detection of tumor necrosis factor-α mRNA induction in ischemic brain tolerance by means of real-time polymerase chain reaction. *J. Cereb. Blood Flow Metab.* **20,** 15–20.

213. Ohtsuki, T., Ruetzler, C. A., Tasaki, K., and Hallenbeck, J. M. (1996) Interleukin-1 mediates induction of tolerance to global ischemia in gerbil hippocampus CA1 neurons. *J. Cereb. Blood Flow Metab.* **16,** 1137–1142.

214. Strijbos, P. J. and Rothwell, N. J. (1995) Interleukin-1β attenuates excitatory amino acid-induced neurodegeneration *in vitro*: involvement of nerve growth factor. *J. Neurosci.* **15,** 3468–3474.

215. Wang, X-K., Yaish-Ohad, S., Li, X., Barone, F. C., and Feuerstein, G. Z. (1998) Use of suppression subtractive hybridization strategy for discovery of tissue inhibitor of matrix metalloproteinase-1 (TIMP-1) gene expression in ischemic tolerance. *J. Cereb. Blood Flow Metab.* **18,** 1173–1177.

216. Blackstock, W.P. and Weir, M.P. (1999) Proteomics: quantitative and physical mapping of cellular proteins. *Trends Biotechnol.* 17, 121–127.

217. Brambrink, A. M., Schneider, A., Noga, H., et al. (2000). Tolerance-Inducing dose of 3-nitropropionic acid modulates bcl-2 and bax balance in the rat brain: a potential mechanism of chemical preconditioning. *J. Cereb. Blood Flow Metab.* **20,** 1425–1436.

218. Sakaki, T., Yamada, K., Otsuki, H., Yuguchi, T., Kohmura, E., and Hayakawa, T. (1995) Brief exposure to hypoxia induces bFGF mRNA and protein and protects rat cortical neurons from prolonged hypoxic stress. *Neurosci. Res.* **23,** 289–296.

219. Cohen, M. V., Baines, C. P., and Downey, J. M. (2000). Ischemic Preconditioning: From Adenosine Receptor to KATP Channel. *Ann. Rev. Physiol.* **62,** 79–109.

220. Dzeja, P. P. and Terzic, A. (1998). Phosphotransfer reactions in the regulation of ATP-sensitive K+ channels. *FASEB J.* **12,** 523–529.

221. Liu, Y., Gao, W. D., O'Rourke, B., and Marban, E. (1996). Synergistic modulation of ATP-sensitive K+ currents by protein kinase C and adenosine. Implications for ischemic preconditioning, *Circ. Res.* **78,** 443–54.

222. Liu, Y., Sato, T., O'Rourke, B., and Marban, E. (1998). Mitochondrial ATP-dependent potassium channels: novel effectors of cardioprotection? *Circulation* **97,** 2463–2469.

223. Grigoriev, S. M., Skarga, Y. Y., Mironova, G. D., and Marinov, B. S. (1999). Regulation of mitochondrial KATP channel by redox agents, *Biochim. Biophys. Acta.* **1410,** 91–96.

224. Tokube, K., Kiyosue, T., and Arita, M. (1996). Openings of cardiac KATP channel by oxygen free radicals produced by xanthine oxidase reaction. *Am. J. Physiol.* **271,** H478–H489.

225. Pain, T., Yang, X. M., Critz, S. D., et al. (2000) Opening of mitochondrial K(ATP) channels triggers the preconditioned state by generating free radicals. *Circ. Res.* **87,** 460–466.

226. Poppe, M., Reimertz, C., Dussmann, H., et al. (2001) Dissipation of potassium and proton gradients inhibits mitochondrial hyperpolarization and cytochrome c release during neural apoptosis. *J. Neurosci.* **21,** 4551–4563.

227. Yu, S. P., Yeh, C. H., Sensi, S. L., et al. (1997) Mediation of neuronal apoptosis by enhancement of outward potassium current. *Science* **278,** 114–117.

228. Debska, G., May, R., Kicinska, A., Szewczyk, A., Elger, C. E., and Kunz, W. S. (2001) Potassium channel openers depolarize hippocampal mitochondria. *Brain Res.* **892,** 42–50.

229. Xu, M., Wang, Y., Ayub, A., and Ashraf, M. (2001) Mitochondrial K(ATP) channel activation reduces anoxic injury by restoring mitochondrial membrane potential. *Am. J. Physiol.* **281,** H1295–H12303.

230. Li, P., Nijhawan, D., Budihardjo, I., et al. (1997) Cytochrome c and dATP-dependent formation of Apaf-1/caspase-9 complex initiates an apoptotic protease cascade. *Cell* **91,** 479–489.

231. Zou, H., Henzel, W. J., Liu, X., Lutschg, A., and Wang, X. (1997) Apaf-1, a human protein homologous to *C. elegans* CED-4, participates in cytochrome c-dependent activation of caspase-3. *Cell* **90,** 405–413.

232. Adrain, C. and Martin, S. J. (2001) The mitochondrial apoptosome: a killer unleashed by the cytochrome seas. *Trends Biochem. Sci.* **26,** 390–397.

233. Bellido, T., Huening, M., Raval-Pandya, M., Manolagas, S. C., and Christakos, S. (2000) Calbindin-D28k is expressed in osteoblastic cells and suppresses their apoptosis by inhibiting caspase-3 activity. *J. Biol. Chem.* **275,** 26328–26332.

234. Dowd, D., MacDonald, P., Komm, B., Haussler, M., and Miesfeld, R. (1992) Stable expression of the calbindin-D28K complementary DNA interferes with the apoptotic pathway in lymphocytes. *Mol. Endocrinol.* **6,** 1843–1848.

235. Antonsson, B. and Martinou, J. C. (2000) The Bcl-2 protein family. *Exp. Cell Res.* **256,** 50–57.

236. Roy, N., Deveraux, Q. L., Takahashi, R., Salvesen, G. S., and Reed, J. C. (1997) The c-IAP-1 and c-IAP-2 proteins are direct inhibitors of specific caspases. *EMBO J.* **16,** 6914–6925.

237. Srinivasula, S. M., Hegde, R., Saleh, A., et al. (2001) A conserved XIAP-interaction motif in caspase-9 and Smac/DIABLO regulates caspase activity and apoptosis. *Nature* **410,** 112–116.

238. Garrido, C., Bruey, J. M., Fromentin, A., Hammann, A., Arrigo, A. P., and Solary, E. (1999) HSP27 inhibits cytochrome c-dependent activation of procaspase-9. *FASEB J.* **13,** 2061–2070.

239. Garrido, C., Mehlen, P., Fromentin, A., et al. (1996) Inconstant association between 27-kDa heat-shock protein (Hsp27) content and doxorubicin resistance in human colon cancer cells. The doxorubicin-protecting effect of Hsp27. *Eur. J. Biochem.* **237,** 653–659.

240. Garrido, C., Ottavi, P., Fromentin, A., et al. (1997) HSP27 as a mediator of confluence-dependent resistance to cell death induced by anticancer drugs. *Cancer Res.* **57,** 2661–2667.

241. Mehlen, P., Kretz-Remy, C., Preville, X., and Arrigo, A. P. (1996) Human hsp27, *Drosophila* hsp27 and human αB-crystallin expression-mediated increase in glutathione is essential for the protective activity of these proteins against TNFα-induced cell death. *EMBO J.* **15,** 2695–2706.

242. Yenari, M. A., Fink, S. L., Sun, G. H., et al. (1998) Gene therapy with HSP72 is neuroprotective in rat models of stroke and epilepsy. *Ann. Neurol.* **44,** 584–591.

243. Sharp, F. R., Massa, S. M., and Swanson, R. A. (1999) Heat-shock protein protection. *Trends Neurosci.* **22,** 97–99.

244. Pringle, A. K., Angunawela, R., Wilde, G. J., Mepham, J. A., Sundstrom, L. E., and Iannotti, F. (1997) Induction of 72 kDa heat-shock protein following sublethal oxygen deprivation in organotypic hippocampal slice cultures. *Neuropathol. Appl. Neurobiol.* **23,** 289–298.

245. Rajdev, S., Hara, K., Kokubo, Y., et al. (2000) Mice overexpressing rat heat shock protein 70 are protected against cerebral infarction. *Ann. Neurol.* **47,** 782–791.

246. Goldberg, M. P., Weiss, J. H., Pham, P. C., and Choi, D. W. (1987) N-methyl-D-aspartate receptors mediate hypoxic neuronal injury in cortical culture. *J. Pharmacol. Exp. Ther.* **243,** 784–791.

247. Patel, M. N., Peoples, R. W., Yim, G. K., and Isom, G. E. (1994) Enhancement of NMDA-mediated responses by cyanide. *Neurochem. Res.* **19,** 1319–1323.

248. Zeevalk, G. D. and Nicklas, W. J. (1991) Mechanisms underlying initiation of excitotoxicity associated with metabolic inhibition. *J. Pharmacol. Exp. Therap.* **257,** 870–878.

249. Guo, Q., Christakos, S., Robinson, N., and Mattson, M. P. (1998) Calbindin D28k blocks the proapoptotic actions of mutant presenilin 1: reduced oxidative stress and preserved mitochondrial function. *Proc. Natl. Acad. Sci. USA* **95,** 3227–3232.

7

Production of Transgenic and Mutant Mouse Models

Alex J. Harper

Summary

Manipulation of the rodent genome by deliberately inserting (transgenic) or removing (knockout) a gene of interest or indeed by selectively breeding animals with a spontaneous or random mutation producing a trait of interest has been developed over several years. Mouse "fanciers" have been selectively breeding interesting mice since the turn of the last century to produce a plethora of different background strains of the common house mouse (*Mus musculus*). Rat (*Rattus norvegicus*) strain development has also proceeded with selective breeding, although the range of strains is more limited. The deliberate and targeted manipulation of the mouse genome has been with us for over two decades, with the rat genome a more recent addition, and yet this technology has been limited to a very narrow range of genes. With the complete mapping of the mouse genome (and the rat genome soon to follow), the powerful techniques of transgenic and knockout rodent production can be applied to the numerous genes whose expression is altered in existing stroke models.

Key Words

Transgenic; knockout; animal; mouse; rat; stroke; ischaemia; mutant; model; vascular; dementia.

1. Introduction

Transgenic models are those in which a protein from the gene of interest is overexpressed. This overexpression can be controlled in a temporal or spatial manner with appropriate use of DNA regulatory sequences. Knockout models are generated by removing the endogenous protein expression from a gene of interest. The first transgenic (*1*) and knockout (*2*) mouse lines followed earlier models of human disease using spontaneous mutants produced from selective breeding programs (*3,4*) or induced mutant models (*5,6*). Mutant models result from breeding that selects for specific traits, whereas transgenic and knockout

From: *Methods in Molecular Medicine, Vol. 104: Stroke Genomics: Methods and Reviews*
Edited by: S. J. Read and D. Virley © Humana Press Inc., Totowa, NJ

models result from the deliberate introduction of precise engineering into an identified gene.

The essential stages of overexpression transgenic model production start with cloning and construction of the relevant cDNA sequence (the construct). A supply of single-celled embryos for injection (harvesting embryos) is required before the DNA is introduced into the embryo (cDNA microinjection). Subsequently the surviving embryos are placed into a recipient mother (transfer of embryos), and finally offspring positive for the transgene are identified and bred on to produce experimental transgenic colonies. The essential stages of making a knockout model are very similar, with a few important differences. The process again starts with cloning and construction of a DNA but using relevant genomic DNA sequences in place of cDNA (the targeted construct). This genomic DNA is introduced into embryonic stem (ES) cells (transfection), and screening takes place to identify the correct recombination of the targeted construct (clone selection). Subsequently the correctly targeted ES cells are introduced into a blastocyst (ES cell microinjection), and surviving embryos are placed into a recipient mother (transfer of embryos) before identifying positive offspring and breeding the appropriate knockout lines.

Animal models of stroke have used nongenetic techniques (7,8) to reproduce the conditions seen in the ischemic human brain. These rodent models attempt to simulate the pathological sequelae that ensue following stroke in the human brain. The production of transgenic animals is routine in many laboratories around the world, and a gene of interest—GOI1—will be used to illustrate how a transgenic or knockout animal may be generated. These methods will help improve our understanding of the role specific genes play in ischemic stroke.

2. Materials

2.1. Microinjection Equipment

1. Microscope for microinjection (Leica, Milton Keynes, UK).
2. Antivibration table (cat. no.702, Leica).
3. Picoinjector (cat . no. P100, Wentworth Laboratories, Sandy, Bedfordshire, UK).
4. Microforge (cat. no. MF-900, Micro Instruments, Oxon, UK).
5. Compressed air supply for the antivibration table.
6. Incubator for 5% CO_2 and 37°C incubation.
7. Standard 4°C fridge with –20°C freezer compartment.
8. M2 media (Sigma-Aldrich, Poole, UK).
9. Micromanipulators (Zeiss, UK).
10. Needle puller/needles (Campden Instruments needle puller; Pre-pulled Needles from Eppendorf).

2.2. Surgical Equipment

1. Dissection equipment including fine scissors, fine forceps, Serafine clip (Fine Science Tools).
2. Surgical staples (9 mm Clay Adams Autoclips).
3. Dissection microscope with large swing-out arm (cat. no. MZ6, Leica).
4. Mouth transfer pipet and transfer capillaries (Sigma).

2.3. Animals

2.3.1. Donor Mice for Overexpression Transgenic Mouse Production

Single-cell embryo donor females are aged 3–6 wk. The precise age depends on the donor strain and health of the animals and will be determined by the quality of supply combined with trial and error. Generally younger or smaller mice give more embryos of poorer quality, and older mice give fewer, better quality embryos. Taking [C57Bl/6 × CBA]hybrid mice as an example, optimum embryo yield was obtained from mice with an average age of 4.5 wk and an average weight of 15.6 g (*see* **Note 1**) Strains of mice respond differently to superovulation (**Table 1**); however, an important consideration is the type of characterization to be carried out on the mice. In the study of vascular dementia, behavioral characterization is often required, and thus a mouse strain that is amenable to specific behavioral analyses is desirable. The inbred FVB/N strain is amenable to the production of transgenic lines, and, being inbred, it is isogenic. However, it has many problems with stereotypic behaviours that make behavioral characterization problematic. If hybrid mice are used, then consideration needs to be given to whether or not an isogenic colony is required. If this is the case, then the colony must be backcrossed to the strain of choice for subsequent generations of breeding. A strain typically used is the C57Bl6. This can be over three to six generations depending on whether speed congenics are used (*see* **Note 2**).

2.3.2. Donor Mice for Knockout Mouse Production

Blastocyst donor females are aged 6–8 wk. An inbred strain of mouse should be used that has proved to be suitable as a host for the ES cell line chosen. Often C57Bl/6 mice are used for this purpose. Superovulation of these donor mice does increase the embryo yield but not in the same numbers as for single-celled superovulation.

2.3.3. Sterile Males

Sterile males are required to mate with mature fertile females in order to generate pseudopregnant dams. Sterile males can be normal healthy hybrids

Table 1
Superovulation Response of Different Mouse Strains

Strain	No. of embryos
C57Bl/6	40–60
BalB/cByJ	40–60
129/SvJ	40–60
CBA/CaJ	40–60
SJL/J	40–60
C58/J	40–60
A/J	≤15
C3H/HeJ	≤15
129/J	≤15
129/ReJ	≤15
DBA2/J	≤15
C57/J	≤15
FVB/N	25
[C57Bl/6 × CBA]hybrid	40–60
C57Bl/6 normal ovulation	8–10

that have been vasectomized or can be a sterile breed of mice (e.g., T/Tw2 from the Jackson Laboratories). Vasectomized male mice maintain performance for up to 1 yr, whereas inbred sterile mice can last for up to 6 mo. In either case a test mating should be carried out once the mice are fit to mate to ensure sterility.

2.3.4. Pseudopregnant Recipient Females

Recipient females are healthy mature mice between 8 and 20 wk of age. Older or younger mice can be used, but as most strains of female mice become fertile at approx 6–8 wk and rapidly lose fertility after 26 wk of age, the percentage of successfully reared pups will decrease. The strain for recipient females should be the same as, or a hybrid of, that used to donate embryos for injection (e.g., C57Bl/6 used for microinjection and [C57Bl/6 × CBA]hybrid used for the recipient female). Recipient females at 0.5 d post-mating are required for single- or two-celled embryo transfer, and 2.5 d post-mating recipient females are required for blastocyst transfers.

2.4. Embryonic Stem Cells

ES cells are derived from the inner cell mass of an expanded blastocyst (*9*). Newly derived ES cells require testing for their ability to integrate into the germline when injected into a host blastocyst as well as their utility in transfec-

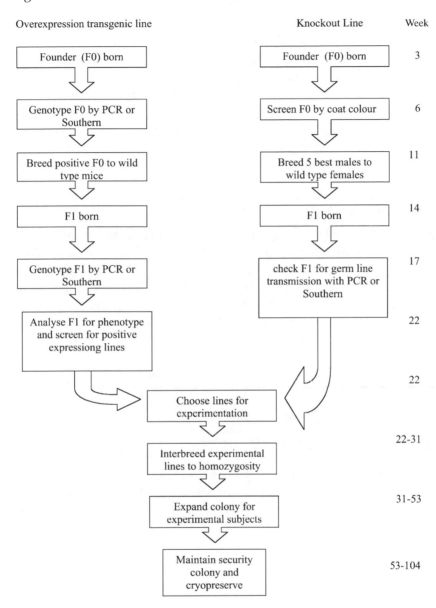

Fig. 1. Diagram of production and breeding timelines.

tion experiments. It is advisable to use proven ES cell lines that are in general use by academic laboratories. Such information can be found in the literature and the relevant lab approached for a supply of ES cells. If possible, a low passage (number of growth cycles since derivation) number should be used, as these provide better germline transmission. The vast majority of ES cell lines are of the 129 background strain of mice, as these have been found to be most amenable to efficient targeted ES cell line production. Other strains of ES cell are available and do offer significant advantages over the 129 strain. For example, if C57Bl/6 ES cells are used and a C57Bl/6 host is used, then the resulting offspring will be 100% C57Bl/6, removing the need for backcrossing to an isogenic population. However, C57Bl6 ES cells are not very amenable to current techniques and are an inefficient method of knockout mouse production.

2.5. Breeding Facility

A standard animal breeding facility with small animal operating suites is ample for the production of transgenic and knockout mice. One important consideration is the health status of such facilities. Animals with an altered genetic makeup may well be more susceptible to infection, and so the highest level of health is required. Full barrier containment procedures are required and should be adhered to in the areas where these animals are bred.

2.6. Gonadotropic Hormones

1. Follicle-stimulating hormone (FSH: Sigma, cat. no. F4021).
2. Luteinizing hormone (LH: Sigma, cat. no. C8554).

2.7 Molecular Biology Equipment

Standard equipment for genetic engineering is required. During the construction of transgenic or knockout animals DNA material for manipulation is obtained from cDNA and genomic libraries. Subsequent manipulations involve the use of mutagenesis, polymerase chain reaction (PCR), and cloning. Additionally, equipment for Southern blot analysis is required for screening the offspring that may carry the genetic manipulation of interest.

3. Methods

3.1. Summary of Process

The process for producing transgenic mice is very similar to that for producing knockout mice.

1. Initially DNA is manipulated to incorporate the key features required.
2. This transgene is introduced into a host embryo, either directly into the nucleus for a transgenic mouse or via an ES cell for a knockout mouse. Knockout mouse production requires a prescreening step to identify positive homologous recombination events within the ES cell.

3. Subsequently these embryos are replaced into a recipient dam, and the resulting offspring are screened for the presence of the transgene. Overexpression transgenic mouse lines with high copy numbers (and hence higher levels of protein) are often selected and bred on to produce experimental colonies.
4. Mice carrying a heterozygous targeted deletion of a gene are interbred to generate mice homozygous for the deletion, which are subsequently bred on for experimentation.
5. There are a number of differences in the breeding regime for overexpression and knockout mouse colonies. These are summarized in **Fig. 1**.

3.2. Expression Cassette Cloning Strategy

3.2.1. Elements Expressed

The manipulated DNA must have certain elements in order to be expressed successfully in the host genome. At the very basic level, a transgene requires a gene of interest and a polyadenylation signal driven by a suitable promoter (**Fig. 2A**). In order to follow the expression pattern of the transgene at a gross tissue level, a reporter gene can be included, separated from the transgene by an internal ribosomal entry sequence element (IRES) *(10)*. This allows two gene products to be produced from one promoter, thus ensuring that the reporter gene will accurately report where the transgene has been expressed (**Fig. 2B**). The expression of a transgene can be temporally regulated by the inclusion of repressor sequences linked to the promoter. This will have the effect of activating gene transcription only when the repressor has been silenced (**Fig. 2C**) by an activator and a trigger. Examples of this activation/repression system include the use of doxycyclin or tetracyclin as the trigger *(11)*. This can be administered in the diet of the mouse to activate the transcription of a gene or, conversely, to stop the repression of a gene knockout, thereby allowing temporal control of the gene expression.

3.2.2. Knockout Targeting Strategy

When engineering the target gene, an important consideration is ensuring that the expression of the endogenous gene is completely removed. If even part of a gene is transcribed, the resulting partial gene product may be translated into a functional protein and have unknown effects possibly entirely unrelated to its original function. To avoid this situation, the endogenous locus of the gene of interest must be genetically engineered to prevent any transcription. This is achieved by manipulating the endogenous locus as close to the transcription initiation site (ATG) as possible. This manipulation should remove at least the first exon of the locus and replace it with a reporter gene, *LacZ* for example. More of the locus can be removed so as to ensure that no endogenous functional gene product is transcribed (**Fig. 3**).

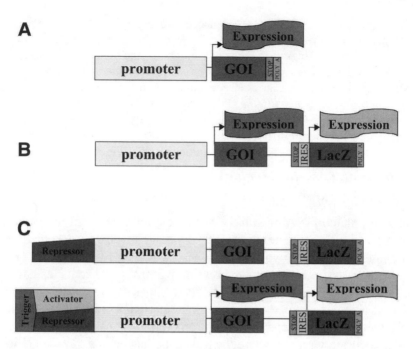

Fig. 2. (**A–C**) Expression vectors.

The example targeting vector in **Fig. 3** includes a stop sequence to prevent further transcription of the endogenous locus, an IRES element to reinitiate transcription of the downstream DNA, a *LacZ* reporter gene and polyadenylation sequence (to signal the end of the primary RNA transcript), and a selectable marker, neomycin. This is one of a number of selectable markers that can be used during targeting vector transfection into ES cells and confers resistance to antibiotic treatment, in this example neomycin. Thus cells that have incorporated the targeting vector correctly will survive when treated with neomycin in tissue culture.

The targeting vector must include regions of homology to the endogenous locus in order to integrate into the host genome. This is achieved by homologous recombination, as shown in **Fig. 4**. The lengths of the homologous arms have been found to be critical to successful integration, with a minimum of 1 kb and a maximum of 4 kb required. Specific sequences of DNA, LoxP (locus of recombination), flank the selectable marker as it is incorporated into the host genome along with the targeting vector. The marker may be expressed in the host animal and so is removed by the addition of cre protein into the transgenic or knockout mouse line (cre causes recombination of the LoxP sites), and the marker is removed *(12)*.

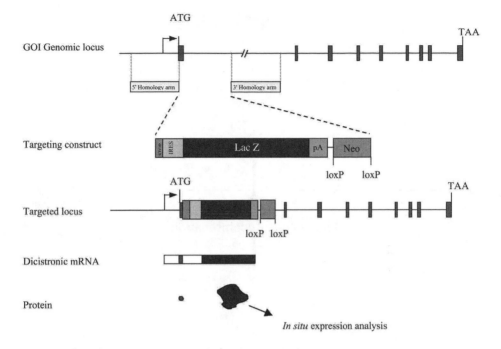

Fig. 3. Knockout targeting strategy. GOI, gene of interest.

3.3. Harvesting Embryos

1. Young females are used to provide the single-celled embryos for microinjection.
2. Females are dosed via the intraperitoneal route with 5 IU of FSH followed 48 h later by 5 IU of LH and are then paired with fertile mature males. The timing of the FSH injection is not critical, but the LH must be dosed before the middle of the second day after FSH administration.
3. Females are checked for vaginal plugs the following morning. In healthy hybrid donor mice, all the paired females will be plugged with an obvious vaginal plug of semen, indicating a successful mating. (*See* **Note 3**)
4. Embryos for injection are recovered from the euthanized donor female by dissecting the coiled oviduct away from the uterus and placing it into a 35-mm Petri dish filled with M2 medium supplemented with 10 mg/mL of hyaluronidase (cat. no. H4272, Sigma). This enzyme degrades the protein holding nutritive cumulus cells next to the embryo.
5. The embryos are gathered in a swelling called the ampulla. The wall of the oviduct is torn at the ampulla, and the embryos are released into the medium.
6. After approx 5 min of incubation, the cumulus cells fall away from the embryo. They are collected with a mouth micropipet and are washed twice in fresh M2. The embryos are then ready for microinjection.

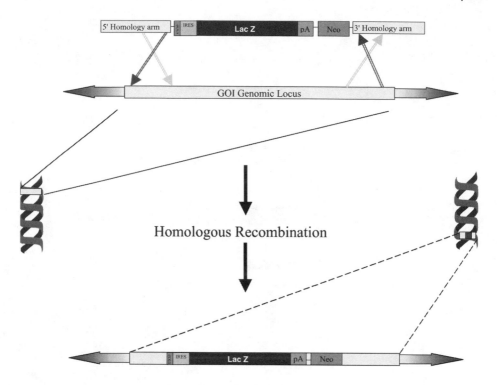

Fig. 4. Knockout targeting vector recombination. GOI, gene of interest.

7. Embryos not used immediately for injection should be stored in an incubator.
8. In order to relocate the embryos easily they are stored in a small volume of M2 medium in a 35-mm Petri dish.
9. Four such microdrops are placed on the Petri dish, one each at the 12, 3, 6, and 9 o'clock positions.
10. These drops of media are overlaid with low-density mineral oil (Sigma, cat. no. M8410) in order to prevent evaporation but allow the diffusion of gases.
11. This dish is stored in an incubator at 37°C and 5% CO_2. All embryos are stored in this Petri dish.
12. Uninjected embryos are stored at 12 o'clock and injected at 3 o'clock; poor or malformed embryos that cannot be injected are stored at 9 o'clock. The 6 o'clock position can be used to store embryos that have survived injection well and still look healthy after at least 1 h of incubation post injection.

3.4. Single-Cell Embryo Injection

1. A microinjection rig (**Fig. 5**) is required to introduce the DNA construct into single-celled mouse embryos.

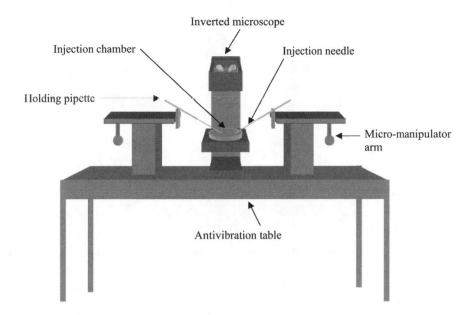

Fig. 5. Microinjection workstation.

2. After the microscope is turned on, the anti-vibration table is filled with compressed air and the picoinjector is turned on, a cavity slide filled with M2 is overlaid with light mineral oil, and the microinjection needle is pulled.
3. After a correctly pulled needle is obtained, it is backfilled with the DNA construct solution (*see* **Note 4**) and loaded onto the needle holder.
4. A holding pipet is mounted onto the opposite holder, and both are lowered into the M2 media under oil.
5. At this point the embryos for injection are introduced into the top part of the chamber. Batches of approx 20 embryos are a manageable number until the operator is proficient.
6. The holding pipet vacuum is turned on, most convieniently by use of a foot pedal, and maneuverd to pick up an embryo from the group. The embryo must be held at its center with the holding pipet resting very gently on the floor of the injection chamber.
7. Then the injection needle is brought into focus opposite the holding pipet (**Fig. 6**).
8. The needle must be in the same focal plane as the male pronucleus of the embryo prior to injection. To achieve this, the needle needs to be moved to a position either north or south of the embryo (in the microscope field of view).
9. The focus is adjusted to the male pronucleus, and then the needle is moved vertically up or down to be in the same focal plane as the pronucleus.
10. The needle is then moved away from, and brought into line with the embryo (the tip will move out of focus but that is acceptable) and then brought forward to inject.

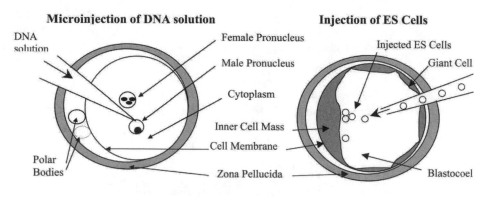

Microinjection of DNA solution

DNA solution
Female Pronucleus
Male Pronucleus
Cytoplasm
Inner Cell Mass
Cell Membrane
Polar Bodies
Zona Pellucida

Injection of ES Cells

Injected ES Cells
Giant Cell
Blastocoel

Fig. 6. Injection of embryos.

11. When the needle pierces the pronucleus, the DNA solution is delivered to the pronucleus by depressing the foot pedal of the picoinjector.
12. Immediately after injection the needle is removed from the embryo, care being taken not to touch the nucleolus.
13. The embryo is moved to the south of the injection chamber, and all the remaining embryos are injected in the same way.
14. If the DNA solution does not flow out of the needle tip it may be because the tip, of the needle has sealed closed when it was pulled. To solve this problem the tip of the needle must be broken on the clean, flat surface of the holding pipet ("tipping the needle"; *see* **Note 5**).
15. After all the embryos are injected in this way, they are removed to recover in the incubator, and another batch is introduced into the injection chamber for injection.
16. This continues until all the embryos are injected. If a needle becomes completely blocked or the tip too big, then the needle is discarded and another backfilled to replace it.

3.5. Blastocyst Injection

Knockout mice are generated by introducing the DNA construct into an embryonic stem cell, which is then injected into the vacuole of a mature blastocyst. Knockout mouse generation is more time-consuming and involves an additional in vitro culture step compared with making transgenic mice, although the injection part is not so difficult. This technology is the only way in which gene expression can be removed, thereby enabling the effect of removing or knocking out a particular gene to be observed.

1. Embryonic cells carrying the targeted locus are prepared in bulk so as to be single celled, with all contaminating debris removed. The ES cells are prepared in tissue culture conditions and stored in culture media in the incubator.
2. Just prior to injection, an aliquot of cells (several hundred) is introduced into a drop of M2 medium in the right-hand cavity of the main injection chamber.

3. Blastocysts are recovered from the donor females at 2.5 d post superovulation. The donor mice must be of the same strain or very closely compatible to the genetic strain of the DNA targeting construct.
4. Blastocyst recovery is achieved by flushing M2 medium through each uterine horn in turn.
5. The entire uterus is dissected out, severing just above the cervix.
6. The ovary and most of the oviduct coils are cut away, and a 25-gage needle is inserted into the top (ovary end) of each uterine horn.
7. With the bottom end of the uterus in a 35-mm Petri dish, a rapid flush of M2 medium is forced into the horn, expelling the embryos into the dish.
8. After flushing from the uterine horns, the blastocysts are washed with M2 and stored in the 12 o'clock position of a 35-mm Petri dish.
9. Just prior to injection, up to 20 blastocysts are placed in the injection holding chamber.
10. An ES cell injection needle is lowered into the drop of M2 to the right of the main injection chamber and is backfilled with M2 media.
11. Up to 20 clean, single ES cells are taken up into the needle.
12. The needle is then raised, moved to the injection chamber (by moving the stage, not the needle) and then lowered.
13. A single blastocyst is held by vacuum on the holding pipet and oriented so that the giant cells are nearest the needle tip.
14. Injection is by rapid movement of the needle tip into the blastocyst and subsequent expulsion of the ES cells. This will cause the blastocyst to collapse. Care should be taken to ensure that the ES cells adhere to the inside wall of the blastocoel and do not follow the tip of the needle as it is withdrawn from the injection site.
15. The needle is moved back to the ES cell chamber and refilled with cells, and the injection procedure is repeated until all the blastocysts are injected.
16. The blastocysts will recover quickly from injection and should have expanded back to their original size within 2 h.

3.6. Transfer of Injected Embryos Into Recipient Female

Protocols for small animal surgery must be adhered to when one is transferring the surviving embryos back into the recipient mothers. The basic surgery for both operations is the same.

1. Initially embryos are loaded into a transfer capillary. Up to 20 single- or two-celled embryos or up to 10 blastocysts are loaded into the capillary preceded by a small bubble and some M2 medium. This is to allow the operator to see when the embryos have been transferred, as they are too small to see with the naked eye.
2. The appropriate post-mating recipient females (*see* **Subheading 2.3.4.**) are anesthetized using a long-acting anesthetic (halothane is ideal or combination Hypnorm/Hypnorvel) and are laid on their belly.
3. A 10-mm incision is made longitudinally in the skin over the spine approximately halfway down the back (**Fig. 7A**).

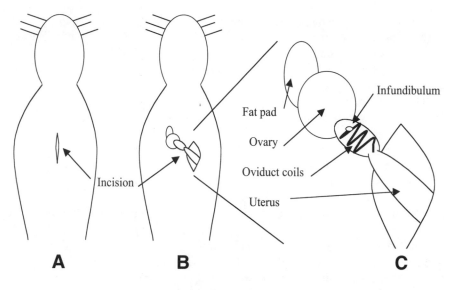

Fig. 7. Surgery.

4. This incision can then be used as a "window" to move around the body wall and search for a cream-colored fat deposit that is attached to the ovary.
5. Once this is found, the body wall is cut with a small 5-mm incision and the fat pad is exteriorized along with the ovary and top of the uterus (**Fig. 7B**).
6. The Serafine clip is used to hold the fat pad out of the body cavity while the binocular microscope if moved into place.
7. This is where the surgery for replacing the different embryos differs. To transfer single- or two-celled embryos, the mouse is examined under low magnification to find the infundibulum (**Fig. 7C**). This opening receives the eggs during natural ovulation and is located at the end of the oviduct coils at the top of the uterus.
8. The injected embryos must be inserted into the infundibulum by hand under low-power magnification of the binocular microscope. This technique requires a steady hand and it is advisable to practise on dead animals with a colored dye prior to live transfer.
9. Transfer is achieved by tearing the bursa near the infundibulum, clasping the bursa with fine forceps to stop the coils from moving, and then gently inserting the tip of a transfer capillary into the infundibulum. This opening is approx 150 μm in diameter, and a good capillary should be approx 100 μm in diameter.
10. The tip of the transfer capillary should be maneuverd around the first oviduct coil bend and the embryos expelled into the ampulla by gently blowing into the mouthpiece of the capillary tubing. The bubble preceding the embryos is used to assess when the capillary has been emptied.
11. The transfer of blastocysts into the recipient mouse is an easier technique. A 27-gage needle is used to make a hole in the uterus wall just before the point where the uterus begins to narrow to the oviduct (**Fig. 7C**).

12. The transfer capillary loaded with blastocysts is inserted into the hole and the embryos expelled again using the bubble before the embryos to assess when the capillary has been emptied.
13. After transfer, the uterine horn, ovary, and fat pad are replaced into the body cavity, and the incision is closed with a suture.
14. Once the same procedure has been repeated on the other uterine horn, the skin incision is closed with a Michel clip, and the animal is allowed to recover from surgery on a heated pad.
15. Saline may be administered to avoid dehydration.

3.7. Colony Expansion

3.7.1. Initial Breeding Strategy for Overexpression Transgenic Mice

1. After identification of the desired mouse line by screening the first founder generation, the colony will need to be expanded in order to generate animals for further characterization and breeding.
2. Initial breeding regimes will be determined by the properties of the gene of interest, for example, sex specificity or copy number (in the case of transgenic animals).
3. The rapid expansion of the colony is best achieved by breeding founder male mice to two wild-type females.
4. This "trio" of mice can be left in the cage to produce litter after litter until enough F_1 generation mice are produced.
5. Gestation in mice is approx 19–21 d, and, when left in the same cage as the male, the female will mate within 12–24 h of parturition (a postpartum mating) and give birth another 19–21 d later.
6. This cycle can be repeated successfully up to six times before breeding performance is affected.

3.7.2. Independently Segregating Transgenes

When a transgene is directly injected into an embryo, multiple copies of the transgene can randomly integrate into the host genome in a concatenation at one integration site or at multiple integration sites. When the mice breed, homologous recombination may separate the transgenes at multiple integration sites, leading to segregation of the transgene. This can result in F_1 offspring having different copies of the transgene and hence different protein expression levels. A Southern blot analysis is required to assess the transgene copy number of different F_1 litters and check for this segregation event. After the F_1 generation, transgene segregation, although not impossible, is far less likely.

3.7.3. Initial Breeding Strategy for Knockout Mice

Founder knockout mice are selected on the basis of the percent contribution of the injected ES cell to the coat color. If the host strain of mouse used is

C57Bl/6 then the coat color of the host will be black. If 129 ES cells are used, then the donor coat color is a light brown. Thus one can assess the percent contribution of the donor ES cells, and hence the proportion carrying the targeted locus, by looking at coat color. A 100% black mouse will not have incorporated many ES cells but will breed vigorously whereas a 100% light brown mouse will have incorporated many ES Cells but will not breed well. Five founder mice with more than 50% of their coat color as light brown should be chosen for breeding onto the F_1 generation. The resulting F_1 generation mice need to be genotyped to assess whether the targeted locus has transmitted through the germline successfully. F_1 mice carrying the targeted mutation can be interbred at this point to generate mice homozygous for the targeted locus. Assuming they are viable, these mice can be analysed to discover if the locus has been targeted correctly and the protein expression is indeed knocked out.

3.7.4. Background Strain

The background strain can have an enormous influence on the phenotype of transgenic and knockout mice. If isogenic donor single-celled embryos are used, then overexpression transgenic mouse lines will be isogenic from birth. However, knockout mouse lines will usually be a mix of C57Bl/6 and 129 strains. When the F_1 generation mouse is a hybrid mix of strains, then more reliable phenotypic data will be obtained from the mouse line being backcrossed to produce an isogenic background. This decision to backcross is made by considering the reasons for making the mouse line. If just one protein is to be investigated in the mouse line, then other proteins from the different strains are less likely to have an impact on the data. However, if many interacting proteins are to be investigated or behavioral observations made, then an isogenic background will produce more reproducible data.

3.8. Summary

Historically mice have been used in transgenic and mutant rodent models, and now rats are becoming more widely employed for this technique. Novel rat transgenic techniques will contribute to the knowledge already acquired from existing nontransgenic rat models of stroke. As an example, the SOD1 rat has been produced *(13)*, following on from the SOD1 mouse *(14)*. Rodent-directed transgenic technologies have allowed researchers to follow the consequences of increasing or decreasing expression of genes selected from the human genome in a live mammalian host. This technique has huge advantages over ex vivo or cell-based technology, as there are many more biochemical interactions among proteins, DNA, and RNA possible in vivo than in other expression systems. As specific genes involved in stroke are identified by existing technologies such as subtractive cDNA cloning *(15,16)* and by newer technologies such as microarray chips *(17)*, it is possible to manipulate these genes and create new models of stroke in which the consequence of very specific gene manipulations can be examined (*see* **Note 6**). With the sequencing of the human genome, it is possible to identify genes as risk factors for stroke *(18)*. Using this information to generate models in which these risk factor genes have been manipulated will allow researchers to follow the consequence of inherited genetic susceptibility to stroke.

4. Notes

1. Different strains of embryos respond differently to culture and microinjection. For example, C57Bl/6 inbred mouse embryos lyse very easily on injection, whereas [C57Bl6 × CBA]hybrid embryos survive very well. Similarly, the inbred embryos will not all proceed to the two-cell stage of development if cultured overnight, whereas the hybrid embryos will
2. Speed congenics is a technique whereby specific allelic variants of the required mouse strain are identified by PCR and appropriate animals selected for breeding. This can reduce the number of generations needed to breed to the selected background from six to three generations (18–9 mo).
3. If very large sterile males are paired with small superovulated females, some females may die because of too much stress and bullying during mating.
4. The needle may clog when it is being back-filled. To clear the capillary, approx 3 mm of capillary can be cut from the needle base in order to provide a clean surface to draw up the DNA solution. The cut is made by scoring with a diamond-tipped pencil and sharply tapping the blunt end of the capillary.
5. If the needle is not pulled accurately, it may be sealed at the tip. In this case the tip of the needle needs to be broken to provide a hole through which the DNA solution can travel. The needle tip is broken by gently dragging the tip across the holding pipet until it breaks. If the hole is visible at high (400×) magnification it is too big, and the needle must be replaced.
6. As with all animal work, the protocols used should be reviewed by a local ethical review board. When one is making novel transgenic or knockout mouse lines, the potential for suffering of these animals is unknown. A wide literature search for related lines should be carried out as a guide to the potential harm that the planned gene manipulation will have. This information can then serve as a guide for the likely level of harm that may be inflicted on the animals as they are produced and hence the desirability to progress with the work.

References

1. Gordon, J. W., Scangos, G. A., Plotkin, D. J., Barbosa, J. A., and Ruddle, F. H. (1980) Genetic transformation of mouse embryos by microinjection of purified DNA. *Proc. Natl. Acad. Sci. USA* **77,** 7380–7384.
2. Gossler, A., Doetschman, T., Korn, R., Serfling, E., and Kemler, R. (1986) Transgenesis by means of blastocyst-derived embryonic stem cell lines. *Proc. Natl. Acad. Sci. USA* **83,** 9065–9069.

3. Letts, V. A., Felix, R., Biddlecome, G. H., et al. (1998) The mouse stargazer gene encodes a neuronal ca2+-channel gamma subunit. *Nat. Genet.* **19,** 340–347.

4. Noebels, J. L., Qiao, X., Bronson, R. T., Spencer, C., and Davisson, M. T. (1990) Stargazer: a new neurological mutant on chromosome 15 in the mouse with prolonged cortical seizures. [Erratum appears in Epilepsy Res. (1992) **11,** 72]. *Epilepsy Res.* **7,** 129–135.

5. Hough, T. A., Nolan, P. M., Tsipouri, V., et al. (2002) Novel phenotypes identified by plasma biochemical screening in the mouse. *Mammal. Genome* **13,** 595–602.

6. Nolan, P. M., Peters, J., Strivens, M., et al. (2000) A systematic, genome-wide, phenotype-driven mutagenesis programme for gene function studies in the mouse. *Nat. Genet.* **25,** 440–443.

7. Davis, M., Mendelow, A. D., Perry, R. H., Chambers, I. R., and James, O. F. (1995) Experimental stroke and neuroprotection in the aging rat brain. *Stroke* **26,** 1072–1078.

8. Sutherland, G. R., Dix, G. A., and Auer, R. N. (1996) Effect of age in rodent models of focal and forebrain ischemia. *Stroke* **27,** 1663–1667.

9. Hogan, B., Beddington, R., Costantini, F., and Lacey, E. (1994) *Manipulating the Mouse Embryo: A Laboratory Manual.* Cold Spring Harbor Laboratory Press, Plain View, NY.

10. Vagner, S., Galy, B., and Pyronnet, S. (2001) Irresistible IRES—attracting the translation machinery to internal ribosome entry sites [Review]. *EMBO Rep.* **2,** 893–898.

11. Burcin, M. M. (1998) A regulatory system for target gene expression. *Front. Biosci.* **3,** C1–C7.

12. Nagy, A. (2000) Cre recombinase: the universal reagent for genome tailoring [Review]. *Genesis* **26,** 99–109.

13. Howland, D. S., Liu, J., She, Y. J., et al. (2002) Focal loss of the glutamate transporter EAAT2 in a transgenic rat model of SOD1 mutant-mediated amyotrophic lateral sclerosis (ALS). *Proc. Natl. Acad. Sci. USA* **99,** 1604–1609.

14. Gurney, M. E., Pu, H. F., Chiu, A. Y., et al. (1994) Motor neuron degeneration in mice that express a human cu,zn superoxide dismutase mutation. *Science* **264,** 1772–1775.

15. Wang, X., Barone, F. C., White, R. F., and Feuerstein, G. Z. (1998) Subtractive cloning identifies tissue inhibitor of matrix metalloproteinase-1 (TIMP-1) increased gene expression following focal stroke. *Stroke* **29,** 516–520.

16. Yokota, N., Uchijima, M., Nishizawa, S., Namba, H., and Koide, Y. (2001) Identification of differentially expressed genes in rat hippocampus after transient global cerebral ischemia using subtractive cDNA cloning based on polymerase chain reaction. *Stroke* **32,** 168–174.

17. Roth, A., Gill, R., and Certa, U. (2003) Temporal and spatial gene expression patterns after experimental stroke in a rat model and characterization of PC4, a potential regulator of transcription. *Mol. Cell. Neurosci.* **22,** 353–364.

18. Rosand, J. and Altshuler, D. (2003) Human genome sequence variation and the search for genes influencing stroke. *Stroke* **34,** 2512–2516.

8

3-Nitropropionic Acid Model of Metabolic Stress

Assessment by Magnetic Resonance Imaging

Toby John Roberts

Summary

3-Nitropropionic acid (3-NPA) is a potent mitochondrial inhibitor that can be administered systemically to create a progressive and localized striatal neurodegeneration mimicking many of the pathological features of Huntington's disease and other forms of metabolic compromise such as cerebral ischemia, carbon monoxide poisoning, and hypoglycemia. Here we describe a method to produce 3-NPA-induced lesions using the systemically administered toxin. We also describe magnetic resonance imaging methods to allow assessment of lesion severity over time within the same animal.

KEY WORDS

3-Nitropropionic acid; Huntington's disease; oxidative stress; magnetic resonance imaging; striatum; caudate putamen; excitotoxicity; neurodegeneration; energy metabolism; succinate dehydrogenase; complex II; mitochondria; dopamine; glutamate; blood–brain barrier.

1. Introduction

In recent years the metabolic toxin 3-nitropropionic acid (3-NPA) has been widely used to provide a preclinical model of Huntington's disease (HD) *(1– 3)*. HD is an autosomal dominant neurodegenerative disease caused by a CAG repeat expansion in the coding region of gene IT15 that encodes for a protein of unknown function called huntingtin *(4)*. This progressive neurodegenerative disease is characterized by selective damage to the striatum during its early stages. There is a growing body of evidence that a chronic deficit in neuronal energy metabolism may be involved in the development of HD *(5–7)*. It is this

From: *Methods in Molecular Medicine, Vol. 104: Stroke Genomics: Methods and Reviews*
Edited by: S. J. Read and D. Virley © Humana Press Inc., Totowa, NJ

fact that has stirred interest in 3-NPA, an irreversible suicide inhibitor of the mitochondrial enzyme succinate dehydrogenase (SDH). SDH is an essential enzyme in the Krebs cycle and forms part of complex II of the mitochondrial electron transport chain. Through this mechanism 3-NPA acts as an acute metabolic stressor that causes cell death both in vitro and in vivo *(8–10)*. When it is administered systemically, a selective loss of striatal neurons occurs, with a preferential loss of the medium-sized γ-aminobutyric acid (GABA)ergic spiny projection neurons that innervate the external globus pallidus and substantia nigra pars reticulata and a relative sparing of the aspiny NADPH-diaphorase–positive interneurons and large cholinergic interneurons *(2,11)*.

This pattern of cell loss is reminiscent of both HD and mild cerebral ischemia *(12,13)*, which points to a selective vulnerability of striatal neurons to metabolic insult. It has been suggested that 3-NPA causes selective striatal damage through a combination of mechanisms unique to the striatum. Some of these mechanisms of striatal damage are thought to be common to other forms of metabolic stress including focal and global ischemia, carbon monoxide poisoning, and hypoglycemia. The large corticostriatal glutamatergic input predisposes the striatum to excitotoxic injury. 3-NPA is thought to cause secondary excitotoxic damage *(14–16)* through a reduction of cellular ATP levels, which leads to depolarization of the cell membrane. The subsequent relief of the voltage-dependent Mg^{2+} blockade of the N-methyl-D aspartate (NMDA) receptors allows basal levels of glutamate to activate the NMDA receptor, causing a sustained increase of intracellular calcium leading to cell death. There is also some evidence that 3-NPA may act directly to increase extracellular levels of glutamate, which may be mediated through a non-energy-dependent inhibition of glutamate synaptic vesicle transporters *(17)* or via a potentiation of glutamate release by elevated dopamine levels *(18)*.

Excitotoxic mechanisms cannot wholly account for the striatal specific toxicity associated with metabolic stress. It is thought that dopamine, which itself is toxic *(19)*, is an important mediator of striatal damage. It has been shown that the nigrostriatal dopaminergic pathway contributes to the toxic effects of 3-NPA in the striatum. Removal of the striatal dopamine input via 6-hydroxydopamine lesioning of the nigrostriatal pathway protects the striatum from 3-NPA induced damage *(20)*. In vivo measurements of dopamine and its metabolites show that 3-NPA causes an elevation in concentration and higher turnover of striatal dopamine *(2,21)*. High levels of dopamine generate free radicals through monoamine oxidase-catalyzed oxidation or via autooxidation, which lead to an enhancement of the oxidative stress already induced by excitotoxic mechanisms and mitochondrial stress. Several studies have demonstrated the neuroprotective effects of free radical scavengers and the enhancement in markers of oxidative stress in this toxin model *(22,23)*. Although excitotoxic and dopamine-mediated mechanisms may be common factors in selective striatal damage caused by other metabolic stressors, it must be noted that 3-NPA may have preference for the striatum owing to at least one other factor. Several groups have demonstrated blood–brain barrier damage localized to the lateral striatal artery. Extravasation of blood-borne proteins, such as albumin and immunoglobulin G, into the striatal parenchyma occurs after 3-NPA administration *(24–26)*. Damage to astrocytic endfeet surrounding the arterial endothelium is in evidence and is thought to be one of the earliest signs of cellular damage in the 3-NPA model *(27)*. Taken together, these data suggest that 3-NP selectivity for the striatum may be partly owing to a regionally enhanced bioavailability as a result of damage to the blood–brain barrier of the lateral striatal artery.

3-NPA provides a very interesting model of metabolic stress. However, one must be aware that the protocol used in this model greatly influences the effect and final outcome of the toxin. The degree and type of damage caused is critically dependent on several factors, as shown by the literature: currently almost no two laboratories use exactly the same methods. Among the factors to take into account are the dose and its route of administration. In general, higher doses or several doses administered over very short periods produce less selective lesions along with a higher mortality rate. Intracranial administration produces more severe neuronal loss with less discrimination between the different neuronal phenotypes lost. Doses administered systemically over chronic periods produce the most selective lesions *(28)*. Animal age is important, with older animals showing much greater sensitivity *(10,29)*. This may be owing to the reduced mitochondrial functioning and enhanced synaptic glutamate levels seen in older animals *(30,31)*. Male animals are more vulnerable to 3-NPA than females, and it has been suggested that estrogen may exert a protective effect, whereas testosterone may exacerbate the damage *(32)*. It has also been shown that the rat strain is also important. In a recent paper by Ouary et al. *(33)*, it was shown that Fischer rats are highly vulnerable to 3-NPA but provide a poor animal model owing to a very high mortality rate. Sprague-Dawley rats, which have been used with this toxin most frequently, show a large interanimal variability and are therefore less than ideal. In contrast, Lewis rats show a much better tolerance to the toxin, and the resultant damage is highly reproducible *(34)*. The reasons for this are not clear, but it is likely that variations in the factors mediating the damage as outlined above between strains may be responsible. The model as described here produces consistent and phenotypically selective damage to the striatum with a very low mortality rate.

To study the evolution of 3-NPA damage, magnetic resonance imaging (MRI) provides an ideal tool that can be used noninvasively over repeated sessions within the same animal (**Fig. 1**). MRI allows the measurement of indirect markers of the lesion progression. This may be achieved through the measure-

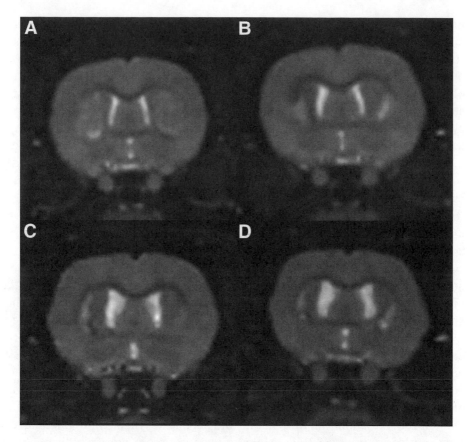

Fig. 1. T2-weighted MRI scans at the level of the striatum at (**A**) 2, (**B**) 4, (**C**) 6, and (**D**) 8 wk after lesioning protocol. Note the bilateral lesions appearing as a hyperintense region across the striatum. The lesion shows a reduction in size over time with the development of cavities within the lateral part of the striatum. The lateral ventricles showed a marked increase in size over time as they expanded to fill the space left by the striatal tissue necrosis.

ment of brain structure sizes and the size of the lesion along with an examination of MR relaxometry parameters that are altered when there is damage to and changes in the effected tissue. Here we also describe basic MRI methods to accomplish such indirect measurements of lesion progression.

2. Materials

2.1. 3-NPA Lesioning

1. 3-NPA (Fluka).
2. Sterile saline.

3. 12-wk-old male Lewis rats (Charles River), 360 ± 20 g in weight allowed access to food (RM1 [p] SDS Diet, Lillico) and water *ad libitum*. Animals are housed with a 12-h on/off light-dark cycle. Temperature is maintained at $20 \pm 1°C$ with humidity of 35–70%.
4. Clamp stand and clamp for holding pH meter and 3-NPA container in place.
5. Accurate balance (± 0.001 g).
6. Fume hood.
7. Sterile container with lid (large enough to accommodate pH probe).
8. Duphalyte (Fort Dodge Animal Health) an injectable supportive maintenance therapy containing vitamin B complex, electrolytes, amino acids, and dextrose.
9. Complan (Heinz).
10. pH meter.
11. 1-mL Syringes with Luer lock to prevent needle from falling off syringe.
12. 25-gage needles.
13. 5 *M* NaOH solution.

2.2. Magnetic Resonance Imaging

1. MRI scanner capable of multiecho spin echo imaging.
2. Equipment for delivery of gaseous anesthetic suitable for a rodent.
3. Rat headholder, nonmagnetic.
4. Sterile saline.
5. Eye gel to prevent drying of animal eyes during scanning.
6. Heating device such as infrared lamp to maintain animal body temperature outside magnet.
7. Thermostatically controlled electric heating blanket and rectal temperature probe (Harvard Apparatus), used to maintain animal body temperature inside the magnet.
8. System for measuring pulse such as blood pressure tail cuff system (Harvard Apparatus).
9. System for measuring respiratory rate such as a fine wound wire coil to be placed on animal abdomen attached to an amplifier system. Breathing movements move the wire in the magnetic field, inducing an electrical current in the wire, which varies with the wire position.

3. Methods

3.1. 3-NPA Lesioning

The 3-NPA lesioning methodology used in our laboratory is based on that of Ouary et al. *(33)* uses 12-wk-old male Lewis rats. This protocol will produce selective striatal lesions in over 80% of subjects. Overt symptomology is usually apparent after 3 d of toxin administration. Symptoms include splayed legs, wobbly gait, intermittent hindlimb paresis, hypoactivity, and poor condition of the fur. There is also a marked weight loss (*see* **Note 1**).

3.1.1. Preparation of the 3-NPA Stock Solution

1. Weigh 3-NPA into a sterilized container of sufficient volume. The total mass of 3-NPA should be equivalent to about 18 mg per animal. **Caution:** This must be done in a fume hood with suitable protective clothing, as 3-NPA is a potent neurotoxin (*see* **Note 2**).
2. Calculate the amount of saline required to give a final 3-NPA concentration of 0.03 g/mL for the mass of 3-NPA weighed out in **step 1**.
3. Add a volume of saline to the 3-NPA less than that calculated in **step 2**. Clamp a pH meter in place to measure the solution pH and adjust the pH to 7.4 by adding 5 *M* NaOH dropwise. Once pH 7.4 is reached, make the solution using sterile saline up to a final concentration of 0.03 g/mL.

3.1.2. Administration of 3-NPA

1. Weigh the animals and calculate the volume of a 0.03 g/mL 3-NPA stock solution needed to give a dose of 42 mg/kg for each animal.
2. Using the volume of stock 3-NPA solution calculated in **step 1**, determine the amount of saline required to make 3-NPA solution up to a total volume of 1 mL i.e., the volume of stock 3-NPA solution calculated in **step 1** subtracted from a total of 1 mL (*see* **Note 3**).
3. Label a 1-mL syringe for each animal using a marker pen. For each subject fill the 1-mL syringe with sterile saline and attach a 25-gage needle (*see* **Note 4**). Remove air bubbles and adjust the volume of saline to that calculated in **step 2**.
4. Finally, fill the rest of the syringe to a total volume of 1 mL by taking up the 0.03 g/mL 3-NPA stock solution into the syringe. This will give a solution of 3-NPA that when injected will provide a dose of 42 mg/kg in a final volume of 1 mL (*see* **Note 5**).
5. Inject the animals intraperitoneally (*see* **Note 6**) taking care not to inject into the lower intestine or other organs. It is suggested that this procedure be carried out with two people to prevent the needle from being kicked out by the animal and risking 3-NPA solution contaminating the experimenter.
6. Repeat this procedure for 5 d, injecting at the same time of day (*see* **Note 7**). 3-NPA should be prepared fresh each day.
7. Weigh animals daily and observe until their weight begins to increase again.

3.1.3. Animal Care

This procedure causes the animals to lose appetite and cease grooming properly, so the general appearance of the animals will deteriorate over the course of lesioning. In general animals should lose between 10 and 15% of their body weight, with weight loss continuing for at least 2 d after the last injection of toxin (**Fig. 2**). It may be necessary to supplement the diet with more palatable food such as Complan, which can be hand-fed, or softened pellets to encourage eating. Subcutaneous injections of vitamin and glucose supplements

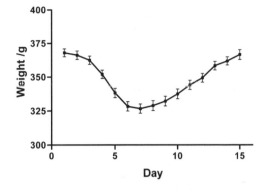

Fig. 2. Decrease in body weight over 15 d starting from the first day of 3-NPA injections. Maximal weight loss occurs around 2 d after the fifth and final injection. Animals return to original body weight approx 2 wk after the first injection.

(Duphalyte) will also aid survival but may interfere with the experimental design. Warming of the animal's environment may also aid recovery.

3.2. Magnetic Resonance Imaging

3.2.1. Magnetic Resonance Image Acquisition

Our laboratory uses conventional MRI techniques available on any research-capable scanner. Animals can be repeatedly imaged depending on the time-point of interest. The 3-NPA lesion is a progressive lesion with changes still occurring weeks to months after toxin treatment.

1. Anesthesia must be used to immobilize and minimize stress to the animal during scanning. We use 2% isoflurane in an O_2/N_2O (30:70% mix) at a flow rate of 1 L/min. Anesthesia is induced at 4% isoflurane over 5–10 min in a clear Perspex chamber with an attached scavenging system.
2. Animals are placed in an MRI-compatible plastic stereotaxic frame that positions the top of the brain parallel to the horizontal axis of the MRI scanner. This allows more direct comparison of MRI images to the Paxinos and Watson histological atlas *(35)* (*see* **Note 8**). Anesthetic is delivered via a plastic face mask.
3. A thermostatically controlled electric heating blanket is placed under the animal to maintain temperature at $37 \pm 0.5°C$.
4. Animals must be monitored for both heart rate and respiratory rate in order to monitor depth of anesthesia while the animals are in the MRI scanner. This is achieved using a blood pressure tail cuff (Harvard Apparatus) to measure the pulse and a home-made winding coil placed on the moving part of the abdomen and connected to a simple amplifier. Readings are displayed using the MP100 system (Biopac Systems) (*see* **Note 9**).

5. A simple, multiecho, spin echo sequence is used to visualize the lesion. This MRI scanning sequence is available on all modern MRI scanners (*see* **Note 10**). The parameters used in our scanner (Unity Inova, Varian) interfaced to a 4.7-T superconducting magnet (Oxford Instruments) are summarized in **Table 1**. Total scan time is not more than 30 min.

6. After MRI scanning, inject the animals sc with 4 mL of saline, 2 mL on either flank (*see* **Note 11**). Animals are kept warm during recovery using an infrared heating lamp.

3.2.2. MRI Analysis: Volume Quantification

From the multiecho spin echo MR images, several quantitative parameters relating to lesion severity can be assessed. We routinely examine the 3-NPA lesion volume as defined by the region of hyperintensity in the striatum in the longest echo time T2-weighted images (echo time [TE] = 66 ms). We also measure total striatal volume from the T2-weighted images and total brain volume from the proton density-weighted images (TE = 22 ms).

1. Convert proprietary format images to Analyze 7.5 format (*see* **Note 12**). This should give separate images for the 22-, 44-, and 66-ms echo time images.

2. Use MRIcro to measure lesion volume, striatal volume, and total brain volume (*see* **Note 13**).

3. Open MRIcro and from the file menu open an analyze format image to be measured (*see* **Note 14**).

4. In the left-hand window is a region of interest (ROI) tool panel. Use the *pen tool with autoclose* to draw around a ROI, e.g., the lesion extent as seen on the T2-weighted (longest echo time) image that will appear hyperintense. This is done by holding down the left mouse button and using the mouse to define the ROI. Once this is complete, release the left mouse button to close the ROI (**Fig. 3**).

5. To count all pixels in the ROI, *the fill* tool in the left-hand ROI panel should be used to fill the inside of each ROI.

6. Repeat **steps 4** and **5** for all areas of the lesion or area being measured through all slices of interest.

7. Once the entire lesion area has been defined, its volume is calculated by using the *ROI min/mean and max* tool in the left-hand ROI panel. This will give the total number of pixels and minimum, mean, and maximum pixel intensities within the entire ROI. The total number of pixels gives the volume of the lesion or area measured. This can be converted to meaningful units by calculating the volume of an individual pixel from the MRI parameters used (**Table 1**), i.e., from the parameters in **Table 1**, pixel volume = (30 mm/96 pixels) × (30 mm/96 pixels) × 0.7 mm = 0.0684 mm^3.

8. The above procedure can be repeated by creating ROIs for the entire brain or any brain structure of interest, e.g., the striatum.

9. Save ROIs using the *save this ROI* button in the ROI panel. Saved ROIs can be used to analyze T2 values as described in **Subheading 3.2.3.** below.

Table 1
Summary of the Main Parameters Used for the Multiecho Spin Echo MRI Sequence

Repetition time(s): 3	Echo times (s): 0.022, 0.044, and 0.066	Number of averages: 4
Sweep width (Hz) /s: 32,746	Dwell time (s): 0.00293	180° Pulse length (ms): 2550
Slice thickness (mm): 0.7	Interslice gap (mm): 0.7	90° Pulse length (ms):2550
Number of slices: 30	Field of view (cm): 3×3	Matrix dimension (pixels): 96 × 96 In plane resolution (mm): 0.312

211

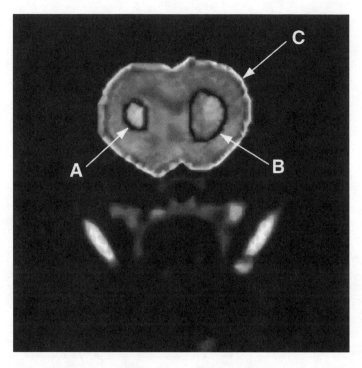

Fig. 3. A T2-weighted image showing the ROIs used to measure the volume of (A) the lesion, (B) the striatum, and (C) the whole brain volume. Measurements are repeated across all slices within which the area of interest occurs.

3.2.3. MRI Analysis: T2 Relaxometry

The advantage of the multiecho spin echo sequence described is that it can also be used to create T2 maps (**Fig. 4**) (*see* **Note 15**). T2 maps are simply a pixel-by-pixel representation of the T2 (or spin–spin) relaxation constant. As it is sensitive to changes in composition of the tissue, it can be used as another indirect measure of lesion progression.

1. ImageJ java-based software (*see* **Note 16**) is used to calculate T2 maps using a multipoint fit.
2. Open the individual 22-, 44-, and 66-ms echo time images (*see* **Note 17**).
3. For each echo time image create a single montage image. (This is a single image with all the slices combined into one slice.) Select the 22-ms echo time image. From the *Image* drop-down window, select *Stacks* and then select *Make Montage*. Enter the columns as 6 and the rows as 5. Enter the first slice as 1, the number of slices as 30, and the increment as 1. Then select OK. This will create a 6 × 5 montage image of the 30 slices in the 22-ms image. Repeat for the 44-ms and 66-ms images (*see* **Note 18**).

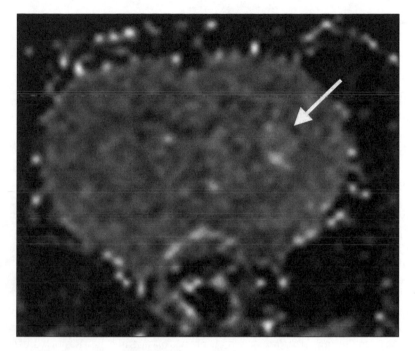

Fig. 4. A T2 map showing a pixel-by-pixel representation of the T2 (spin–spin) relaxation constant. Note the brighter area in the striatum where the lesion occurs, indicating a higher T2 constant (arrow).

4. Close all images except for the three montage images. From the *Image* drop-down menu, select *Stacks*, then select *Convert Images to Stack*. This will combine all three montage images into a single image of three slices. The three slices will contain a montage of the 22-, 44-, or 66-ms echo time images respectively (*see* **Note 18**).

5. To calculate the T2 values from the three echo times, select the newly created stack image. Convert the image to 32-bit by selecting the *Image* drop-down menu, and then select *type* and choose 32-bit. Repeat this and select 16-bit (*see* **Note 19**).

6. Select the stack image and from the *plugin* drop-down menu select the *MRI Analysis Calculator* (*see* **Note 20**). Choose T2 calculation and select OK. Choose the montage stack image created in **step 4** as the T2 image stack on which to perform the calculation. Enter the TE values for each slice in seconds (in this case 0.022, 0.044, and 0.066 s). The other two options are thresholds for the calculation and can be left as the default values. Choose OK. The message stating that an attempt will be made to convert images to 32-bit should be ignored. Choose OK. A progress bar in the main tool window will appear, and the T2 map will be created.

7. To convert the T2 map montage image back to a single image per slice, select the T2 map montage image created in **step 6**. From the *plugin* drop-down menu,

select the *Stack Maker* plugin tool. In the Stack Maker window enter the number of images per row and the number of images per column, which will be 6 and 5, respectively, in this case. Then select OK. This will then create a single image from the 6 × 5 image montage (*see* **Note 21**).

8. Save the T2 map image created in **step 6** (*see* **Note 22**).
9. To measure the T2 value in a given ROI, use MRIcro. Open the T2 map image in MRIcro. From the ROI drop-down menu, open the previously saved ROI corresponding to the subject and area to be measured e.g., the striatum.
10. To calculate the mean T2 in this ROI, click on the *ROI min/mean and max* button in the ROI panel on the left-hand side. The mean T2 value is shown along with the number of pixels and minimum and maximum pixel intensities.

4. Notes

1. In our experience, the maximal percentage weight loss is a very reliable indicator of the severity of the lesion, as indicated by the lesion volume measured by T2-weighted MRI scans (**Fig. 5**).
2. **Caution:** 3-NPA is a potent neurotoxin in humans as well as animals. Cases of human poisoning have been reported in China as the result of consumption of moldy sugar cane *(36)*. 3-NPA is a mycotoxin produced by various species of fungus including *Arthrinium*, which grows on sugar cane. Reported effects on humans include putaminal necrosis accompanied by dystonia. Appropriate precautions should be taken (including protective clothing such as face mask, eye protection, gloves (two pairs), and lab coat), and all handling other than animal injections should be carried out in a fume hood.
3. All injections are given in a total volume of 1 mL.
4. When filling syringes with sterile saline, do not use the same needle that will be used to inject the animal, as this will blunt the needle, making it more difficult to inject the animal and increasing the likelihood of the experimenter becoming contaminated with 3-NPA.
5. Filling first with saline to the required volume reduces the amount of handling of syringes containing 3-NPA solution and also allows removal of air bubbles and adjustment of volumes before any toxin is in the syringe.
6. Injections may be given subcutaneously for a slower release of toxin; however, this is slightly more difficult to achieve safely. Many experiments have also delivered 3-NPA via a subcutaneously implanted osmotic pump (o.p.). Several publications show this gives a more consistent lesion. However, it appears that, at least in our hands, a dose of 42 mg/kg delivered by o.p. for 5 d (the same total dose and time-course as in the method described here) does not produce a lesion. We have used up to 60 mg/kg delivered by o.p. with no apparent lesion. Of the publications using o.p. delivery in the Lewis rat, there seems to be a large variation in dose between groups to achieve similar lesion damage. This may be because of a difference in the susceptibility of the animals, as they come from different sources, or even differences in the type of osmotic pump used *(33,34)*.

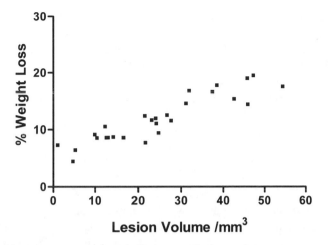

Fig. 5. Maximum percentage weight loss vs lesion volume as measured from T2-weighted (TE = 66 ms) images. A strong correlation ($R^2 = 0.883$) exists between weight loss and lesion size.

7. 3-NPA injections should be carried out preferably in the early morning, as this will allow the animals more time to recover before they begin nocturnal feeding. To accommodate this, it may be necessary to adjust timing of the 12-h light dark cycle.
8. The stereotaxic head frame system consists of a tooth bar and ear bars that can be laterally adjusted to accommodate rats of varying size. The tooth bar sits 3 mm below the ear bars (which are tapered plastic screws) to allow a flat head position. Most MRI research units will have some form of head-holder system available. A frame can be made from plastics available in any reasonably well-equipped machine shop.
9. Anesthesia monitoring is particularly important during the first 7 d after the 3-NPA injection procedure as the animals are weakened and more sensitive to anesthesia so should be kept at as low a level of anesthesia as is practicably possible.
10. Because the multiecho spin echo imaging sequence contains a user-defined number of echoes at varying echo times, we get both proton density (with contrast dependent on the density of protons—usually from water and fat) and T2 (with contrast dependent on MR relaxation parameters) weighted scans within a single scan acquisition.
11. Subcutaneous saline injections post scanning improve recovery by preventing dehydration. This is particularly important if the animals are to be scanned with short recovery intervals.
12. Analyze 7.5 is a widely used digital image file format developed by the Mayo Clinic, Rochester, MN (http://www.mayo.edu/bir/).
13. Most MRI laboratories will have locally available software that may provide a more efficient method to do this kind of analysis. MRIcro (Chris Rorden, Univer-

sity of Nottingham, UK) is described here because it is freely available, will operate on a variety of computing platforms, and is easy to use. It can be obtained from the University of Nottingham, at the following website: http://www.psychology.nottingham.ac.uk/staff/cr1/index.html.

14. If no local conversion program is available, MRIcro can read most image formats simply by inputting parameters for the image size, image data offset, and image bit type. To do this, open MRIcro and within the header information panel input the number of pixels into the x and y values. For the MRI acquisition described here $x = 96$ and $y = 96$. Input the number of slices to z, in this case 30. Finally, input the image data offset. This is usually 0 if the header information is held in a separate file from the image. When this is not the case, the offset is the number of bits before the image information is encoded. Once these parameters are inputted, open the *import* drop-down window and choose *open foreign*. Choose the file containing the image data and open it. The image should then appear in the main viewing window. To convert the data to Analyze 7.5 format, set up the correct header parameters, and then from the *import* drop-down window choose *Convert foreign to Analyze*. From the number of files window, keep the defaults and click on select. Choose the data to be converted and click open. This will prompt for a file name. Finally choose save to create the Analyze image.

15. Most MRI laboratories will have in-house or commercial software that will be able to do these calculations and measurements. They may be more automated and therefore save time. The measurements described are typical of those that might be measured in any structural MRI study, and the methods described here are used because they are freely available and will run on virtually any contemporary computer system.

16. ImageJ can be obtained from the Research Services Branch (RSB) of the National Institute of Mental Health (NIMH) at the following site: http://rsb.info.nih.gov/ij/download.html. ImageJ can be used on any computer format that is Java-enabled. Follow the instructions in the readme file in order to install it. Various image manipulation tools are available as separate plugins including an MRI Analysis Calculator (Karl Schmidt, Harvard University, Cambridge, MA), a plugin called stack Maker (Wayne Rasband, NIH) that is used in the whole brain T2 calculation method described here, and a plugin for loading and saving Analyze format images (Guy Williams, University of Cambridge, Cambridge, UK). These should all be downloaded and installed by placing the compiled java plugin (.class) file in the /Imagej/plugins directory. Another useful plugin for separating the individual echo times called stack splitter (Patrick Kelly and Harvey J. Karten, University of San Diego, San Diego, CA) is available. As most multiecho spin echo image datasets are produced as a single image file (i.e., all echo time images in the same image), this plugin is useful for separating out the individual echo times so they can be analyzed separately. All these plugins can be obtained from http://rsb.info.nih.gov/ij/plugins/.

17. To open an image in ImageJ, use the *File* drop-down menu, and use the *import* function. Choose the RAW data option and open the image. A window will open

asking for the image header parameters. These include the image bit number, signed 8- or 16-bit for Analyze 7.5, the width and height of the imported image in number of pixels, and the offset to the image data. As the Analyze 7.5 format images have a separate header file, this offset will be 0. The number of images also needs to be input that corresponds to the number of slices. Finally, the gap between images must be entered as 0 for Analyze 7.5 format images, i.e., the number of bytes separating each slice. It is also possible to define byte type, which is defined by the computing format the images were generated on, i.e., little-endian for Intel-based machines or big-endian for Sun Microsystem-based machines. Alternatively, the plugin described in **Note 16** for loading and saving Analyze images may be used. Simply open the plugin drop-down menu, and use the Analyze Reader plugin.

18. Creation of the montage is necessary because the MRI Analysis plugin described in **Note 16**. will only do calculations for a single slice at a time. Therefore it is necessary to put all the images from a given echo time into a single slice to perform the calculation on every slice in one go. The MRI analysis tool requires that each of the different echo time images collected be in the same image with a separate slice for each echo time. This is because, as described above in **Note 16**, most multiecho spin echo datasets will be in the form of a single-image file, with a separate slice for each echo time collected.

19. The MRI Analysis plugin works with 32-bit datasets. A quirk in the 16-bit to 32-bit conversion of the images by the MRI Analysis plugin software requires Analyze 7.5 16-bit images to be converted from 16-bit to 32-bit and back to 16-bit manually before a correct T2 calculation will occur.

20. The MRI Analysis Calculator plugin allows calculation of T2 maps from multiecho spin echo datasets. It may be downloaded and installed as described in **Note 16**.

21. It is necessary to convert the T2 map montage image back to a single image of 30 slices if the ROIs created in **Subheading 3.2.2.** are to be used to calculate the T2 in the area defined by the ROI. Stack maker is a plugin that allows the conversion to be carried out. It can be downloaded and installed as described in **Note 16**.

22. Images can be saved in a raw data format, but to be read in MRIcro, they must be changed to 16-bit as described in **Subheading 3.2.3., step 6**. To save as raw data open the *File* drop-down menu in ImageJ and choose *Save As* and then choose the *Raw Data* format. These images may then be opened or converted by MRIcro to analyze format data as described in **Note 14**. Alternatively, the image can be saved directly as Analyze 7.5 format using the plugin for loading and saving Analyze data described in **Note 16**. Use the plugin drop-down menu and use the Analyze Writer plugin.

References

1. Hamilton, B. F. and Gould, D. H. (1987) Nature and distribution of brain lesions in rats intoxicated with 3-nitropropionic acid: a type of hypoxic (energy deficient) brain damage. *Acta Neuropathol.* **72,** 286–297.

2. Beal, M. F., Brouillet, E., Jenkins, B. G., et al. (1993) Neurochemical and histologic characterization of striatal excitotoxic lesions produced by the mitochondrial toxin 3-nitropropionic acid. *J. Neurosci.* **13,** 4181–4192.
3. Brouillet, E., Guyot, M. C., Mittoux, V., et al. (1998) Partial inhibition of brain succinate dehydrogenase by 3-nitropropionic acid is sufficient to initiate striatal degeneration in rat. *J. Neurochem.* **70,** 794–805.
4. The Huntington's Disease Collaborative Research Group (1993) A novel gene containing a trinucleotide repeat that is expanded and unstable on Huntington's disease chromosomes. *Cell* **72,** 817–818.
5. Gourfinkel-An I., Vila, M., Faucheux, B., et al. (2002) Metabolic changes in the basal ganglia of patients with Huntington's disease: an in situ hybridization study of cytochrome oxidase subunit I mRNA. *J. Neurochem.* **80,** 466–476.
6. Jenkins, B. G., Koroshetz, W. J., Beal, M. F., Rosen, B. R. (1993) Evidence for impairment of energy metabolism in vivo in Huntington's disease using localized 1H NMR spectroscopy. *Neurology* **43,** 2689–2695.
7. Beal, M. F. (2000) Energetics in the pathogenesis of neurodegenerative diseases. *Trends Neurosci.* **23,** 298–304.
8. Behrens, M. I., Koh, J., Canzoniero, L. M., Sensi, S. L., Csernansky, C. A., and Choi, D. W. (1996) 3-Nitropropionic acid induces apoptosis in cultured striatal and cortical neurons. *Neuroreport* **6,** 545–548.
9. Pang, Z. and Geddes, J. W. (1997) Mechanisms of cell death induced by the mitochondrial toxin 3-nitropropionic acid: acute excitotoxic necrosis and delayed apoptosis. *J. Neurosci.* **17,** 3064–3073.
10. Brouillet, E., Jenkins, B. G., Hyman, B. T., et al. (1993) Age-dependent vulnerability of the striatum to the mitochondrial toxin 3-nitropropionic acid. *J. Neurochem.* **60,** 356–359.
11. Reynolds, G. P., Pearson, S. J., Heathfield, K. W. (1990) Dementia in Huntington's disease is associated with neurochemical deficits in the caudate nucleus, not the cerebral cortex. *Neurosci. Lett.* **113,** 95–100.
12. Ferrante, R. J., Kowall, N. W., Beal, M. F., Richardson, E. P. Jr., Bird, E. D., and Martin, J. B. (1985) Selective sparing of a class of striatal neurons in Huntington's disease. *Science* **230,** 561–563.
13. Calabresi, P., Centonze, D., Pisani, A., et al. (1998) Striatal spiny neurons and cholinergic interneurons express differential ionotropic glutamatergic responses and vulnerability: implications for ischemia and Huntington's disease. *Ann. Neurol.* **43,** 586–597.
14. Alexi, T., Hughes, P. E., Faull, R. L., Williams, C. E. (1998) 3-Nitropropionic acid's lethal triplet: cooperative pathways of neurodegeneration. *Neuroreport* **9,** R57–R64.
15. Lee, W. T., Shen, Y. Z., Chang, C. (2000) Neuroprotective effect of lamotrigine and MK-801 on rat brain lesions induced by 3-nitropropionic acid: evaluation by magnetic resonance imaging and in vivo proton magnetic resonance spectroscopy. *Neuroscience* **95,** 89–95.

16. Nishino, H., Hida, H., Kumazaki, M., et al. (2000) The striatum is the most vulnerable region in the brain to mitochondrial energy compromise: a hypothesis to explain its specific vulnerability. *J. Neurotrauma* **17,** 251–260.

17. Tavares, R. G., Santos, C. E., Tasca, C. I., Wajner, M., Souza, D. O., Dutra-Filho, C. S. (2001) Inhibition of glutamate uptake into synaptic vesicles from rat brain by 3-nitropropionic acid in vitro. *Exp. Neurol.* **172,** 250–254.

18. Godukhin, O. V., Zharikova, A. D., and Budantsev, A. Y. (1984) Role of presynaptic dopamine receptors in regulation of the glutamatergic neurotransmission in rat neostriatum. *Neuroscience* **12,** 377–383.

19. Ben-Shachar, D., Zuk. R., Glinka, Y. (1995) Dopamine neurotoxicity: inhibition of mitochondrial respiration. *J. Neurochem.* **64,** 718–723.

20. Maragos, W. F., Jakel, R. J., Pang, Z., Geddes, J. W. (1998) 6-Hydroxydopamine injections into the nigrostriatal pathway attenuate striatal malonate and 3-nitropropionic acid lesions. *Exp. Neurol.* **154,** 637–644.

21. Johnson, J. R., Robinson, B. L., Ali, S. F., Binienda, Z. (2000) Dopamine toxicity following long term exposure to low doses of 3-nitropropionic acid (3-NPA) in rats. *Toxicol. Lett.* **116,** 113–118.

22. Fontaine, M.A., Geddes, J.W., Banks, A., Butterfield, D.A. (2000) Effect of exogenous and endogenous antioxidants on 3-nitropionic acid-induced in vivo oxidative stress and striatal lesions: insights into Huntington's disease. *J. Neurochem.* **75,** 1709–1715.

23. Binienda, Z., Simmons, C., Hussain, S., Slikker, W. Jr., Ali, S. F. (1998) Effect of acute exposure to 3-nitropropionic acid on activities of endogenous antioxidants in the rat brain. *Neurosci. Lett.* **251,** 173–176.

24. Reynolds, D. S. and Morton, A. J. (1998) Changes in blood-brain barrier permeability following neurotoxic lesions of rat brain can be visualised with trypan blue. *J. Neurosci. Methods* **79,** 115–121.

25. Hamilton, B. F. and Gould, D. H. (1987) Correlation of morphologic brain lesions with physiologic alterations and blood-brain barrier impairment in 3-nitropropionic acid toxicity in rats. *Acta Neuropathol.* **74,** 67–74.

26. Nishino, H., Shimano, Y., Kumazaki, M., Sakurai, T. (1995) Chronically administered 3-nitropropionic acid induces striatal lesions attributed to dysfunction of the blood-brain barrier. *Neurosci. Lett.* **186,** 161–164.

27. Nishino, H., Kumazaki, M., Fukuda, A., et al. (1997) Acute 3-nitropropionic acid intoxication induces striatal astrocytic cell death and dysfunction of the blood-brain barrier: involvement of dopamine toxicity. *Neurosci. Res.* **27,** 343–55.

28. Borlongan, C. V., Nishino, H., Sanberg, P. R. (1997) Systemic, but not intraparenchymal, administration of 3-nitropropionic acid mimics the neuropathology of Huntington's disease: a speculative explanation. *Neurosci. Res.* **28,** 185–189.

29. Bossi, S. R., Simpson, J. R., Isacson, O. (1993) Age dependence of striatal neuronal death caused by mitochondrial dysfunction. *Neuroreport* **4,** 73–76.

30. Shigenaga, M. K., Hagen, T. M., and Ames, B. N. (1994) Oxidative damage and mitochondrial decay in aging. *Proc. Natl. Acad. Sci. USA* **91,** 10771–10778.

31. Price, M. T., Olney, J. W., and Haft, R. (1981) Age-related changes in glutamate concentration and synaptosomal glutamate uptake in adult rat striatum. *Life Sci.* **28,** 1365–1370.
32. Nishino, H., Nakajima, K., Kumazaki, M., et al. (1998) Estrogen protects against while testosterone exacerbates vulnerability of the lateral striatal artery to chemical hypoxia by 3-nitropropionic acid. *Neurosci. Res.* **30,** 303–312.
33. Ouary, S., Bizat, N., Altairac, S., et al. (2000) Major strain differences in response to chronic systemic administration of the mitochondrial toxin 3-nitropropionic acid in rats: implications for neuroprotection studies. *Neuroscience* **97,** 521–530.
34. Blum, D., Gall, D., Cuvelier, L., Schiffmann, S. N. (2001) Topological analysis of striatal lesions induced by 3-nitropropionic acid in the Lewis rat. *Neuroreport* **12,** 1769–72.
35. Paxinos, G. and Watson, C. (1997) *The Rat Brain in Stereotaxic Coordinates.* Academic, San Diego.
36. Ludolph, A. C., He, F., Spencer, P. S., Hammerstad, J., and Sabri, M. (1991) 3-Nitropropionic acid-exogenous animal neurotoxin and possible human striatal toxin. *Can. J. Neurol. Sci.* **18,** 492–498.

III

CLINICAL PARADIGMS

9

Practicalities of Genetic Studies in Human Stroke

Ahamad Hassan and Hugh S. Markus

Summary

Considerable evidence suggests genetic factors are important in the pathogenesis of multi-factorial stroke. However, studies identifying the underlying genes have been largely disappointing. This chapter reviews the different approaches and their relative merits. It is likely stroke is a polygenic disorder and that underlying genes may interact within environmental risk factors. Stroke itself is a syndrome caused by a number of different pathologies, which may result from different genetic predispositions. Therefore accurate stroke subtyping is likely to be important in identifying genetic associations. Previous studies have suffered from small sample size, lack of adequate phenotyping, and poor case-control matching.

The most popular approach to identifying genes in human polygenic ischemic stroke has been the candidate gene approach. The relative merits of this approach are discussed. More recently this has been extended to famly-based association studies. Linkage-based approaches have been used less although current studies are implementing the affected-sibling-pair method. In the future genome-wide association studies are likely to become more widely used.

Key Words

Cerebrovascular disease; genetics; polygenic stroke.

1. Introduction

Many separate lines of evidence suggest that genetic factors are important in the pathogenesis of cerebrovascular disease. These include twin *(1,2)* and family history studies *(3,4)* and studies of animal models of stroke *(5,6)*. The identification and characterization of genes that are involved in the pathogenesis of stroke have recently been highlighted as a "field of need"*(7)*. It is hoped that research will eventually lead to improved assessment of individual disease risk and reveal new therapeutic pathways. To date no single gene has been unequivocally demonstrated in human stroke, partly because of the limitations of widely used techniques, such as case control studies. Attention has turned toward refining these techniques, including better stroke phenotyping, and

From: *Methods in Molecular Medicine, Vol. 104: Stroke Genomics: Methods and Reviews*
Edited by: S. J. Read and D. Virley © Humana Press Inc., Totowa, NJ

adopting alternative approaches such as the use of intermediate phenotypes, family-based studies, and genomic strategies. Novel techniques have certain theoretical advantages but may be difficult to implement. Ethical frameworks, with respect to acquisition and storage of human DNA, apply to all genetic techniques. In this chapter we provide an overview of the methods and practicalities of genetic studies in ischemic human stroke.

2. Genetic Disease Models of Ischemic Stroke

Before discussing the practicalities of genetic research, some background as to how genetic factors could be operating in the disease will be helpful. Most stroke geneticists consider that a disease model similar to that adopted in other complex traits such as hypertension and diabetes is a useful one. The principle features of this model are as follow:

1. Stroke is a polygenic disorder reflecting the influence of several genetic loci modulating different pathophysiological processes. The traditional view is that there may be many hundreds of genetic variants, each conferring equally small amounts of disease risk. However, studies in animal models *(5,6)*, and more recently in humans *(8)*, have supported the alternative notion that a few major stroke genes may confer the bulk of genetic risk.
2. Genetic factors may express their influence only under certain conditions. This is a phenomenon that is referred to as variable penetrance and is responsible for the complex patterns of inheritance seen in polygenic disorders such as stroke. Variable penetrance can arise for several reasons. For example, the presence of several genes may be required to increase the risk of disease in an additive manner (a gene dose effect), or a gene may first have to interact with another risk factor such as an environmental trigger (e.g., smoking) or another gene (epistatic interaction). Gene–environment or gene–gene interactions of this kind are usually synergistic, with the net result being a multiplicative increase in disease risk.
3. Genetic factors may be comparatively more important in early-onset disease. It is hypothesized that early-onset disease has a more homogenous substrate, with genetic factors having comparatively more influence than environmental factors *(9)*. There is anecdotal evidence for this theory from other polygenic diseases, as well as direct evidence for stroke more recently from twin *(10)* and family history studies *(3)*.

Cerebral infarction is the end result of a number of complex pathophysiological processes. If one accepts that stroke is a polygenic condition, it is also helpful to consider at what level in the stroke pathway genetic influences might be operating to increase disease risk. Genes could be acting at the level of conventional risk factors such as hypertension, either by predisposing to the risk factors themselves, or by modulating their effects on end organs. Alternatively, genetic factors could be operating independently of these known risk factors by directly participating in the processes that lead to atherosclerosis, plaque rupture, and thromboembolism. Finally, genetic factors could influence the response of brain tissue to ischemia, for example, by determining collateral blood flow, or neuronal susceptibility to ischemia.

3. Difficulties in Identifying Human Stroke Genes

Although disease models facilitate our understanding of stroke genes, identifying individual molecular factors may be difficult, for several reasons. First, stroke is usually a late-onset disease. This means that techniques relying on available family members for genetic comparison are difficult to perform, as often the relevant individuals, such as parents or siblings, will have died. Second, stroke may exhibit the phenomenon of locus heterogeneity, i.e., many different disease alleles may predispose to the same phenotype. Third, stroke is a phenotypically heterogeneous disorder, and different stroke subtypes have different underlying disease mechanisms, including those at a molecular genetic level. Therefore the individual contribution of a single gene may be diluted if stroke is considered as a single entity.

Another factor that makes molecular genetic studies of stroke difficult is incomplete gene penetrance. This can lead to loss of statistical power if the end result is subclinical disease in control subjects. Confounding is another potential problem: coexistent risk factors such as hypertension and diabetes may share some susceptibility genes with ischemic stroke. Therefore it may be difficult to dissect out the role of an individual gene in one disease.

4. Strategies for Detecting Genes in Human Polygenic Ischemic Stroke
4.1. The Candidate Gene Approach

Candidate gene studies involve first identifying a molecular variant or polymorphism within a functionally relevant gene (candidate gene) and then confirming an association with disease. These methods require an *a priori* understanding of stroke pathophysiology and therefore lack the ability to identify novel genes. However, a large number of potential candidate genes have been described with potentially functional polymorphisms, which has made this a practical and popular approach *(11)*. Approaches falling into this category include case control studies, intermediate phenotype, and family-based association studies.

4.1.1. Case Control Studies

The current mainstay of genetic studies of stroke in humans has been the use of case control association studies. This involves comparing the frequency of polymorphisms among stroke cases and unaffected individuals, all of whom are unrelated. In most case control studies, DNA has been prospectively collected from patients presenting to stroke units or outpatient clinics while they have been undergoing investigation or treatment. Therefore detailed accurate phenotyping is possible. A number of different control groups have been used for case control association studies including hospital and spouse controls. Probably the best controls to use are those randomly selected from family practice or local electoral registers, stratified for age and sex, as other methods of control recruitment are more likely to introduce bias.

Most case control studies provide an estimate of the risk conferred by the polymorphism in disease, expressed as the odds ratio (OR). Statistical techniques such as logistic regression can then be used to determine whether a variant is an independent risk factor for stroke or whether there are interactions with other vascular risk factors such as hypertension and smoking, or even other genetic polymorphisms. A statistically significant association does not necessarily imply a causal role of the polymorphism, as it can also reflect linkage disequilibrium (close proximity between polymorphism and actual disease-causing locus), or statistical (type I) error. One of the most important causes of type I error is population stratification, which arises because of marked differences in the frequency of genetic polymorphisms within different ethnic populations. Other important causes are multiple hypothesis testing, which can lead to association through chance, and publication bias in favor of studies with positive findings. Another type of statistical error arises through missed association (type II error). Potential causes of this type of error include small sample sizes and poor selection of phenotype.

Bias owing to population stratification can be reduced by collecting DNA in defined populations and following them up longitudinally. Most often a nested case control design is performed in which cases developing stroke during the follow-up period are matched with individuals who did not develop stroke. Although biases are less of a problem, this type of approach has two main disadvantages. First, the total number of strokes is usually small, and second, accurate phenotyping of strokes can be difficult since individuals present with stroke at an unpredictable time and location and may be investigated in different ways. For example, in the Physicians Health Study *(12)* 14,916 men were followed up for 12 yr, but 338 cases of stroke occurred. Differentiating ischemic stroke from hemorrhagic stroke was difficult, and ischemic stroke subtyping was not feasible.

4.1.2. Refining the Case Control Approach: Stroke Phenotyping and Stroke Subtypes

An important issue recognized in case control genetic studies is selection of stroke phenotype. For example, many early studies used predominantly elderly

Table 1
Role of Family History as a Risk Factor According to Different Pathogenic Stroke Subtypes[a]

Stroke subtype	No. of cases	OR (95% CI) Univariate	OR (95% CI) Multivariate
		Family history of early-onset stroke (≤65 yr old)	
All ischemic strokes	944	1.69 (1.25–2.29)**	1.38 (1.00–1.90)*
Large artery stroke	262	2.24 (1.49–3.36)**	1.67 (1.08–2.66)*
Small vessel disease	232	1.93 (1.25–2.97)*	1.49 (0.94–2.37)
Cardioembolic	118	0.67 (0.30–1.50)	0.60 (0.26–1.39)
Undetermined etiology	296	1.27 (0.82–1.97)	1.11 (0.70–1.77)

[a]The risk of ischemic stroke conferred by a family history of young stroke in a first-degree relative is shown for all ischemic stroke and also specific ischemic stroke subtypes. Multivariate analysis = adjusted for age, sex, arterial hypertension, diabetes mellitus, serum cholesterol, and smoking status. OR, odds ratio.
   *, $p < 0.05$.
   **, $p < 0.001$.

subjects, in whom genetic influences may be less important than environmental influences *(3,10)*. These studies were probably not sufficiently powerful to detect genes that confer small amounts of risk. A number of later studies have focused primarily on individuals with early-onset disease *(13,14)*, although often the potential advantage of this has been offset by smaller sample sizes.

Stroke is a phenotypically heterogenous disorder, with different mechanisms underlying each specific subtype. Genetic factors that influence ischemic sensitivity may be common to all stroke subtypes. However, there are likely to be genes that are specific to certain stroke subtypes, for example, by predisposing to certain forms of vessel injury. By analogy in the autosomal dominant disease cerebral autosomal dominant arteriopathy with subcortical infarctions and leukoencephalopathy (CADASIL) mutations in the *Notch 3* gene result in recurrent lacunar stroke secondary to cerebral small vessel disease *(15)* but do not predispose to other stroke subtypes such as large vessel disease or cardioembolism. Similarly, certain genetic polymorphisms may be specific risk factors for certain subtypes of multifactorial stroke. Family history studies have also been consistent with the notion of genetic heterogeneity between different subtypes. Intuitively, one might suspect that the strongest genetic influences would be detected in stroke owing to large or small vessel disease, as these are more likely to reflect discrete pathological entities, whereas cardioembolic or undetermined stroke subtypes may reflect a large number of disease processes.

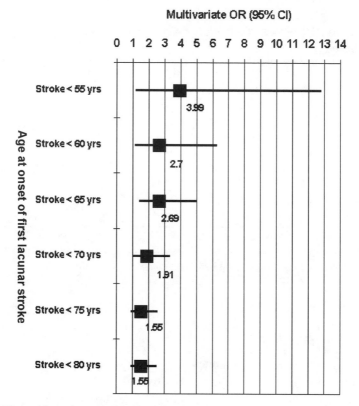

Fig. 1. The odds ratio associated with a family history of young stroke (≤65 yr) as a risk factor for lacunar stroke (small vessel disease) presenting at different ages. OR, odds ratio. Bars represent 95% CI for OR.

This hypothesis was confirmed in a study where excess familial clustering of stroke was found in small vessel and large vessel disease related strokes *(16)*. We have found a very similar pattern when we examined family history of early onset stroke, in a case control study comparing 1000 consecutive ischaemic stroke patients and 800 community controls (**Table 1**). In addition, within large and small vessel disease strokes, there was evidence for an interaction with age, with genetic factors predominating in cases of early-onset disease (**Fig. 1**). Therefore genetic case control studies could be more effective if they focused on phenotypes with a narrow pathological substrate such as large or small vessel disease stroke. Further evidence for this is provided by studies that have focused on the genetic contribution to the respective intermediate phenotypes for stroke owing to large and small vessel disease (*see* **Subheading 4.1.3.** below).

Recent advances in neuroimaging have made it easier to identify the underlying disease mechanisms and appropriately classify stroke in many cases, although in up to 30% the mechanism may be undetermined. Stroke subtype classification systems that are frequently used include the TOAST (Trial of Org 10172 in Acute Stroke Treatment) classification *(17)* or the Oxford Community Stroke Project Classification *(18)*, although the latter system is primarily a clinical classification rather than a pathological one and therefore less precise and not as well suited for genetic association studies.

4.1.3. Using Intermediate Phenotypes

As stroke is the end result of a number of pathologically different processes, it is possible that many genes, each conferring a small amount of risk, are involved in influencing the end phenotype. Conventional case control studies may not be sufficiently powerful to detect the contribution of an individual disease allele, and one logical step is to use intermediate phenotypes. Intermediate phenotypes represent specific components of the disease process. As the number of genes involved is likely to be less than that for ischemic stroke, it is hypothesized that the overall contribution of each allele to the intermediate phenotype is greater and therefore easier to detect. A further advantage of intermediate phenotypes is that often it is possible to express the intermediate phenotype as a quantitative trait or continuous variable, rather than the presence or absence of disease. This greatly increases the statistical power of studies. Equally importantly, the ability to detect "subclinical disease" avoids the reduction in power owing to incomplete penetrance in case control studies. As well as using cross sectional methods, it is also possible to follow progression of the intermediate phenotype longitudinally, thus providing a prospective element to the study.

A number of intermediate phenotypes have been used for human stroke (**Fig. 2**). Common carotid artery intima medial thickness (IMT), determined by ultrasonography, has been widely used as a marker for early carotid atherosclerosis, whereas the presence and size of carotid plaque has been used as an estimate of more advanced disease. Carotid ultrasound has been widely used in this context to determine the role of a variety of conventional and novel risk factors such as chronic infection and inflammation *(19)*. Estimates of IMT and extent of carotid plaque may therefore be useful intermediate phenotypes for large vessel stroke. Silent white matter hyperintensities on T2-weighted MRI can be considered an intermediate phenotype for small vessel disease stroke *(20,21)*. Family and twin studies are consistent with the notion that genetic factors strongly influence these intermediate phenotypes (22–24), and there have been several studies testing for association between these intermediate phenotypes and candidate genes.

Animal models have also been used to represent intermediate phenotypes of human stroke (5,6). For example, sensitivity to cerebral ischemia can be determined by measuring infarct volumes following experimental middle cerebral artery (MCA) artery occlusion or stroke susceptibility estimated based on latency to stroke. The phenotypic variance of these traits has been found to be determined by relatively few genetic loci, and there are candidate genes relevant to human stroke that lie on syntenic chromosomes in humans (25). However, one should remain cautious about using these approaches, as it may not be possible to extrapolate data from these highly inbred animal strains to complex human stroke.

4.1.4. Family-Based Association Studies

Family-based association studies have been used with some success in several late-onset polygenic diseases (26,27). These methods were originally developed to detect very close linkage between a marker and a disease, by comparing marker frequencies between affected and unaffected different family members. Subsequently these techniques have also been used in candidate gene studies. Genetic association in these circumstances cannot be explained by population stratification and is most likely to be a result of the variant being either directly responsible for the disease or being in linkage disequilibrium with a variant that is. The TDT (28) is one variant of this approach; it relies on the availability of both living unaffected parents and an affected living offspring, for genotyping. If one parent is heterozygous for a disease allele, then we would expect this allele to be transmitted in 50% of cases to the offspring simply by chance. If there is distortion of the transmission frequency to the offspring, then the variant is associated with the disease.

This method has a potential benefit in that it can allow the differential effects of maternal vs paternal transmission to be detected. However, in stroke and other late-onset polygenic diseases, it is rare to have both available parents who are alive. Therefore recruitment of sufficient numbers of parent offspring trios would be extremely difficult. Variants of the transmission disequilibrium test (TDT) have been developed, which may facilitate recruitment. As individuals with stroke frequently have unaffected living siblings, one method that could be used is based on genotyping the affected individual and the unaffected living siblings. This is known as the sibling transmission disequilibrium test (S-TDT) (29). If the allele in question is associated with disease, we would expect the frequency of a disease allele to be greater among affected siblings

Table 2
Sample Sizes Required, and Number of Ischemic Stroke Patients Requiring Screening, to Achieve Sufficient Cases/Families for Case Control, TDT, and S-TDT Methodology

Odds ratio associated with study of genetic variant	Case control study		TDT study		S-TDT study	
	Sample size	Strokes needing screening	Sample size	Strokes needing screening	Sample size	Strokes needing screening
1.5	1727	3421	1727	132,160	2581	12,808
2	414	820	414	31,680	617	3062
2.5	219	433	219	16,758	327	1622
3	144	285	144	11,018	215	1066

[a]Four levels of risk associated with the genetic variant under study are shown. Illustrative sample size calculations are based on individuals with young onset disease ($\leq$65 yr). We have assumed that 50% of living parents and siblings can be recruited, and that for the family-based studies at least one of the available parents was heterozygous for the allele being tested (30). TDT, transmission disequilibrium test; S-TDT, sibling TDT.

compared with unaffected siblings. Although families of this type can be easily recruited, they tend to be less genetically informative. Therefore this method is not as powerful as the case control approach or the TDT. A further problem is that unaffected siblings may have subclinical disease. This leads to overmatching and additional loss of statistical power.

In a recent study *(30)* we estimated sample sizes (**Table 2**) for the different association-based approaches to investigating ischemic stroke and the number of patients that would need to be screened to obtain these sample sizes. The estimates were based on studies of individuals with early-onset cerebrovascular disease (≤65 yr) presenting to our stroke service. For association studies, assuming an odds ratio of 2 and allele frequency of 0.1, the following sample sizes would be required: case control methodology 820, TDT methodology 31680, and S-TDT 3062. The sample size estimates would be much higher if older patients were being recruited, or if less important alleles of lower frequency were being investigated. In terms of recruitment rate and amount of genotyping required, the standard case control study, despite some of the disadvantages mentioned, would seem the most efficient. The estimates emphasize that for TDT and S-TDT approaches, studies are likely to be multicenter. This is particularly the case if individual stroke subtypes are to be studied, which will require even larger number of stroke patients to be screened.

4.2. Genomic Approaches to Studying Ischemic Stroke

An important limitation of association-based studies using candidate genes is that they rely on existing knowledge of the pathophysiology of stroke and lack the potential to identify novel stroke genes. In contrast, with genome-wide screening techniques, a hypothesis is not required concerning the identity of the gene, which is identified according to its physical position on the genetic maps now becoming available *(31,32)*. Therefore linkage techniques have the potential for novel gene discovery. Techniques falling into this category include linkage analysis and genome-wide association studies.

4.2.1. Linkage Approaches

Linkage-based approaches rely on whether fragments of genome are transmitted within families along with the phenotype according to certain patterns of inheritance. If a marker is linked to the disease, it will cosegregate accordingly. In most polygenic diseases, the inheritance pattern is more complex and cannot be predicted. However, one would still expect alleles that are linked to the disease to be more frequently shared between related affected family members than that predicted by chance. By using a framework of polymorphic markers spanning the genome and computational methods, it is possible to determine extent of allele sharing across the genome and locate a chromosomal region containing the gene responsible for disease.

Although linkage approaches have been very successful for monogenic diseases, there has been less success in the polygenic diseases. One reason for this is that linkage studies may have less power than association studies to detect genes of low relative risk (relative risk < 3) *(33)*. If there are many genes, each with a small relative risk (locus heterogeneity), linkage studies may not be a suitable approach in ischemic stroke. However, complex patterns of inheritance can be produced by disorders involving few genes with high relative risks but weak penetrance, and if this scenario is applicable to stroke, a linkage study might be an efficient tool *(34)*.

Currently the linkage-based method remains the only approach that is amenable to genome-wide screening strategies, although genome-wide association studies (*see* **Subheading 4.2.2.**) may be implemented in the future. Furthermore, population stratification is less of a problem with linkage studies than with conventional or genome-wide association studies. Linkage approaches that are currently being tested in stroke include the affected sibling pair approach and the genealogical approach. We will consider each of these approaches in turn.

4.2.1.1. The Affected Sibling Pair Method

This is the simplest form of linkage study. It is based on the premise that under random segregation, two affected siblings would be expected to share an identical allele (identity by descent [IBD]) 50% of the time *(9)*. In the case of a marker being linked to the disease, we would expect the IBD proportion to deviate significantly from that predicted by chance.

The power of an affected sibling pair methodology to detect linkage depends on the magnitude of IBD anticipated. One of the most important determinants of this is the relative risk conferred by a disease locus, which is in turn dependent on the overall genetic contribution in the disease (sibling relative risk), and the number of different disease loci *(35)*. The sibling relative risk (λ) is defined as the increased risk of stroke conferred by an affected individual to a sibling. In our study based on reported family history data, we found that the highest degree of sibling clustering of stroke was found among individuals concordant for early-onset disease. We estimated the sibling relative risk ratio to be approx 3, with a 95% CI of 1.5–6.0 *(30)*. This estimate of risk is similar to the levels of risk quoted in ischemic heart disease *(36)* but is lower than that seen in other polygenic diseases such as type 1 diabetes mellitus ($\lambda = 16$) *(37)*, in which the affected sibling pair approach has been attempted. For such estimates of sibling risk, sample size estimates are provided in **Table 3**. It can be

Table 3
Estimated Number of Affected Sib Pairs Required for a Linkage Study Based on a Sibling Relative Risk of 3.08, and Different Number of Anticipated Stroke Genes[a]

Predicted stroke loci	IBD	ASPs needed
1	0.67	208 (164–1157)
3	0.58	953 (512–8256)
5	0.55	2446 (1157–18,580)
10	0.53	6800 (4642–74,334)
20	0.51	61,230 (18,580–351,310)

[a]The numbers given are based on detecting significant linkage in a genome wide screen. ASP, affected sibling pair; IBD, expected proportion of alleles identical by descent. Figures in parentheses are sample sizes based on the 95% confidence intervals for the estimate of sibling relative risk 1.45–6.54 *(30)*.

appreciated that affected sibling pair studies are unlikely to be successful if there are many stroke genes, each conferring small effects. However, if there were a few major stroke genes, an affected sibling approach seems practical.

An important consideration centers around the population prevalence of affected stroke sibling pairs. As stroke is a late-onset disease, it may be that affected siblings are no longer alive. The author's own experience indicates that 9% of stroke patients reported a sibling history of stroke *(30)*. However, in only one-half was the affected sibling alive and potentially available for genotyping. Another study using a similar interview-based approach reported higher sibling concordance rates of 20%, with 11% alive at the time of sampling *(38)*. These findings suggest that large numbers of stroke cases would have to be screened even to obtain conservative sample size estimates for a sibling pair approach. As an illustration, we estimated that if DNA was required from 1000 living sibling pairs both concordant for young stroke, then almost 60,000 stroke cases would have to be screened to obtain this number of affected sib pairs. If only 50% of affected siblings were willing to participate, this figure would be nearly 120,000. Therefore a large multicenter approach is required, as exemplified by the Siblings With Ischaemic Stroke (SWISS) study. This study has the aim of recruiting 300 affected sibling pairs from 50 centers in the United States and Canada *(39)*.

4.2.1.2. THE GENEALOGICAL APPROACH

An alternative approach to the affected sibling pair method involves tracking genetic information through large extended families with affected and unaffected members. This type of linkage approach is particularly suited to

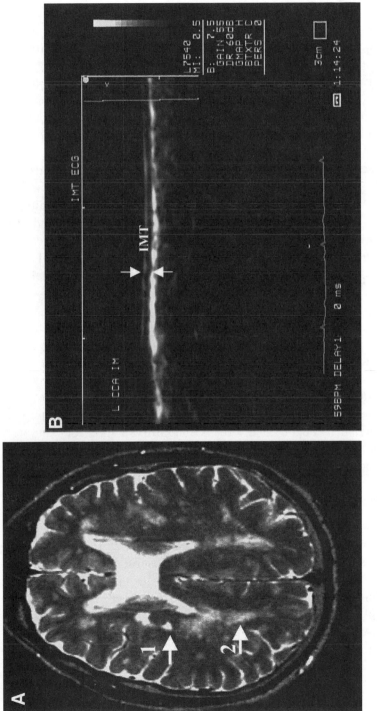

Fig. 2. (A) Axial T2-weighted MRI scan demonstrating imaging evidence of small vessel disease. Both discrete lacunar infarcts (arrow 1) and confluent white matter hyperintensities (arrow 2) can be seen. (B) Carotid ultrasound demonstrating intima medial thickness (IMT; arrows), defined as the distance between the lumen–intima and media–adventitia interfaces.

populations with extensive genealogical data and to populations that have been isolated over many centuries. These populations are likely to be more genetically homogenous. There may be fewer different stroke genes within the gene pool, i.e., reduced locus heterogeneity, leading to the possibility of variants with high relative risk. Furthermore overrepresentation of alleles among disease populations is more likely to be a causal phenomenon because of the possibility of a founder effect. Recently a genealogical approach was adopted in Iceland to provide strong evidence for the existence of a novel stroke gene *(8)*. Nearly all Icelanders are descendants of Norwegian Vikings and the Irish settlers of the 9th century. Iceland is also unique in that genealogical records exist including all 270,000 living Icelanders and many of their ancestors. A population-based list of all living stroke individuals was compared with genealogical data to construct 179 extended families containing multiply affected and unaffected individuals, who were asked to provide DNA following informed consent. Genome-wide screening subsequently revealed strong evidence of linkage to chromosome 5q12 *(8)*, where a gene concerned with endothelial proliferation was found to account for much of the relative risk of stroke in this population (K. Steffansson, personal communication, 2001). Interestingly, this locus was found to contribute equally to all stroke subtypes, with the possible exception of hemorrhagic stroke. This study would suggest that relatively few stroke genes may be present that account for the bulk of stroke genetic risk and that may be important in the expression of different disease subtypes. However, although this study may provide novel insights into stroke pathogenesis, it should be remembered that the findings apply to a unique and isolated population. Furthermore, many countries lack the social infrastructure (for example, a system of integrated health care provision) or political will needed for this type of approach to be successful.

4.2.2. Genome-Wide Association Studies

Because linkage studies have only had modest success in most polygenic diseases, some authors have proposed that a genome-wide single-nucleotide polymorphism case control study would provide a useful alternative to a linkage study *(40,41)*. It is predicted that association techniques have a better chance of finding disease-causing alleles with low relative risk but share the advantage of linkage studies in not requiring an *a priori* hypothesis. A screening approach would involve systematically comparing the frequency of bi-allelic markers (single-nucleotide polymorphisms [SNPs]) across the genome in order to find association (linkage disequilibrium) with disease. The power of genome-wide association studies depends on disease allele frequencies and the extent of linkage disequilibrium between marker and disease alleles. Linkage disequilibrium normally extends for short chromosomal distances, typically 5–500 kb, and it has been estimated that a useful marker framework would contain between 60,000 and 500,000 *(41,42)* evenly spaced SNPs. This approach requires much higher levels of significance to avoid type I error, typically 8.3 × 10^{-7} and 1 × 10^{-7} using a Bonferroni correction. At this level of marker density, there is still a possibility that a locus could go undetected because of weak or absent linkage disequilibrium between a disease allele and adjacent SNPs. However, the power of genome-wide association studies could be increased by using more markers or focusing on coding polymorphisms that are more likely to be directly involved in disease pathogenesis *(33)*.

High-throughput methods have facilitated identification and cataloguing of sequence variation *(43)*, which in turn has made genome-wide association studies a realistic prospect. Many hundreds of thousands of SNPs have been released into the public domain through public/private collaborations such as the SNP Consortium (http://brie2.cshl.org). Currently the cost per genotype using these techniques remains prohibitive, and studies of a few hundred individuals would be likely to cost many millions, given the number of SNPs that would have to be typed. Furthermore, current genotyping methods are probably still not sufficiently rapid. However, once some of these practical restraints are removed, genome-wide association studies could be applied to existing DNA stroke databases, which so far have been used exclusively for single-gene association studies.

4.3. Ethical Issues in Stroke Genetic Research

Ethical frameworks, with respect to acquiring and storage of human DNA, apply no matter which genetic technique is to be used. All studies need to be approved by the hospital research ethics committee. The study protocol should include obtaining voluntary informed consent from the individual and ensuring that data confidentiality is maintained. Clinical information should be kept on secure protected databases and DNA samples coded, so that a particular individual cannot be identified. Family-based or linkage research involves separate issues such as privacy of other family members, e.g., an affected sibling. Once a researcher has identified an appropriate family structure from the proband history, it may not be ethically appropriate to "cold contact" other family members asking for their participation. One solution that has been adopted is the principle of proband-initiated contact *(44)*, whereby the affected individual informs other family members of the study and they can then decide whether or not to contact the researcher.

4.4. Conclusions

Human stroke is a complex genetic disorder, and identifying the underlying genetic factors to date has been problematic. Progress is likely to involve the use of a number of different genetic approaches with selection of appropriate sample sizes in each case. This will require large multicenter collaborations for some types of study. The most powerful studies are likely to be those that embrace the concept of individual stroke mechanisms, through the use of either intermediate phenotypes or detailed stroke subtyping.

Acknowledgments

We would like to thank Paula Jerrard Dunne for her analysis of family history in relation to stroke subtype and Professor Pak Sham for helpful discussions. Ahamad Hassan was funded by a Stroke Association grant.

References

1. Brass, L. M., Isaacsohn, J. L., Markings, K. R., and Robinette, C. D. (1992) A study of twins and stroke. *Stroke* **23,** 221–223.
2. Bak, S., Gaist, D., Sindrup, S. H., Skytthe, A., and Christensen, K. (2002) Genetic liability in stroke: a long-term follow-up study of Danish twins. *Stroke* **33,** 769–774.
3. Jousilahti, P., Rastenyte, D., Tuomilehto, J., Sarti, C., and Vartiainen, E. (1997) Parental history of cardiovascular disease and risk of stroke. A prospective follow-up of 14371 middle-aged men and women in Finland. *Stroke* **28,** 1361–1366.
4. Liao, D., Myers, R., Hunt, S., et al. (1997) Familial history of stroke and stroke risk. The Family Heart Study. *Stroke* **28,** 1908–1912.
5. Jeffs, B., Clark, J. S., Anderson, N. H., et al. (1997) Sensitivity to cerebral ischaemic insult in a rat model of stroke is determined by a single genetic locus. *Nat. Genet.* **16,** 364–367.
6. Rubattu, S., Volpe, M., Kreutz, R., Ganten, U., Ganten, D., and Lindpaintner, K. (1996) Chromosomal mapping of quantitative trait loci contributing to stroke in a rat model of complex human disease. *Nat. Genet.* **13,** 429–434.
7. Boerwinkle, E., Doris, P. A., and Fornage, M. (1999) Field of needs: the genetics of stroke. *Circulation* **99,** 331–333.
8. Gretarsdottir, S., Sveinbjornsdottir, S., Jonsson, H. H., et al. (2002) Localization of a susceptibility gene for common forms of stroke to 5q12. *Am. J. Hum. Genet.* **70,** 593–603.
9. Lander, E. S. and Schork, N. J. (1994) Genetic dissection of complex traits. *Science* **265,** 2037–2048.
10. Brass, L. M., Page, W. F., and Lichtman, J. H. (1998) Stroke in twins III: a follow-up study. *Stroke Suppl.* **29,** 256.
11. Hassan, A. and Markus, H. S. (2000) Genetics and ischaemic stroke [Review]. *Brain* **123,** 1784–1812.

12. Zee, R. Y., Ridker, P. M., Stampfer, M. J., Hennekens, C. H., Lindpaintner, K. (1999) Prospective evaluation of the angiotensin-converting enzyme insertion/ deletion polymorphism and the risk of stroke. *Circulation* **99,** 340–343.

13. De Stefano, V., Chiusolo, P., Paciaroni, K., et al. (1998) Prothrombin G20210A mutant genotype is a risk factor for cerebrovascular ischemic disease in young patients. *Blood* **91,** 3562–3565.

14. Longstreth, W. T. Jr., Rosendaal, F. R., Siscovick, D. S., et al. (1998) Risk of stroke in young women and two prothrombotic mutations: factor V Leiden and prothrombin gene variant (G20210A). *Stroke* **29,** 577–580.

15. Chabriat, H., Vahedi, K., Iba-Zizen, M. T., et al. (1995) Clinical spectrum of CADASIL: a study of 7 families. Cerebral autosomal dominant arteriopathy with subcortical infarcts and leukoencephalopathy. *Lancet* **346,** 934–939.

16. Polychronopoulos, P., Gioldasis, G., Ellul, J., et al. (2002) Family history of stroke in stroke types and subtypes. *J. Neurol. Sci.* **195,** 117–122.

17. Adams, H. P. Jr., Bendixen, B. H., Kappelle, L. J., et al. (1993) Classification of subtype of acute ischemic stroke. Definitions for use in a multicenter clinical trial. TOAST. Trial of Org 10172 in Acute Stroke Treatment. *Stroke* **24,** 35–41.

18. Bamford, J., Sandercock, P., Dennis, M., Burn, J., Warlow, C. (1991) Classification and natural history of clinically identifiable subtypes of cerebral infarction. *Lancet* **337,** 1521–1526.

19. Crouse, J. R. III and Thompson, C. J. (1993) An evaluation of methods for imaging and quantifying coronary and carotid lumen stenosis and atherosclerosis. *Circulation* **87,** II17–II33.

20. Kobayashi, S., Okada, K., Koide, H., Bokura, H., and Yamaguchi, S. (1997) Subcortical silent brain infarction as a risk factor for clinical stroke. *Stroke* **28,** 1932–1939.

21. Mantyla, R., Aronen, H. J., Salonen, O., et al. (1999) Magnetic resonance imaging white matter hyperintensities and mechanism of ischemic stroke. *Stroke* **30,** 2053–2058.

22. Carmelli, D., DeCarli, C., Swan, G. E., et al. (1998) Evidence for genetic variance in white matter hyperintensity volume in normal elderly male twins. *Stroke* **29,** 1177–1181.

23. Duggirala, R., Gonzalez Villalpando, C., O'Leary, D. H., Stern, M. P., and Blangero, J. (1996) Genetic basis of variation in carotid artery wall thickness. *Stroke* **27,** 833–837.

24. Zannad, F., Visvikis, S., Gueguen, R., et al. (1998) Genetics strongly determines the wall thickness of the left and right carotid arteries. *Hum. Genet.* **103,** 183–188.

25. Read, S. J., Parsons, A. A., Harrison, D. C., et al. (2001) Stroke genomics: approaches to identify, validate, and understand ischemic stroke gene expression. *J. Cereb. Blood Flow Metab.* **21,** 755–778.

26. Rogus, J. J., Moczulski, D., Freire, M.B., Yang, Y., Warram, J. H., Krolewski, A. S. (1998) Diabetic nephropathy is associated with AGT polymorphism T235: results of a family-based study. *Hypertension* **31,** 627–631.

27. Niu, T., Yang, J., Wang, B., et al. (1999) Angiotensinogen gene polymorphisms M235T/T174M: no excess transmission to hypertensive Chinese. *Hypertension* **33,** 698–702.

28. Spielman, R. S., McGinnis, R. E., and Ewens, W. J. (1993) Transmission test for linkage disequilibrium: the insulin gene region and insulin-dependent diabetes mellitus (IDDM). *Am. J. Hum. Genet.* **52,** 506–516.

29. Spielman, R. S. and Ewens, W. J. (1998) A sibship test for linkage in the presence of association: the sib transmission/disequilibrium test. *Am. J. Hum. Genet.* **62,** 450–458.

30. Hassan, A., Sham, P. C., and Markus, H. S. (2002) Planning genetic studies in human stroke: sample size estimates based on family history data. *Neurology* **58,** 1483–1488.

31. Lander, E. S., Linton, L. M., Birren, B., et al. (2001) Initial sequencing and analysis of the human genome. *Nature* **409,** 860–921.

32. Venter, J. C., Adams, M. D., Myers, E. W., et al. (2001) The sequence of the human genome. *Science* **291,** 1304–1351.

33. Risch, N. J. (2000) Searching for genetic determinants in the new millennium. *Nature* **405,** 847–856.

34. Gulcher, J. R., Kong, A., and Stefansson, K. (2001) The role of linkage studies for common diseases. *Curr. Opin. Genet. Dev.* **11,** 264–267.

35. Risch, N. (1990) Linkage strategies for genetically complex traits. II. The power of affected relative pairs. *Am. J. Hum. Genet.* **46,** 229–241.

36. Pohjola-Sintonen, S., Rissanen, A., Liskola, P., and Luomanmaki, K. (1998) Family history as a risk factor of coronary heart disease in patients under 60 years of age. *Eur. Heart J.* **19,** 235–239.

37. Weeks, D. E. and Lathrop, G. M. (1995) Polygenic disease: methods for mapping complex disease traits. *Trends Genet.* **11,** 513–519.

38. Meschia, J. F., Brown, R. D. Jr., Brott, T. G., Hardy, J., Atkinson, E. J., and O'Brien, P. C. (2001) Feasibility of an affected sibling pair study in ischemic stroke: results of a 2-center family history registry. *Stroke* **32,** 2939–2941.

39. Meschia, J. F., Brown, R. D. Jr., Brott, T. G., Chukwudelunzu, F. E., Hardy, J., Rich, S. S. (2002) The Siblings With Ischemic Stroke Study (SWISS) Protocol. *B.M.C. Med. Genet.* **3,** 1.

40. Risch, N. and Merikangas, K. (1996) The future of genetic studies of complex human diseases. *Science* **273,** 1516–1517.

41. Collins, F. S., Guyer, M. S., and Charkravarti, A. (1997) Variations on a theme: cataloging human DNA sequence variation. *Science* **278,** 1580–1581.

42. Kruglyak, L. (1999) Prospects for whole-genome linkage disequilibrium mapping of common disease genes. *Nat. Genet.* **22,** 139–144.

43. Syvanen, A. C. (2001) Accessing genetic variation: genotyping single nucleotide polymorphisms. *Nat. Rev. Genet.* **2,** 930–942.

44. Worrall, B. B., Chen, D. T., Meschia, J. F. (2001) Ethical and methodological issues in pedigree stroke research. *Stroke* **32,** 1242–1249.

10

Evaluation of the Interactions of Common Genetic Mutations in Stroke

Zoltán Szolnoki

Summary

Stroke is a common entity. It is the third leading cause of death and the leading cause of adult disability in the developed world. More than 110 heritable disorders, more than 175 genetic loci, and more than 2050 unique mutations predisposing to stroke are known. Although ischemic stroke can result from merely one gene defect (and a number of clearly defined mendelian hereditary disorders do lead to stroke), the interaction of unfavorable genetic factors such as the Leiden V, methylenetetrahydrofolate reductase (MTHFR) 677TT, apolipoprotein E (ApoE) 4, and angiotensin-converting enzyme (ACE) D/D genotypes, which alone are not major risk factors, can in specific patterns exert a synergistic effect on certain clinical risk factors. This chapter discusses how to evaluate these interactions and the interpretation of findings.

Key Words

Gene interactions; stroke; genetic risk; Leiden V; apolipoprotein E (ApoE); angiotensin-converting enzyme (ACE); methylenetetrahydrofolate reductase (MTHFR).

1. Introduction

At a population level, the common sporadic form of ischemic stroke is underpinned by both environmental and genetic risk factors (1). Typically, in clinical practice, environmental and clinical risk factors such as hypertension, diabetes mellitus, smoking, alcohol drinking, and other factors, are usually considered to be more important than genetic factors. However, it is the interplay of both environmental and genetic factors that leads to the development of ischemic stroke. Indeed, a complex network of interactions between these genetic factors and clinical risk factors has been demonstrated (2–4; see **Note 1**). When present alone, the unfavorable genetic polymorphisms and common

From: *Methods in Molecular Medicine, Vol. 104: Stroke Genomics: Methods and Reviews*
Edited by: S. J. Read and D. Virley © Humana Press Inc., Totowa, NJ

genetic mutations, such as the Leiden V, prothrombin G20210A, and methylenetetrahydrofolate reductase 677TT (MTHFR C677T) mutations, as well as the apolipoprotein E (ApoE) and angiotensin-converting enzyme I/D (ACE I/D) polymorphisms, have not been proved to be major genetic risk factors for ischemic stroke. However, in a specific combination pattern, these unfavorable genetic mutations and polymorphisms can constitute a specific risk factor for certain stroke subtypes (3; see **Note 2**). The Leiden V mutation alone or in different combination patterns with the ACE D, ApoE 4, and MTHFR 677T alleles will predispose to large vessel infarction (3). The ACE D/D genotype combined with the MTHFR allele or the ApoE 4 allele will predispose to small vessel infarction (3).

In addition to these interactions, unfavorable genetic factors such as the Leiden V and MTHFR 677TT mutations and the ApoE 4 and ACE D/D genotypes, which alone are not major risk factors, can modify and exert a synergistic effect on certain clinical risk factors (4). It has been demonstrated that the Leiden V mutation in combination with hypertension or diabetes mellitus can increase the risk of ischemic stroke (4). Additionally, synergistic effects between the ACE D/D and MTHFR 677TT genotypes and alcohol drinking or smoking have been reported by Szolnoki et al. (4). Futhermore, the presence of the ApoE 4 allele can greatly facilitate the unfavorable effects of hypertension, diabetes mellitus, smoking, or drinking on the incidence of ischemic stroke (4). These data reveal that common unfavorable mutations and polymorphisms such as the Leiden V, MTHFR C677T, ApoE, and ACE I/D polymorphisms can play important roles in the development of ischemic stroke. Indeed, the clustering of unfavorable frequent genetic factors and clinical risk factors in one person can multiply the relative risk of ischemic stroke (4).

For example, leukoaraiosis, a neuroimaging entity that is observed in certain computed tomography (CT) scans, can reflect a broad public health problem caused by a cognitive impairment ranging from mild slowness of thinking to full-blown subcortical dementia (5–11). The term leukoaraiosis refers to bilateral patchy or diffuse areas of hypodensity in CT scans, or to hyperintensity in T2-weighed magnetic resonance imaging (MRI) scans; it involves only the white matter, and there are no changes in the adjacent ventricles or sulci (12,13). One-fourth of subjects aged 65 yr or over are affected by some degree of white matter changes (5). Aging, hypertension, an earlier stroke event, diabetes mellitus, and cardiac disease appear to be the most important risk factors for leukoaraiosis (13,14). Ischemic demyelination and small vessel disease have been assumed to be underlying pathological processes in the evolution of this entity (5,13,15–17). These well-known vascular risk factors, however, do not fully explain the incidence of leukoaraiosis (5,13,14,17–21). The exact etiological factors and pathomechanism of leukoaraiosis remain to be identified.

Table 1
Recommended Genotype Identification in Primary or Secondary Ischemic Stroke or Leukoaraiosis Prevention[a].

Status of patients	Recommended genotype identification
Hypertension or diabetes mellitus	Leiden V
Smoker or heavy drinker	MTHFR C677T and ACE I/D
Hypertension or diabetes mellitus or smoker or heavy drinker	ApoE
Large vessel infarction defined by MRI	Leiden V, MTHFR C677T, ACE I/D, ApoE
Small vessel infarction defined by MRI	MTHFR C677T, ACE I/D, ApoE
Presence of leukoaraiosis defined by MRI	MTHFR C677T, ACE I/D, ApoE
Altered circadian rhythm characterized by a lack of nocturnal physiological fall in blood pressure, or wide daily fluctuation of the blood pressure or a hypotensive crisis	MTHFR C677T, ACE I/D, ApoE
Presence of a cognitive deficiency with or without focal neurological signs	MTHFRC677T, ACE I/D, ApoE

[a]According to the rating scale of Fazekas et al. *(24)*, leukoaraiosis is defined as irregular periventricular hyperintensities extending into the deep white matter in the T2-weighed MRI scans (periventricular hyperintensities grade 3) and deep white matter hyperintense signals with the initial confluence of foci or with large confluent areas in the T2-weighed MRI scans (deep white matter hyperintense signals grades 2–3). We disregard punctiform lesions or caps and pencil-thin periventricular linings because they represent normal anatomic variants *(25)*. ACE, angiotensin-converting enzyme; ApoE, apolipoprotein E; MTHFR, methylenetetrahydrofolate reductase.

One possibility that must be explored is that of a genetic susceptibility. It has been noted that the co-occurrence of the ACE D/D and MTHFR 677TT genotypes can yield an independent genetic risk for the evolution of leukoaraiosis *(22)*. Likewise, it has been demonstrated that either of the two unfavorable genotypes in combination with the ApoE 2 or 4 allele can predispose to the development of leukoaraiosis *(23)*.

Knowledge of common genetic risk factors (and their interaction) may give rise to a new preventive strategy in clinical practice and provide an answer to the question of the differing degrees of susceptibility to ischemic stroke or leukoaraiosis in patients with hypertension or diabetes mellitus or patients who smoke or drink alcohol excessively (*see* **Note 3**). When the above data are taken into account, a new methodological protocol for medical practice may be recommended in order to facilitate more specific and effective primary and secondary stroke prevention (**Table 1**).

Table 2
Sequences and Concentrations of the Primers for Leiden V
and MTHFR C677T Mutations and ApoE Genotypes

Primer	Sequences	Concentration (μM/L)
Leiden V for	TTA CTT CAA ggA CAA AAT ACC TgT AAA gCT	0.7
Leiden V rev	CAT gAT CAg AgC AgT TCA AC	0.7
MTHFR for	Agg gAg CTT TgA ggC TgA CCT gAA	0.4
MTHFR rev	ACg ATg ggg CAA gTg ATg CCC Atg	0.4
ApoE for	TCC Aag gAg CTg CAg gCg gCg CA	1.0
ApoE rev	ACA gAA TTC gCC CCg gCC Tgg TAC ACT gCC A	1.0

ApoE, apolipoprotein E; MTHFR, methylenetetrahydrofolate reductase.

2. Materials

1. For the Leiden V and the MTHFR C677T mutations, a total volume of 10 μL polymerase chain reaction (PCR) mix contains 1 μL of DNA (40–80 ng), 1 μL reaction buffer (GeneAmp 10X PCR Buffer II, Perkin-Elmer), 200 μM of each dNTP, 0.6 μL of 25 mM MgCl$_2$ stock solution, and 0.5 U *Taq* polymerase.
2. For the ApoE polymorphism, a total volume of 10 μL PCR mix contains 1 μL of DNA (40–80 ng), 1 μL reaction buffer (GeneAmp 10X PCR Buffer II, Perkin-Elmer), 200 μM of each dNTP, 0.6 μL of 25 mM MgCl$_2$ stock solution, 0.5 U *Taq* polymerase, and 1 μL dimethyl sulfoxide (DMSO).
   The sequences and concentrations of the primers for the Leiden V and MTHFR C677T mutation and ApoE genotypes are given in **Table 2**.
3. For the ACE I/D polymorphism, a total volume of 10 μL PCR mix contains 1 μL of DNA (40–80 ng), 1 μL of reaction buffer (LightCycler DNA master hybridization probes 10X buffer; Roche Diagnostics), 1 μL of 25 mM MgCl$_2$ stock solution, and 0.5 μL of DMSO (Sigma, cat. no. D-8418).
   The sequences and concentrations of the primers for the ACE I/D polymorphism are given in **Table 3**.

3. Methods
3.1. Isolation of DNA

Genomic DNA can be extracted with the QIAamp DNA Blood Mini Kit (Qiagen) according to the manufacturer's instructions from 200 μL of peripheral blood anticoagulated with EDTA.

3.2. The PCR Technique

The Leiden V and MTHFR C677T mutations and the ACE I/D and ApoE polymorphisms should be examined by means of the PCR technique. The PCR

Table 3
Sequences and Concentrations of the Primers for the ACE I/D Polymorphism

Primer	Sequence	Concentration (μM/L)
ACE for	CTg gAg ACC ACT CCC ATC CTT TCT	0.2
ACE rev	GAT gTg gCC ATC ACA TTC gTC AgA T	0.2
ACE det	CgT gAT ACA gTC ACT TTT Atg-FLU	0.2
ACE anch	LCRed640-ggT TTC gCC AAT TTT ATT CCA gCT CTg-P	0.2

ACE, angiotensin-converting enzyme.

technique for the Leiden V mutation *(26)*, the MTHFR C677T mutation *(27)*, and the ApoE polymorphism *(28)* should be carried out in accordance with the modification of the original descriptions.

1. After an initial denaturation step of 3 min at 94°C, the PCR program comprises 35 cycles of 94°C/60 s, 52°C/60 s, and 72°C/60 s for the Leiden V mutation; 35 cycles of 94°C/30 s, 71°C/30 s, and 72°C/60s for the MTHFR C677T mutation; and 35 cycles of 94°C/30 s, 65°C/30 s, and 72°C/30 s for the ApoE polymrphism.
2. The final extension step lasts for 7 min at 72°C to ensure a complete extension of all PCR products.
3. The amplified PCR fragments are digested with 3 U of restriction enzyme *Hind*III, *Hin*fI, and *Hha*I, for the Leiden V and MTHFR C677T mutations and the ApoE polymorphism, respectively.
4. The digestion products are resolved on an agarose gel.
5. The digestion patterns are shown in **Table 4**.

It has been suggested that the ACE I/D polymorphism should be examined by a PCR method developed with a view to decreasing the examination time and providing a better detection of heterozygotes *(29)*. This method includes a fluorescent probe melting point analysis performed with fluorescently labeled oligonucleotide hybridization probes on a LightCycler™ instrument.

1. This PCR method is performed in disposable capillaries (Roche Diagnostics).
2. The PCR conditions are as follows: initial denaturation at 95°C for 60 s, followed by 40 cycles of denaturation (95°C/0 s, 20°C/s), annealing (61°C/10 s, 20°C/s), and extension (72°C/15 s, 20°C/s).
3. The melting curve analysis consists of one cycle at 95°C/10 s, 40°C/10 s, and then an increase of the temperature to 65°C at 0.2°C/s.
4. The fluorescence signal (F) is monitored continuously during the temperature ramp and then plotted against the temperature (T).
5. These curves are transformed to derivative melting curves [($-dF/dT$) vs T]. The I/I, I/D, and D/D genotypes give peaks at 61°C; 53°C and 61°C; and 53°C, respectively.

Table 4
The Digestion Patterns for Leiden V, MTHFR C677T Mutations,
and APOE Polymorphism

Mutation	Enzyme	Band (bp)
Leiden V normal	*Hin*dIII	222
Leiden V heterozygous		30 + 192 + 222
Leiden V mutant		30 + 192
MTHFR normal	*Hin*fI	147
MTHFR heterozygous		51 + 96 + 147
MTHFR mutant		51 + 96
ApoE E2/E2	*Hha*I	83 + 91
ApoE E3/E3		35 + 48 + 91
ApoE E4/E4		35 + 48 + 72
ApoE E2/E3		35 + 48 + 83 + 91
ApoE E2/E4		35 + 48 + 72 + 83 + 91
ApoE E3/E4		35 + 48 + 72 + 91

ApoE, apolipoprotein E; MTHFR, methylenetetrahydrofolate reductase.

3.3. Assessment of the Results

The genotypes and genotype patterns found can indicate an increased risk of certain ischemic stroke subtypes or leukoaraiosis. The assessment of the clinical impact of the genotypes found should be guided by the original publications *(2–4,22,23)*. A specific combination pattern of mutations and polymorphisms can be a specific genetic risk factor for a specific subgroup of stroke or leukioaraiosis.

The Leiden mutation in combination with hypertension or diabetes mellitus is a predisposing factor for the development of ischemic stroke or large vessel infarction *(4)*. The homozygous MTHFR 677TT and ACE D/D genotypes can facilitate the unfavorable effects of smoking or drinking on the incidence of ischemic stroke *(4)*. The presence of at least one ApoE 4 allele in combination with hypertension or diabetes mellitus or smoking or drinking can increase the relative risk of ischemic stroke *(4)*.

The ACE D/D genotype alone or in combination with the MTHFR 677T or ApoE 4 allele or with both is a risk factor for small vessel infarction *(2,3)*. The Leiden V mutation alone or in different combination patterns with the ACE D, ApoE 4, and MTHFR 677T alleles predisposes to large vessel infarction *(2,3)*. The ApoE 4 allele alone means a general, minor genetic risk factor for ischemic stroke overall *(3)*. The MTHFR 677T allele alone is not a risk factor for any stroke subtype. In the different specific predisposition gene combinations, however, both the ApoE 4 and MTHFR 677T alleles can increase the relative risk of a given stroke subgroup (large and small vessel infarction) *(3)*.

Patients with a homozygous MTHFR C667T mutation combined with an ACE D/D genotype have an increased risk of leukoaraiosis with or without cerebral infarction *(22)*.

The homozygous MTHFR 677TT mutation and ACE D/D genotype in combination with the ApoE 4/4 + 4/3 or ApoE 2/2 +2/3 genotypes can contribute significantly to the development of leukoaraioasis *(23)*.

4. Notes

1. A frequent clinical entity such as ischemic stroke is affected by various genetic and environmental factors. The common unfavorable genetic factors alone play very little part in the development of ischemic stroke. Specific unfavorable polygenetic patterns alone or in combination with other common classic clinical factors, however, can play an important role in the evolution of a specific ischemic subgroup. The presence of a specific clustering of unfavorable genetic polymorphisms and mutations should be taken into careful consideration as a genetic risk factor in clinical practice. If a common unfavorable genetic factor, such as the MTHFR 677TT or ACE D/D genotype, is uprooted from the clinical and other genetic context, it is not possible be able to draw a correct conclusion concerning the relative risk of an ischemic stroke event. Accumulation of genetic knowledge will lead to a new approach to specific stroke prevention and therapy. As a first step, after it has been elucidated whether the discovered genotype pattern, alone or in combination with the classic clinical risk factors, can enhance the risk of stroke, a new individual complex preventive strategy should be created by taking much stricter preventive measures and organizing a stricter medical follow-up. Concerning the interactions between the unfavorable genetic factors and clinical risk factors, the data are far from complete *(3,4)*. The more unfavorable genetic factors are involved in the complex system of interactions, the more interactions will become known. The recently published data may only provide the first approach to the complex assessment of genetic risk in ischemic stroke *(2–4,22,23)*.

2. Knowledge of unfavorable genetic patterns can also draw attention to new subclassifications of the ischemic stroke groups *(2,3)*. The recently published data demonstrate that the final vascular cerebral pathology classification of ischemic stroke, such as large or small vessel infarcts (measured via MRI scans), may be reassessed taking into account the specific genetic risk factors and the patterns of unfavorable genetic polymorphisms and mutations *(2–4)*.

3. It should also be noted that the occurrences of common unfavorable genetic mutations and polymorphisms such as the MTHFR C677T, Leiden V, ACE I/D, and ApoE genotypes can vary widely in different geographic areas. The genetic protocol recommended here is based on associations found in a European Caucasian population. These associations should be confirmed in other geographic areas.

References

1. Saver, J. L. and Tamburi, T. (2000) Genetics of cerebrovascular disease, in *Neurogenetics* (Stefan M. P., ed.), Oxford University Press, Oxford, pp. 403–431.
2. Szolnoki, Z., Somogyvári, F., Kondacs, A., Szabó, M., and Fodor, L. (2001) Evaluation of the roles of the Leiden V mutation and ACE I/D polymorphism in subtypes of ischaemic stroke. *J. Neurol.* **248**, 756–761.
3. Szolnoki, Z., Somogyvári, F., Kondacs, A., Szabó, M., and Fodor, L. (2002) Evaluation of the interaction of common genetic mutations in stroke subtypes. *J. Neurol.* **249**, 1391–1397.
4. Szolnoki, Z., Somogyvári, F., Kondacs, A., et al. (2003) Evaluation of the modifying effects of unfavourable genotypes on classical clinical risk factors for ischaemic stroke. *J. Neurol. Neurosurg. Psychiatry* **74**, 1615–1620.
5. van Gijn, J. (1998) Leukoaraiosis and vascular dementia. *Neurology* **51(suppl. 3)**, 3–8.
6. Steingart, A., Hachinski, V. C., Lau, C., et al. (1987) Cognitive and neurologic findings in demented patients with diffuse white matter lucencies on computed tomographic scan (leuko-araiosis). *Arch. Neurol.* **44**, 36–39.
7. Steingart, A., Hachinski, V. C., Lau, C., et al. (1987) Cognitive and neurologic findings in subjects with diffuse white matter lucencies on computed tomographic scan (leuko-araiosis). *Arch. Neurol.* **44**, 32–35.
8. Hachinski, V. C., Potter, P., and Merskey, H. (1987) Leuko-araiosis. *Arch. Neurol.* **44**, 21–23.
9. Junque, C., Pujol, J., and Vendrell, P. (1990) Leuko-araiosis on magnetic resonance imaging and speed of mental processing. *Arch. Neurol.* **47**, 151–156.
10. Liu, C. K., Miller, B. L., and Cummings, J. L. (1992) A quantitative MRI study of vascular dementia. *Neurology* **42**, 138–143.
11. Ylikoski, R., Ylikoski, A., Erkinjuntti, T., Sulkava, R., Raininko, R., and Tilvis, R. (1993) White matter changes in healthy elderly persons correlate with attention and speed of mental processing. *Arch. Neurol.* **50**, 818–824.
12. Hachinski, V. C., Potter, P., and Merskey, H. (1986) Leuko-araiosis: an ancient term for a new problem. *Can. J. Neurol. Sci.* **13**, 533–534.
13. Pantoni, L. and Garcia, J. H. (1997) Pathogenesis of leukoaraiosis, a review. *Stroke* **28**, 652–659.
14. Inzitari, D., Diaz, F., and Fox, A. (1987) Vascular risk factors and leuko-araiosis. *Arch. Neurol.* **44**, 42–47.
15. Wiszniewska, M., Devuyst, G., Bogousslavsky, J., Ghika, J., and van Melle, G. (2000) What is the significance of leukoaraiosis in patients with acute ischemic stroke? *Arch. Neurol.* **57**, 967–973.
16. Spolveri, S., Baruffi, M. C., and Cappelletti, C. (1998) Vascular risk factors linked to multiple lacunar infarcts. *Cerebrovasc. Dis.* **8**, 152–157.
17. Leys, D., Englund, E., Del Ser, T. (1999) White matter changes in stroke patients. Relationship with stroke subtype and outcome [Review]. *Eur. Neurol.* **42**, 67–75.

18. Iijima, M., Ishino, H., Seno, H., Inagaki, T., Haruki, A. (1993) An autopsy case of Binswanger's disease without hypertension and associated with cerebral infarction in the terminal stage. *Jpn. J. Psychiatr. Neurol.* **47,** 901–907.
19. Loizou, L. A., Jefferson, J. M., and Smith, W. T. (1982) Subcortical arteriosclerotic encephalopathy (Binswanger's type) and cortical infarct in a young normotensive patient. *J. Neurol. Neurosurg. Psychiatry* **83,** 423–439.
20. Ma, K. C., Lundberg, P. O., Lilja, A., and Olsson, Y. (1992) Binswanger's disease in the absence of chronic arterial hypertension: a case report with clinical, radiological and immunhistochemical observation on intracerebral blood vessels. *Acta Neuropathol. (Berl.)* **83,** 434–439.
21. Henon, H., Godefroy, O., Lucas, C., Pruvo, J. P., and Leys, D. (1996) Risk factors and leukoaraiosis in stroke patients. *Acta Neurol. Scand.* **94,** 137–144.
22. Szolnoki, Z., Somogyvári, F., Kondacs, A., Szabó, M., and Fodor, L. (2001) Evaluation of the roles of common genetic mutations in leukoaraiosis. *Acta Neurol. Scand.* **104,** 281–287.
23. Szolnoki, Z., Somogyvári, F., Kondacs, A., et al. (2003) Specific APO E genotypes in combination with the ACE D/D or MTHFR 677TT mutation yield an independent genetic risk of leukoaraiosis. *Acta Neurol. Scand.* **109,** 222–227.
24. Fazekas, F., Chawluk, J. B., Alavi, A., Hurtig, H. I., and Zimmerman, R. A. (1987) MR signal abnormalities at 1.5 T in Alzheimer's dementia and normal aging. *AJR Am. J. Roentgenol.* **149,** 351–356.
25. Fazekas, F., Kleinert, R., and Offenbacher, H. (1993) The pathologic correlate of incidental MRI white matter signal hyperintensities. *Neurology* **43,** 1683–1689.
26. Greengard, J. S., Xu, X., Gandrile, S., and Griffin, J. H. (1995) Alternative PCR method for diagnosis of mutation causing activated protein C resistant Gln506-factor V. *Thromb. Res.* **80,** 441–443.
27. Clark, Z. E., Bowen, D. J., Whatley, S. D., Bellamy, M. F., Collins, P. W., and McDowell I. F. (1998) Genotyping method for methylenetetrahydrofolate reductase (C677T thermolabile variant) using heteroduplex technology. *Clin. Chem.* **44,** 2360–2362.
28. Crook, R., Hardy, J., and Duff, K. (1994) Single-day apolipoprotein E genotyping. *J. Neurosci. Methods* **53,** 125–127.
29. Somogyvári, F., Szolnoki, Z., Márki-Zay, J., and Fodor, L. (2001) Real-time PCR assay with fluorescent hybridization probes for exact and rapid genotyping of the angiotensin-converting enzyme gene insertion/deletion polymorphism. *Clin. Chem.* **47,** 1728–1729.

IV

ASSESSING DIFFERENTIAL EXPRESSION

11

Technologies of Disease-Related Gene Discovery Using Preclinical Models of Stroke

Xinkang Wang

Summary

Stroke is a clinically defined neurological syndrome characterized by rapidly progressing symptoms and signs of focal loss of cerebral function. The initiation, propagation, and maturation of ischemic stroke are associated with de novo expression of multiple genes in endogenous brain tissues and infiltrated inflammatory cells. This chapter provides an overview for the use of state-of-the-art molecular biological approaches to investigate de novo gene expression in animal models of focal stroke, including subtractive cDNA library screening, mRNA differential display, suppression subtractive hybridization, representational difference analysis, serial analysis of gene expression, and microarrays. Identification of stroke-related gene expression will facilitate the understanding of the molecular basis of stroke pathogenesis and may provide a novel therapeutic intervention of the disease.

Key Words

Brain ischemia; gene expression; mRNA; stroke; subtraction.

1. Introduction

Ischemic stroke is a leading cause of death and disability in developed countries. It is commonly the outcome of obstruction of blood flow in a major cerebral vessel (usually the middle cerebral artery), which, if not resolved within a short, will lead to a core of severely ischemic brain tissue that may not be salvageable. The dynamic changes in brain tissue following ischemic stroke observed in preclinical models are illustrated in **Fig. 1**. Ischemic stroke leads to depletion of energy (ATP, phosphocreatine) and hence to membrane voltage reduction causing ionic fluxes across the cell membrane. The increase in extra-

From: *Methods in Molecular Medicine, Vol. 104: Stroke Genomics: Methods and Reviews*
Edited by: S. J. Read and D. Virley © Humana Press Inc., Totowa, NJ

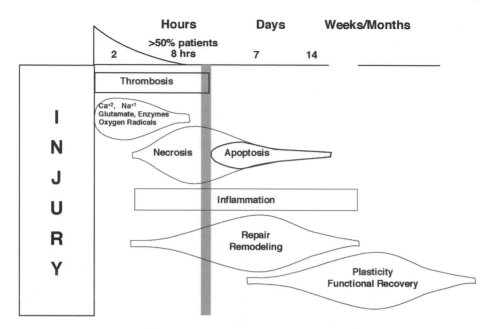

Fig. 1. Dynamic pathophysiological changes following focal stroke.

cellular potassium can reach levels sufficient to release neurotransmitters such as glutamate and aspartate and to stimulate sodium/calcium channels coupled to glutamate receptors, leading to cytotoxic edema. The calcium overload causes further mitochondrial damage and impairment in ATP production as well as extensive breakdown of cellular phospholipids, proteins, and nucleic acids owing to activation of phospholipases, proteases, and endonucleases. Free radicals are also produced during ischemia and contribute to membrane lipid peroxidation, protein and nuclear DNA toxic changes, and cellular injury (i.e., necrosis and apoptosis).

Brain ischemia involves a robust leukocyte and macrophage infiltration and accumulation in the lesions and results in considerable tissue remodeling including gliosis, necrosis/apoptosis and neovascularization (1,2). Various studies have demonstrated that de novo expression of genes, as well as their protein synthesis and actions, play an essential role in these pathophysiological changes. Thus, identification and manipulation of these genes' expression and functions may provide novel therapeutic opportunities for the treatment of ischemic stroke.

A large number of genes have been identified for their induced expression following focal stroke using various molecular biological approaches. Based on their temporal expression profile, these genes can be categorized into several waves of de novo gene expression (**Fig. 2**). Transcription factors such as c-fos, c-jun, and egr-1 (zif268) represent the first wave of gene induction, which are upregulated within minutes after ischemic insult *(3)*. The second wave consists of heat shock proteins (Hsps), the mRNA of which is usually expressed at 1–24 h. The third wave is largely comprised of inflammatory mediators including cytokines, chemokines, and adhesion molecules; their mRNA expression usually starts 1–3 h post-injury, peaks around 12 h, and then decreased to a basal level 24 h after focal ischemia. The last wave of gene expression consists largely of tissue remodeling proteins such as transforming growth factor-β, osteopontin, and metalloproteinases. Overall, these different waves of gene induction and their known/potential functions correlate with the dynamic pathophysiological changes after ischemic brain injury.

Various technologies have been successfully applied in a number of laboratories to identify stroke-related gene expression in preclinical models of stroke in the last decade, including subtractive cDNA library screening *(4,5)*, mRNA differential display *(6,7)*, suppression subtractive hybridization (SSH; *8–10*), representational difference analysis (RDA; *11*), serial analysis of gene expression (SAGE; *12*) and microarrays *(13–15)*. The experimental procedures for some techniques, including subtractive cDNA library screening, mRNA differential display, and SSH are described in this chapter; others have been covered in detail in other chapters.

1.1. Subtractive cDNA Library Screening

Conventional subtractive cDNA library is constructed by subtracting two RNA sources, such as normal and diseased tissues, of which the identical mRNA species are removed by means of various hybridization approaches *(4,5)*. Subtractive cDNA library screening is in general combined with a differential hybridization strategy using probes made from two different sources. Using a subtractive cDNA library followed by differential screening in gerbil, Abe et al. *(4)* identified a Hsp70-like gene upregulated 8 h after transient forebrain ischemia. Similarly, using the subtractive cDNA library strategy, Wang et al. *(5)* identified tissue inhibitor of matrix metalloproteinase-1 (TIMP-1) and other genes induced in ischemic cortex in response to permanent occlusion of the middle cerebral artery (MCAO) in rats.

1.2. mRNA Differential Display

mRNA differential display is a PCR-based technique comparing all poly(A) mRNA between experimental and control, or disease and normal populations *(16)*. This method consists of two basic steps, including RT using a set of 3'-anchored primers and PCR amplification of cDNA fragments using arbitrary (upstream) primers and 3'-anchored primers. The amplified cDNA fragments

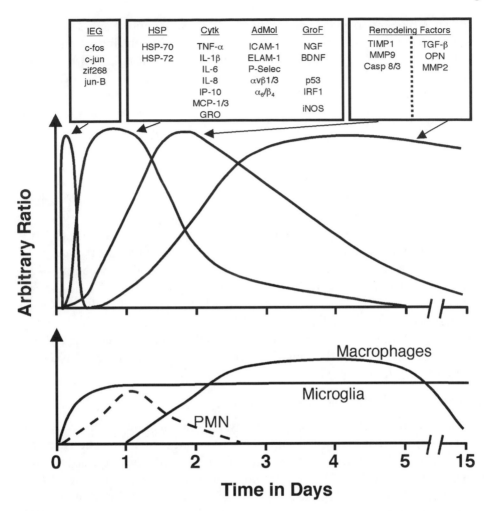

Fig. 2. Gene expression in rat ischemic cortex after middle cerebral artery occlusion. Waves of ischemic gene expression include early response genes/transcription factors (wave 1), heat shock proteins (Hsps; wave 2), proinflammatory mediators (wave 3), proteinases, proteinase inhibitors, and delayed remodeling proteins (wave 4). The leukocyte wave, including early polymorphonuclear (PMN) neutrophils and later monocytes/macrophages infiltration, as well as microglia activation, is illustrated in the lower panel. Astrocyte activation (gliosis) occurs following leukocyte infiltration in the ischemic brain tissue. BDNF, brain-derived neurotrophic factor; ELAM, endothelial leukocyte adhesion molecule; ICAM, intercellular adhesion molecule; IEG, immediate-early gene; IL, interleukin; MMP9, matrix matalloproteinase 9; NGF, nerve growth factor; P-Selec, P-selectin; TGF-β, transforming growth factor-β; TIMP1, tissue inhibitor of metalloproteinase 1; TNF, tumor necrosis factor.

are resolved by electrophoresis and allow for side-by-side comparison and thus identification of both up- and down-regulated genes. A number of reports have shown its utility in stroke research *(6,7)*.

1.3. Suppression Subtractive Hybridization

SSH is a PCR-based technology that combines cDNA subtraction with suppression PCR amplification, which allows us to amplify differentially expressed genes selectively *(17)*. Total cellular RNA isolated from two sources of tissues (such as diseased and normal) is reverse-transcribed to generate cDNA. Specially designed adapters are ligated onto restriction enzyme-digested DNA fragments (usually the diseased samples), followed by two rounds of hybridization and suppression PCR to enrich disease-related genes of interest. A number of genes have been identified in the brain after focal stroke using this technique *(8–10)*.

1.4. Representational Difference Analysis

RDA is also a PCR-based subtractive hybridization technology that was initially used for investigation of genome differences *(18)* and was later adapted to cDNA *(19)*. RDA is performed as a simultaneous subtraction of two cDNA populations in both directions. One can monitor the progress of the subtraction by observing differences emerging from two directions of subtraction in successive rounds. RDA can yield million-folds of amplification of differentially expressed sequences and thus can be used to identify as little as a single copy of particular genomic DNA fragment or cDNA. Using RDA technology, Bates et al. *(11)* have identified a number of genes that were induced in the brain following permanent MCAO (pMCAO) in the rat.

1.5. Serial Analysis of Gene Expression

SAGE technology was initially developed for the quantitative and simultaneous analysis of a large number of transcripts based on the 3' signature SAGE tags of 10 bases unique to each transcript *(20)*. This method has been successfully used for comparison of gene expression between a pair of RNA samples to identify differentially expressed genes. Recently, SAGE Genie, a set of tools for processing SAGE data, has been developed *(21)* to provide a very powerful tool for archiving and analyzing the expression profile for any given gene in a biological context. Using SAGE technology, Trendelenburg et al. *(12)* successfully studied differentially expressed genes in a mouse model of focal cerebral ischemia.

1.6. Microarrays

Microarray technology *(22)* has received much attention in recent years for large scale gene expression analysis using either oligonucleotides or cDNA arrays. This technology has been successfully adopted in many laboratories/ institutions, including its application in gene expression profiling in a preclinical model of stroke *(13–15)*.

2. Materials

2.1. Subtractive cDNA Library Screening

A representative procedure of this subtractive cDNA library strategy used in our laboratory for the cloning of tissue inhibitor of matrix metalloproteinase-1 *(5)* is described here.

1. Poly(A)+ RNA isolated from rat cortex 2 and 12-h after MCAO.
2. Oligotex-dT (QIAGEN).
3. Superscript II RNase H$^-$ reverse transcriptase and reaction buffer (Gibco-BRL).
4. ^{32}[P]-α-dATP (Amersham).
5. RNase inhibitor (Boehringer Mannheim).
6. $(dA)_{30}(dG)_{10}$ Oligodeoxynucleotide (Sigma); lZAP vector (Stratagen).

2.2. mRNA Differential Display

As an example, a gene that encodes adrenomedullin, a member of the calcitonin gene-related peptide (CGRP) family, was isolated in the brain after pMCAO in rats using the mRNA differential display approach:

1. $T_{12}MN$ oligonucleotides (M or N = G, A, T, or C).
2. T/A cloning kit (Invitrogen).
3. MMLV reverse transcriptase (Gibco-BRL).
4. AmpliTaq DNA polymerase (Perkin-Elmer).
5. RNAmap kit (GenHunter); [^{33}P]-α-dATP (Amersham).
6. Total cellular RNA isolated from ischemic and nonischemic cortical tissues.

2.3. Suppression Subtractive Hybridization

1. Clontech PCR-Select cDNA Subtraction Kit (Clontech, Palo Alto, CA).
2. T/A cloning kit (Invitrogen).
3. CR-Select Differential Screening Kit (Clontech).
4. QIAquick PCR Purification kit (QIAGEN).
5. *RsaI*, *EagI*, and *SmaI* restriction enzymes.
6. poly(A)+ mRNA from ischemic cortex 24-h post pMCAO or from nonischemic cortex.

3. Methods

3.1. Substractive cDNA Library Screening

1. For reverse transcription (RT) reaction, 20 µg poly(A)+ RNA isolated from 2 and 12-h nonischemic cortex (10 µg each) in 75 µL water was mixed with 25 µL of 10% (w/v) oligotex-dT and heated at 70°C for 5 min followed by rapid cooling on ice.

2. After addition of 100 µL of 2X TMK buffer (100 m*M* Tris-HCl, pH 8.3, 200 m*M* KCl, and 20 m*M* MgCl$_2$), the mixture was incubated at 37°C for 20 min, and then microcentrifuged for 10 min at room temperature.

3. The precipitate containing mRNA/oligotex/dT complex was dissolved in 400 µL RT buffer (80 µL 5X first strand buffer [Gibco-BRL], 40 µL 0.1 *M* dithiothreitol [DTT], 2 µL each of dNTPs [0.5 m*M*], 1 µL of 1:20 diluted ^{32}[P]-α-dATP, 300 U RNase inhibitor, 10,000 U Superscript II RNase H$^-$ reverse transcriptase) and incubated at 37°C for 90 min.

4. The reaction mixture was then heated at 90°C for 3 min and rapidly cooled on ice.

5. The RNA dissociated from the cDNA/oligotex/dT was removed by microcentrifugation.

6. The precipitate was washed with 400 µL TE followed by centrifugation and dissolved in 100 µL TE containing 100 µg (dA)$_{30}$(dG)$_{10}$ oligodeoxynucleotide.

7. The suspension was heated at 65°C for 5 min, and 20 µL of 3 *M* NaCl was added and incubated at 37°C for 10 min to block the free oligo(dT) residues on the oligotex/dT.

8. The excess (dA)$_{30}$(dG)$_{10}$ was removed by centrifugation and saved for later uses.

9. The precipitate was dissolved in 200 µL of 1.25X hybridization buffer (12.5 m*M* Tris-HCl, pH 7.5, 125 m*M* NaCl, 1.25 m*M* EDTA, 0.125% sodium dodecyl sulfate (SDS), 2 µg oligo(dT)$_{12-18}$, and 4 µg poly(A)+ RNA isolated from 2- and 12-h ischemic cortex in 50 µL water was added and hybridized at 55°C for 20 min.

10. The reaction mixture was centrifuged at room temperature for 10 min and the supernatant (containing the subtracted mRNA) was save at 4°C.

11. The precipitate was dissolved in 400 µL TE, heated at 90°C for 3 min, cooled on ice, and then centrifuged at 4°C for 10 min to removed the RNA.

12. The cDNA/oligotex/dT precipitate was dissolved again in TE containing (dA)$_{30}$(dG)$_{10}$ and the subtractive hybridization was repeated for an additional six times.

13. The subtractive mRNA was used for cDNA library construction.

14. The first and second strand of cDNA was synthesized using a cDNA cycle kit for RT-polymerase chain reaction (PCR) (Invitrogen) according to the manufacturer's specification.

15. An *Eco*RI adapter was ligated onto the cDNA, and fractions of the cDNA larger than 400 bp were collected, digested with EcoRI enzyme, and ligated into λZAP vector (*Eco*RI sites) (Stratagene).

16. A panel of subtracted clones from this cDNA library was further analyzed by differential Southern hybridization.

17. Briefly, miniprep DNA was digested with *Eco*RI to release the cDNA insert from the plasmid and analyzed by agarose gel electrophoresis.

18. Southern hybridization was carried out using a probe generated by RT reaction with poly(A)+ mRNA isolated from either 2- and 12-h ischemic cortex (as the ischemic probe) or the nonischemic cortex (as the nonischemic probe).

19. Superscript II RNase H⁻ reverse transcriptase (Gibco-BRL) was used for this labeling reaction in the presence of both ^{32}[P]-α-dATP and ^{32}[P]-α-dCTP (Amersham).

20. The positive clones (e.g., TIMP-1) were identified and further confirmed by Northern analysis (*5*; *see* **Note 1**).

3.2. mRNA Differential Display

1. Total cellular RNA was reverse-transcribed to yield the first strand cDNA primed with $T_{12}MN$ oligonucleotides, which allows all the mRNAs having a poly (A) tail to be reverse-transcribed. Typically, this RT reaction was divided into four subgroups each using a different $T_{12}MN$ primer with G, A, T, or C at the last base of the 3'-end.

2. Amplification was carried out using a 5' decamer arbitrary primer (e.g., 5'-GACCGCTTGT-3') and a 3' primer (e.g., $T_{12}NA$) in the presence of [^{33}P]-α-dATP.

3. The PCR reaction was carried out for 40 cycles as follows: 94°C 30 s for denaturing, 40°C 2 min for annealing, 72°C 30 s for extension, followed by one cycle for extension at 70°C for 10 min.

4. The amplified cDNA fragments were resolved by electrophoresis and subjected to autoradiographic analysis.

5. Following mRNA differential display, the bands of interest were excised from the dried sequencing gel, released by heating at 95°C for 10 min in 100 μL TE (10 m*M* Tris, pH 7.5, and 1 m*M* EDTA) and ethanol precipitated in the presence of glycogen.

6. The isolated bands were reamplified using the same sets of primers as in the original PCR.

7. The recovered DNA band was served as a probe to confirm mRNA expression by means of Northern blot analysis and/or was subcloned into a pCRII vector for DNA sequencing analysis.

8. Based on the sequence information, the identity of the differentially expressed genes was determined by searching a computer database (e.g., GenBank).

9. If the sequence represented an unknown sequence, a cDNA library was screened using this DNA as a probe to obtain the full length cDNA clone.

10. In the case of adrenomedullin, after it was subcloned into pCRII vector, the DNA sequence was determined and found to be an unidentified gene; therefore, the full-length cDNA was isolated by screening the rat stroke cDNA library (*6*).

11. Adrenomedullin was demonstrated to play a role in exacerbating focal brain ischemic damage when the synthetic peptide was intracerebroventricularly administered in the brain (*6*).

3.3. Suppression Subtractive Hybridization

1. SSH was carried out using a Clontech PCR-Select cDNA Subtraction Kit (*9*).

2. Three micrograms of poly(A)+ mRNA from ischemic cortex 24 h post pMCAO was used for a tester or that from normal brain cortex as a driver for cDNA synthesis.

3. After first and second rounds of hybridization between the tester and driver cDNA, the hybridization mixture was PCR amplified in a volume of 25 μL, containing 1 μL of the diluted subtraction mixture, 1 μL PCR primer 1 (10 μ*M*), 10X PCR reaction buffer, 0.5 μL dNTPs mix (10 m*M*) and 50X Advantage cDNA Polymerase Mix (Clontech).

4. Amplification condition were as follows: incubation at 75°C for 5 min to extend the adaptors followed by 94°C for 25 s in a Perkin-Elmer GeneAmp PCR System 9700, and then followed by 27 cycles at 94°C 10 s, 66°C 30 s, 72°C 1.5 min.

5. The primary PCR mixture was diluted 10-fold and 1 μL was used in a secondary PCR with nested primers for the primer 1 and primer 2R for 10 cycles at 94°C 10 s, 68°C 30 s, 72°C 1.5 min.

6. The subtraction efficiency was evaluated by measuring the levels of two housekeeping genes (G3PDH and rpL32) in the subtracted cDNA pool using PCR.

7. Thereafter, products of the selective secondary PCR were cloned into pCR2.1 vector using the T/A cloning kit and transformed into an *E. coli*, IVNαF'.

8. Screening of the subtracted cDNA sample was carried out using the PCR-Select Differential Screening Kit against randomly selected bacterial colonies cultured overnight in a 96-well plate and blotted onto the 96-well Bio-Dot apparatus (Bio-Rad).

9. The membranes were treated with 0.5 *M* NaOH, 1.5 *M* NaCl for 4 min, 0.5 *M* Tris-HCl also for 4 min, and then washed for 30 min in 0.2X SSC and 0.2% SDS at 63°C.

10. Prehybridization was carried out at 42°C in a buffer containing 5X SSPE, 50% formamide, 5X Denhardt's solution, 2% SDS, 200 μg/mL salmon sperm DNA, 100 μg/mL Poly A, 10 mL of 0.3 μg/mL of oligonucleotides corresponding to the nested primers and complementary sequences.

11. The probes were generated using a random priming method in the presence of ([^{32}P]-α-dATP using either the subtracted ischemic template or normal rat cortex template.

12. The adaptor in the subtracted templates was removed by a a QIAquick PCR Purification kit after *RsaI*, *EagI*, and *SmaI* restriction enzyme (the sites present in the adaptor sequences) digestions.

13. Hybridization was performed overnight with 1–2 × 10^6 cpm/mL. The membranes were washed at room temperature for 15 min with 2X SSC, followed by 66°C with 2X SSC and 0.5% SDS for 3X 15 min, and 0.2X SSC and 0.1% SDS for 10–15 min. The hybridization results were analyzed using autoradiography (*see* **Note 2**).

4. Notes

1. Subtractive cDNA library screening is a reliable but not sensitive method to detect less abundantly expressed genes. Thus, to detect a low-copy number of transcripts, alternative methods using PCR-based technologies should be considered. The limitation and advantage of each technology are briefly summarized in **Table 1**.

Table 1
Pros and Cons of Gene Discovery Technologies

Technology	Pro	Con
Subtractive library	Reliable	Insensitive
Differential display	Sensitive, simple and rapid for identifying a lead	High incidence of false positives
Suppression subtractive hybridization (SSH)	Very sensitive and reproducible	High technique skills to ensure the subtraction
Representational differential analysis (RDA)	Very sensitive	High technique skills to avoid potential contamination
Serial analysis of gene expression (SAGE)	High throughput	Labor-intensive; requires bioinformatics
Microarray	High throughput	Expensive, limited to known sequences, insensitive to low copy number of transcripts

2. mRNA differential display and SSH are PCR-based technologies. These techniques are in general very sensitive for detection of a low-copy number of transcripts, but often they also cause high incidence of false positive (such as mRNA differential display) or require more sophisticated technical skills (such as SSH and RDA). Therefore, each of these technologies needs further refinement to better serve the purpose of gene discovery. In fact, significant improvement has been made in almost all of these technologies since their initial description. For example, SAGE was developed to profile gene expression based on limited sequence information (10-base 3' signature for each transcript) for gene discovery; now following the development of SAGE Genie it has become a more powerful and attractive tool for archiving and analyzing gene expression profiles and taking advantage of the completion of the human genome sequence. While microarray is powerful and has high throughput, it is limited to the study of known sequences (genes) laid on a chip; in addition, it has to deal with a complex cDNA probe (that consists of as many as 10,000 different mRNA species, ranging from a few to thousands of copies per cell), which makes it very difficult for this technique to detect genes that only express a low-copy number of transcripts. To overcome these inherited problems, for example, the combination of SSH with microarray may be adopted to provide a more attractive and powerful technology to identify differentially expressed genes within a given biological context.

References

1. Clark, R. K., Lee, E. V., Fish, C. J., et al. (1993) Development of tissue damage, inflammation and resolution following stroke: an immunohistochemical and quantitative planimetric study. *Brain Res. Bull.* **31,** 565–572.
2. Garcia, J. H., Liu, K. F., Yoshida, Y., Chen, S., and Lian, J. (1994) Brain microvessels: factors altering their patency after the occlusion of a middle cerebral artery (Wistar rat). *Am. J. Pathol.* **145,** 728–740.
3. Wang, X. and Feuerstein, G. Z. (2000) Role of immune and inflammatory mediators in CNS injury. *Drug News Perspect.* **13,** 133–140.
4. Abe, K., Sato, S., Kawagoe, J., Lee, T. H., and Kogure, K. (1993) Isolation and expression of an ischaemia-induced gene from gerbil cerebral cortex by subtractive hybridization. *Neurol. Res.* **15,** 23–28.
5. Wang, X., Barone, F. C., White, R. F., and Feuerstein, G. Z. (1998) Subtractive cloning identifies tissue inhibitor of matrix metalloproteinase-1 (TIMP-1) increased gene expression following focal stroke. *Stroke* **29,** 516–520.
6. Wang, X., Yue, T. L., Barone, F. C., et al. (1995) Discovery of adrenomedullin in rat ischemic cortex and evidence for its role in exacerbating focal brain ischemic damage. *Proc. Natl. Acad. Sci. USA* **92,** 11,480–11,484.
7. Tsuda, M., Imaizumi, K., Katayama, T., et al. (1997) Expression of zinc transporter gene, ZnT-1, is induced after transient forebrain ischemia in the gerbil. *J. Neurosci.* **17,** 6678–6684.
8. Wang, X., Yaish-Ohad, S., Li, X., Barone, F. C.m and Feuerstein, G. Z. (1998) Use of suppression subtractive hybridization strategy for discovery of increased tissue inhibitor of matrix metalloproteinase-1 gene expression in brain ischemic tolerance. *J. Cereb. Blood Flow Metab.* **18,** 1173–1177.
9. Wang, X., Li, X., Yaish-Ohad, S., Sarau, H. M., Barone, F. C., and Feuerstein, G. Z. (1999) Molecular cloning and expression of the rat monocyte chemotactic protein-3 gene: a possible role in stroke. *Mol. Brain Res.* **71,** 304–312.
10. Yokota, N., Uchijima, M., Nishizawa, S., Namba, H., and Koide, Y. (2001) Identification of differentially expressed genes in rat hippocampus after transient global cerebral ischemia using subtractive cDNA cloning based on polymerase chain reaction. *Stroke* **32,** 168–174.
11. Bates, S., Read, S. J., Harrison, D. C., et al. (2001) Characterisation of gene expression changes following permanent MCAO in the rat using subtractive hybridisation. *Mol. Brain Res.* **93,** 70–80.
12. Trendelenburg, G., Prass, K., Priller, J., et al. (2002) Serial analysis of gene expression identifies metallothionein-II as major neuroprotective gene in mouse focal cerebral ischemia. *J. Neurosci.* **22,** 5879–5888.
13. Soriano, M. A., Tessier, M., Certa, U., and Gill, R. (2000) Parallel gene expression monitoring using oligonucleotide probe arrays of multiple transcripts with an animal model of focal ischemia. *J. Cereb. Blood Flow Metab.* **20,** 1045–1055.
14. Jin, K., Mao, X. O., Eshoo, M. W., et al. (2001) Microarray analysis of hippocampal gene expression in global cerebral ischemia. *Ann. Neurol.* **50,** 93–103.

15. Schmidt-Kastner, R., Zhang, B., Belayev, L., et al. (2002) DNA microarray analysis of cortical gene expression during early recirculation after focal brain ischemia in rat. *Mol. Brain Res.* **108,** 81–93.

16. Liang, P. and Pardee, A. B. (1992) Differential display of eukaryotic messenger RNA by means of the polymerase chain reaction. *Science* **257,** 967–971.

17. Diatchenko, L., Lau, Y. F., Campbell, A. P., et al. (1996) Suppression subtractive hybridization: a method for generating differentially regulated or tissue-specific cDNA probes and libraries. *Proc. Natl. Acad. Sci. USA* **93,** 6025–6030.

18. Lisitsyn, N., Lisitsyn, N., and Wigler, M. (1993) Cloning the differences between two complex genomes. *Science* **259,** 946–951.

19. Hubank, M. and Schatz, D. G. (1994) Identifying differences in mRNA expression by representational difference analysis of cDNA. *Nucleic Acids Res.* **22,** 5640–5648.

20. Velculescu, V. E., Zhang, L., Vogelstein, B., and Kinzler, K. W. (1995) Serial analysis of gene expression. *Science* **270,** 484–487.

21. Boon, K., Osorio, E. C., Greenhut, S. F., et al. (2002) An anatomy of normal and malignant gene expression. *Proc. Natl. Acad. Sci. USA* **99,** 11,287–11,292.

22. Schena, M., Shalon, D., Davis, R. W., and Brown, P. O. (1995) Quantitative monitoring of gene expression patterns with a complementary DNA microarray. *Science* **270,** 467–470.

12

Quantitative Analysis of Gene Transcription in Stroke Models Using Real-Time RT-PCR

David C. Harrison and Brian C. Bond

Summary

Many researchers have sought to study changes in gene expression in preclinical models of stroke. These range from in vitro models of ischemia, neuronal death, and regeneration to in vivo animal models aimed at replicating pathologies and regenerative processes typical of the clinical situation. In all such models, changes in gene expression occur, which may be assessed by measuring the abundance of the mRNA transcribed from particular genes of interest. The advent of real-time reverse-transcriptase polymerase chain reaction (RT-PCR) has vastly improved the sensitivity and accuracy of mRNA detection and is now the method of choice in many studies. Although this is a relatively simple and rapid technique, it has a number of pitfalls, especially in experimental design and data analysis. In this chapter we describe a detailed experimental protocol for real-time RT-PCR detection of mRNA transcripts, as used in the rat permanent middle cerebral artery occlusion model. We also discuss methods for analysis and interpretation of the resulting data.

KEY WORDS

RT-PCR; statistical analysis; principle components analysis; analysis of covariance.

1. Introduction

One of the most useful means of gaining insights into the physiological and pathological processes taking place in stroke models is the analysis of gene expression. There are many published reports demonstrating changes in gene expression, corresponding to the complex series of events that unfold following the initial ischemic injury (reviewed in **ref. 1**). These include genes associated with a generalized response to injury, such as the immediate early genes,

From: *Methods in Molecular Medicine, Vol. 104: Stroke Genomics: Methods and Reviews*
Edited by: S. J. Read and D. Virley © Humana Press Inc., Totowa, NJ

which may be upregulated in the immediate aftermath of ischemia. Later changes include responses typical of inflammatory events, such as upregulation of cytokines, and also expression of genes associated with apoptosis. Later still, changes indicative of neuronal regeneration and remodeling take place, including the expression of growth factors.

The rationale for studying gene expression in stroke models has typically fallen into one of two categories: (1) studies that aim to measure and characterize the expression of a particular gene and its protein to answer a specific hypothesis regarding its role in the disease process, and (2) as confirmation of hits from so-called gene-fishing exercises in which the mRNA of ischemic and nonischemic tissues is compared to identify changes in expression arising from treatment.

Methods available for the study of gene expression have developed significantly over the last decade, and real-time polymerase chain reaction (PCR) and reverse transcription (RT)-PCR *(2,3)* in particular have allowed greater sensitivity, accuracy, and throughput in gene expression analysis. Both types of study mentioned above have benefited from these developments. The first scenario has enabled the study of expression of an individual gene to be undertaken in ever-increasing detail *(4)*. In the second type of study, confirmation of hits and their wider characterization over a detailed time-course can be undertaken with greater throughput and accuracy *(5)*.

Nevertheless, there are limitations to this technology: detailed spatial resolution is not possible, and gene expression at the level of the mRNA transcript may not reliably predict changes in functional protein expression. Consequently, real-time RT-PCR may be viewed as an intermediate step in a cascade of studies on the understanding of stroke pathology *(1)*.

This methodology is applicable to any in vivo model of stroke, although studies on nonrodent species may be limited by the lack of gene sequence information. With minor modifications to RNA isolation steps (*see* **Note 1**), in vitro stroke models (including cellular and organotypic preparations) may also be studied by these methods.

As in many other scientific areas, the rigorous application of appropriate statistical methods and the careful choice of graphics may enhance the experimenter's understanding of expression data produced by real-time RT-PCR while ensuring that the resulting conclusions are valid. Good analyses often increase the sensitivity in identifying treatment effects by minimizing the variability, and thoughtful design will eliminate bias from scientific methods.

The methods used in this chapter have been used to quantify changes in expression of a number of genes in the permanent middle cerebral artery occlusion (MCAO) model of ischemic stroke in the rat *(4–6)*. To provide examples of statistical methods, we have carried out further analysis on data that have been published previously on suppressor of cytokine signaling-3 (SOCS-3), leukemia-inhibitory factor (LIF), the nerve growth factor-inducible gene VGF glyceraldehyde-3-phosphate dehydrogenase (GAPDH), and cyclophilin *(5)*. In addition, we have included additional data (on Fas-associated death domain [FADD]) derived from the same experiment. The study was carried out as previously described *(5)* and involved MCAO and sham-operated rats taken at four time points (3, 6, 12, and 24 h, four rats per group and time-point) following surgery. Tissue samples were taken from the region defined by the maximal extent of the lesion from the left (ipsilateral) cerebral cortex and the corresponding area of the right (contralateral) cerebral cortex. The subsequent methodology is described in this chapter.

2. Materials

2.1. Reagents

1. Liquid nitrogen.
2. TRIzol reagent (Invitrogen).
3. PCR grade water (RNAse and DNAse free).
4. Chloroform, isopropanol, and 75% ethanol (all RNAse and DNAse free).
5. Superscript II reverse transcriptase kit (Invitrogen).
6. Oligo(dT)$_{12-18}$ primer (Invitrogen).
7. TaqMan Universal 2X Master Mix (Applied Biosystems).
8. Oligonucleotides, including TaqMan probes.
9. Genomic DNA.

2.2. Equipment

1. Thermo-Fast 96 PCR detection plate (ABgene).
2. Adhesive PCR foil seal (ABgene).
3. ABI Prism Optical adhesive covers (Applied Biosystems).

2.3. Hardware

1. Tissue homogenizer.
2. Hydra-96 (Robbins Scientific).
3. ABI 7700 sequence detection system (Applied Biosystems).

2.4. Software

1. Primer and probe design: Primer Express v1.5 (Applied Biosystems).
2. Univariate statistical analysis: Genstat v6.
3. Multivariate statistical analysis: SIMCA-P v9.
4. Graphics: Statistica v6.

3. Methods

3.1. Experimental Design

A number of issues need to be addressed when designing the experiment. These issues include: how much and where to replicate, choice of tissue samples, whether to pool samples, and dealing with large studies that cannot be carried out in one session or require multiple plates.

3.1.1. Replication

The choice of where and how much to replicate in an experiment is important in a study, as it will contribute to the power of the eventual statistical analysis. The experimenter will usually have several replication options: at the animal level, at the sample level, or at the measurement level. Treatment will usually be applied at the animal level and so the decision on the number of animals is the most important. If possible, this should be assessed by a sample size calculation based on the variability expected, the power desired, and the size of the effect you are wanting to identify *(8)*. However, in practice, resources are usually limited, and this dictates the number of animals used. Needless to say, the more animals used, the more sensitive the analysis will be. Finally, replication at the measurement level, i.e., the number of RT reactions from a single RNA sample, is not so critical. RT to RT variability is generally much smaller than other sources of variability, and replication at this level in the past has been necessary because of the large number of reaction failures. As the methodology has become more reliable, there is less need to replicate RTs, and if an experiment needs to be reduced in size, we would recommend reducing RT replication.

3.1.2. Tissue Sampling

The choice of tissues to sample will vary according to the aims of the study and the model being used. Studies of focal ischemia in which a spatially defined unilateral lesion is present have typically compared either brain hemispheres *(9)* or a more restricted area defined by the extent of the lesion *(4)*. The inclusion of tissue that is unaffected by treatment may have the effect of "diluting" treatment effects and reducing sensitivity. On the other hand, the extent of a lesion may vary between animals and over time, and this may compromise reproducibility.

3.1.3. Tissue Pooling

If the region to be sampled is very small, it may be necessary to pool RNA from animals in the same experimental group to ensure a sufficient sized sample for extracting mRNA. This should be avoided if possible; if it is performed the experimenter should try to minimize the numbers pooled. Variability between pools will very likely be smaller than the variability between animals, but this could be offset by the smaller number of pools. How deceptive this is will depend on the relationship between the variability of the pools and the number of pools.

3.1.4. Large Studies

If the number of samples to assay is large, then samples may have to be distributed across multiple plates and assayed over a period of time. Although it may be practically easier to assay one treatment per experimental session or per plate, this can cause bias in the results, leading to false conclusions. Therefore, it is important to ensure balance of treatments across plates and sessions to ensure that treatments are not biased by possible plate or day effects and are also taken into account when performing an analysis to increase sensitivity.

3.2. Primer and Probe Design

There are two basic methods for detection of an accumulating PCR product in real time, those that rely on a fluorophore that binds nonspecifically to double-stranded DNA, such as SYBR green (*see* **Note 2**), and those that use an oligonucleotide that anneals specifically within the amplicon. The latter include the Taqman probes (as used in the method described here), fluorescence resonance energy transfer (FRET) probes, and molecular beacons. SYBR green can be used in conjunction with standard PCR primers, which can be designed with any available primer design software, or "by eye" by someone with sufficient expertise (*see* **Note 2**). A TaqMan probe and primers can be designed using Primer Express Software. However, with the increasing number of publications in which TaqMan probes have been employed, primer and probe sequences are becoming more abundant in the literature. In addition, ready-made primer and probe "sets" for commonly studied inducible genes such as cytokines are becoming more widely available. An initiative is also currently under way to create an internet database of real-time PCR primer and probe sequences *(10)*.

3.3. RNA Isolation

Isolation of high-quality RNA is arguably the single most critical step to obtaining good real-time RT-PCR data. In this study we used TRIzol reagent (Invitrogen), a phenol chloroform-based method, although other methods for RNA isolation have given equally satisfactory results (*see* **Note 1**).

1. Immediately following euthanasia, dissect the required tissues and snap freeze in liquid nitrogen. Frozen tissue samples may be stored at –80°C.

2. Remove tissue samples from the freezer, weigh, and add TRIzol reagent using 20 μL per mg of tissue. Homogenize immediately. At this stage the homogenate may be divided into aliquots (e.g., 1 mL) and stored at –80°C.

3. Thaw 1 aliquot, containing 1 mL of homogenate, incubate at room temperature for 5 min, and add 200 μL of chloroform. Shake vigorously for 15 s. Incubate at room temperature for 5 min. Centrifuge at 12 000g, 4°C for 15 min to separate phases.

4. Transfer the aqueous phase (500–600 μL) to a fresh 1.5-mL Eppendorf tube, add a further 500 μL of TRIzol, and repeat **step 3**.

5. Transfer the aqueous phase to a fresh 1.5-mL Eppendorf tube, add 500 μL of isopropanol, and mix thoroughly. Incubate at room temperature for 15 min. Centrifuge at 12,000g, 4°C for 10 min.

6. Remove the supernatant, add 1 mL of 75% ethanol, and shake the tube vigorously. Centrifuge at 7500g, 4°C for 5 min.

7. Repeat **step 6**.

8. Remove the supernatant completely, but do not allow the pellet to dry, as this may make the RNA pellet difficult to dissolve.

9. Resuspend the pellet in PCR-grade water. (We recommend using 25 μL for brain RNA derived from a 1 mL TRIzol aliquot. This should give an RNA concentration of approx 2 μg/μL.)

10. Following RNA isolation, the RNA concentration may be determined by A_{260} measurement. It may also be useful to run approx 1 μg of RNA on a 1% agarose/TBE gel. This should reveal distinct bands corresponding to 18S and 28S ribosomal RNA, which give a useful indication of RNA integrity.

3.4. Reverse Transcription

The RT step may be performed separately from the PCR (two-step RT-PCR) or within a single reaction tube (single-step RT-PCR). The method described here is the two-step method. This has the advantage that many genes may be studied in the product of a single RT reaction, including control housekeeping genes, which may be used to correct for discrepancies in RNA quality or RT efficiency. cDNA is more stable than RNA and can be stored for long periods at –20°C. Prior to this step, RNA may be polyA-enriched or DNAse-treated to exclude any signal from genomic DNA (*see* **Note 3**). If not, it may be necessary to include extra "minus-RT" reactions that omit the RT enzyme to control for genomic DNA contamination.

1. Pipet between 10 and 1000 ng of each RNA sample (*see* **Note 4**) into wells of a 96-well plate. Add 0.5 μg oligo(dT) primer (*see* **Note 5**) and PCR grade water to a total volume of 10 μL. Heat to 90°C for 2 min and then rapidly cool on ice.

2. Add remaining reagents: 2 μL 0.1 *M* dithiothreitol, 2 μL 10 m*M* dNTP mix, 200 U RNAse inhibitor, 40 U RT enzyme (except for no-RT controls), PCR grade water, total volume 10 μL, and incubate at 42°C for 1 h.

3. Stop the reaction by heating to 95°C for 2 min.
4. Add 60 μL of PCR-grade water to each well, mix, and distribute between twenty 96-well plates using a Hydra-96 robot (4 μL per well) or multichannel pipet.
5. Seal plates with foil covers, and store at –20°C until further use.

3.5. Real-Time PCR

1. Thaw a plate containing cDNA aliquots from the RT reaction.
2. Mix the following reagents, multiplied by the number of reactions: 12.5 μL 2X quantitative PCR master mix (Applied Biosystems), 0.5 μL each PCR primer (stock concentration 10 μM), 0.5 μL TaqMan probe (stock concentration 5 μM), and 7 μL water.
3. Add a series of calibration standards (if required; *see* **Subheading 3.6.**), 4 μL per well, to spare wells on the plate.
4. Add 21 μL of reaction mixture to each cDNA sample and calibration standard.
5. Seal the plate with an optical cover (Applied Biosystems), and place in the ABI instrument.
6. Set up the instrument according to plate layout, PCR cycling conditions, and reporter dye system, and run the PCR for 40 cycles.

3.6. Calibration

The primary output from real-time PCR is the threshold cycle—defined as the cycle at which the fluorescent signal reaches an arbitrary threshold value. This can be set automatically by the instrument, or manually, but the operator should ensure that it is within the exponential phase of the PCR (**Fig. 1A**). If a series of serial dilutions of a PCR template are subjected to a real-time PCR, this will result in a regular shift in the threshold cycle corresponding to each dilution (**Fig. 1B**). Assuming that the PCR is running at maximal efficiency, a shift of one cycle corresponds to a twofold dilution of template. It is thus possible to plot the threshold cycle against template quantity (log scale) to obtain a standard curve (**Fig. 1C**).

There are various materials one can use as standards to calibrate a real-time RT-PCR plate. In the ideal situation, one would use standards containing a known copy number of a molecule identical to that which is being measured, i.e., cDNA encoding the test gene. In reality, this is not possible, and so a compromise must be struck by using either known quantities of a different molecule (genomic DNA, plasmid DNA), which may behave differently under PCR conditions, or cDNA for the test gene in a set of serial dilutions such that relative concentrations are known. If calibration standards are included on the reaction plate, the software will automatically calculate copy numbers for wells marked as "unknown" from the threshold cycles.

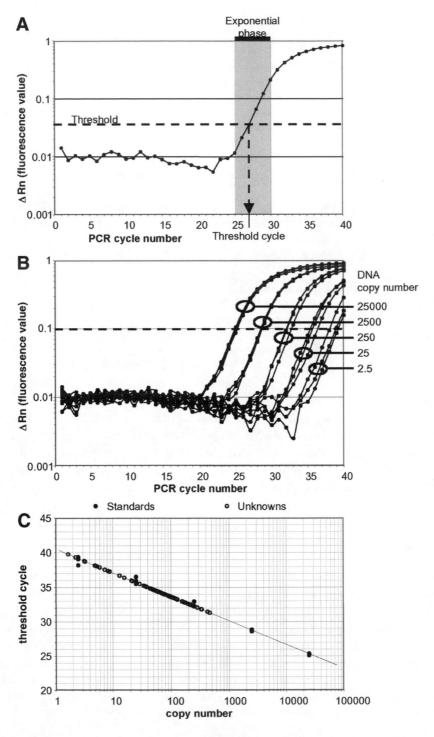

3.6.1. Genomic DNA

This can only be employed when the amplicon lies entirely within one exon of the gene; however, it is readily available and may be used for many different genes. To determine copy number, one must assume that there is one pair of copies per genome. Mammalian genomes typically contain approx 3.0×10^9 bp with a total molecular weight of 1.9×10^{12} Daltons. This corresponds to approx 600,000 copies of each gene per microgram of genomic DNA.

3.6.2. Plasmid DNA

This avoids the constraint on primer design that arises with genomic DNA but obviously may not be available for every gene to be tested. Copy numbers can be worked out roughly from the size of the plasmid plus the insert by using an average molecular weight of 635 Daltons/bp.

When using this method, it is extremely important to avoid any contamination of other reaction components with plasmid DNA.

3.6.3. cDNA

Complementary DNA, reverse-transcribed from the RNA of a tissue or cell line that abundantly expresses the gene of interest, is a useful option when genomic DNA cannot be used and plasmid DNA is unavailable. A set of serial dilutions of the cDNA can be made, and an arbitrary copy number can be ascribed according to the relative amounts.

3.6.4. Non-Use of Standards

If no suitable standards are available, another option is to assume a twofold increase in copy number with every PCR cycle. This would correspond to a slope of -3.32 in the standard curve (**Fig. 2C**), i.e., it would require 3.32 cycles for a 10-fold increase in PCR product to occur. To calculate a copy number, Q from a threshold cycle, C_t, the following formula can be employed, where Cn is the number of PCR cycles in the reaction:

$$Q = 10^{(C_n - C_t / 3.32)}$$

Fig. 1. *(facing page)* (**A**) Increase in reporter dye fluorescence over the course of a PCR, indicating the exponential phase of the PCR, arbitrary threshold and threshold cycle. (**B**) Serial 10-fold dilutions of template result in regular shifts of the threshold cycle of approx 3.32 cycles. (**C**) Standard curve. Threshold cycle is plotted against template copy number (logarithmic scale) for the calibration standards (filled circles). This results in a linear line of best fit. When the unknown samples (open circles) are fitted to this line, a copy number can be calculated from the threshold cycle.

3.7 Standardization and Use of Reference Genes

Expression levels of mRNA are commonly measured as a ratio of test to reference gene (often referred to as housekeepers). The reasons for using reference genes are several:

1. Reference genes, such as β-actin, GAPDH, and cyclophilin, are assumed to act as steady-state controls across treatment groups; the assumption is that housekeeping genes are found in all cells and tissues and that they are unchanged in the presence of treatment *(11,12)*. We wish to remove any systematic variation between samples by using a gene we believe is unaffected by treatment and whose variation in mRNA level only reflects the technical and quality variation between the samples. It has been shown that reference genes in this and other models may fail the criteria for steady-state controls *(9)*. We wish to remove any systematic variation between samples by using a gene we believe is unaffected by treatment and whose variation in mRNA level reflects the technical and quality variation between the samples only.
2. By comparing test mRNA with internal reference mRNA, the component of variation common to both can be removed from the analysis (*see* **Subheading 3.8.**). The mRNA for reference genes, as for test genes, often reflects the global changes. Conclusions on mRNA levels for test genes will vary according to the choice of reference gene to which they are normalized. Therefore, assessment of reference genes is important in deciding on their use in the analysis, and the following interpretation will critically depend on this choice and how the adjustment is performed.

3.8. Data Analysis and Statistical Methods

3.8.1. Log Transformation of Data

Investigation of a wide variety of mRNAs has consistently revealed that genes expressed at low levels have a small variability and highly expressed genes have a large variability. Similarly, within a gene when expression levels change with treatment, the variability correspondingly changes in proportion. Additionally, the relationship between different gene expression levels is often nonlinear in nature. However, once the data are transformed to the log scale, the variability is homogeneous, regardless of the magnitude of expression, and the relationship between genes is linearly related, all of which are required for valid parametric statistical analysis. This should not be surprising, as the raw interpolated data are calculated on a log scale (**Fig. 1C**). Strong correlations are often seen between the log-transformed mRNA transcript numbers, regardless of which genes are assessed relative to one another. An example of the correlations between five genes from the study is presented in **Fig. 2**.

3.8.2. Analysis of Covariance

Expressing data as a ratio to adjust for an uncontrollable variable is a simple and widely used method in many areas of science *(13)*. It is commonly used to express real-time PCR data. However, taking a ratio of test to reference mRNA can lead to bias. A number of assumptions need to be made when analyzing ratios. Many of these assumptions have been shown to be not viable *(9)*. An alternative statistical method to correct for a reference gene is the Analysis of Covariance (ANCOVA) which is widely used to adjust for the effect of independent biological variables *(8,14)*. ANCOVA has a number of advantages over the analysis of ratios. It allows the experimenter to model the relationship between test and reference genes precisely and hence improves the sensitivity of the analysis. Treatment means can be viewed in the original units rather than as a percentage of reference mRNA, and orders of magnitude between treatment groups can be estimated with realistic asymmetrical confidence intervals.

The precise level of a specific mRNA change and the probability of obtaining that result varies according to the choice of reference mRNA to which it is normalized. This indicates that some reference genes explain more of the nonspecific variability than others, leading to a greater improvement in the statistical sensitivity.

When a reference mRNA has been affected by treatment, cautious interpretation of results is advisable, as the adjustments will not only remove any differences in mRNA quality, but may also remove or introduce a treatment effect on the test mRNA. If these methods are used without knowing how treatment has affected the reference mRNA, then mistakes can easily be made. False negatives occur when both test and reference are affected in the same way, as one cancels out the effect of the other. False positives occur when the reference but not the test gene mRNA is affected by treatment; for example, the reference decreases, the test remains stable, and the differential is claimed to be an increase in test gene. We recommend analyzing individual reference mRNA levels with analysis of variance (ANOVA) prior to their use as covariates. Once examined, test gene mRNA differences in treatment groups can be interpreted as the effect over and above the effect seen in the reference. If there are no differences in the test gene mRNA, then this should be interpreted as an effect similar to that seen in the reference mRNA.

3.8.2.1. Example 1

Log FADD against log GAPDH was plotted (**Fig. 3C**), indicating that the two genes are correlated, and that adjustment for log GAPDH levels could help

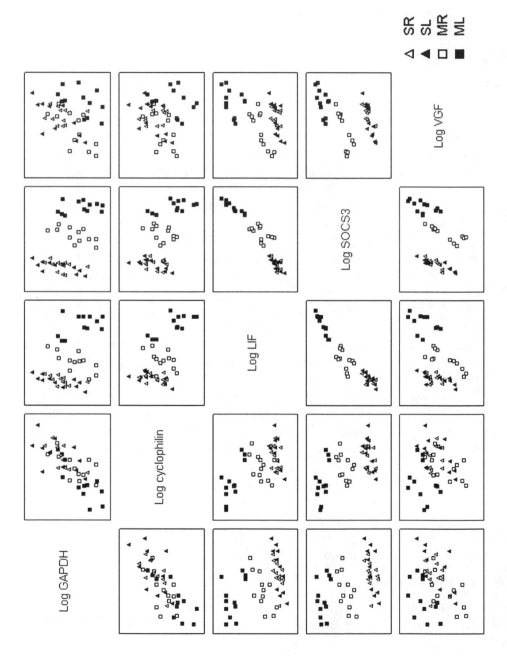

276

in improving the sensitivity of the analysis of log FADD. Additionally, it appears that the 24-h subgroup of MCAO animals has elevated log FADD on both the left and right hemispheres. This should be investigated statistically by looking at the treatment by time interaction in an ANOVA or ANCOVA.

ANOVA on the individual genes indicates no statistically significant effects ($p < 0.05$) in either log FADD (**Fig. 3A**) or log GAPDH. GAPDH is a reference gene, and this confirms the assumption we made that log GAPDH is not affected by treatment. In the ANCOVA of log FADD using log GAPDH as a covariate, we find the treatment by time interaction is now statistically significant ($p = 0.035$). The use of log GAPDH as a covariate has adjusted the FADD levels for the relatively low values of GAPDH seen in the 24-h MCAO right hemisphere (**Fig. 3B**) and additionally decreased the residual mean square, improving the sensitivity of the statistical tests.

3.8.3 Principal Components Analysis

Univariate ANCOVA analysis allows probability values to be given to the mRNA changes of one transcript referenced to another. Principal component analysis (PCA) is a statistical method that gives an overview of the complete dataset. PCA acts as a useful aid to ANCOVA by assessing all the mRNA transcripts and treatment groups in a single analysis. Its goal is to seek out latent relationships, or principal components, between the set of genes under investigation *(15)*. Thus PCA has been recommended for quantifying mRNA as a way of separating the non-specific variation from that caused by treatment *(11,12,16)*.

A PCA loadings plot allows an assessment of which mRNA transcripts cluster together, reflecting similar behavior under the different experimental treatments. A component that has all positive or all negative loadings for genes represents a weighted average of the gene set and pulls out a common trend from the dataset, often reflecting the systematic variation between samples often associated with reference genes. A PCA scores plot indicates which samples have similar profiles across the set of measured genes. If samples within a treatment have been affected in a similar way, then they will cluster together. An assessment can be made of which genes have been affected by the treat-

---

Fig. 2. *(facing page)* Matrix plot of five genes measured in the cortex of rats from an MCAO study on a log10 scale, demonstrating the linear relationship between genes and homogeneous variability on the log scale. SR, sham, right (contralateral) hemisphere; SL, sham, left (ipsilateral) hemisphere; MR, MCAO, right hemisphere; ML, MCAO, left hemisphere.

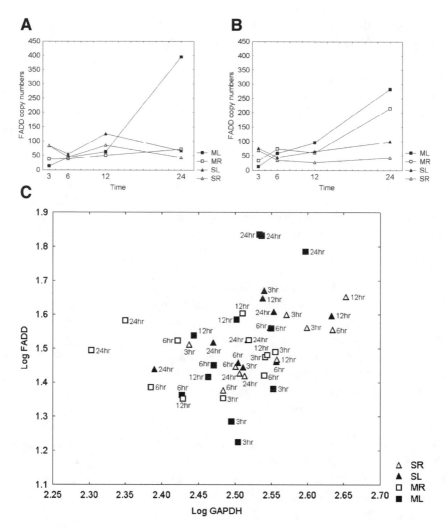

Fig. 3. **(A)** Fas-associated death domain (FADD) mRNA: unadjusted treatment group geometric mean RNA copy numbers plotted over time. **(B)** FADD mRNA, adjusted using glyceraldehyde-3-phosphate dehydrogenase (GAPDH) mRNA copy numbers: treatment group geometric means plotted over time. **(C)** Log FADD vs log GAPDH individual data points plotted by treatment and time. SR, sham, right (contralateral) hemisphere; SL, sham, left (ipsilateral) hemisphere; MR, MCAO, right hemisphere; ML, MCAO, left hemisphere. Note, in plot **(C)**, the high levels of the MCAO treatment group at 24 h for FADD, taking into account their relatively low GAPDH levels. Before adjustment for GAPDH via the covariate analysis (plot A), the MCAO right treatment group at 24 h is low and comparable to levels in the sham groups. After adjustment for GAPDH (plot B), the MCAO right treatment group mean at 24 h has been adjusted upward and is now similar to the MCAO left group mean at 24 h.

ment by looking for mRNA transcripts with large loadings in the same direction on the loadings plot as the treatment effect in the scores plot.

3.8.3.1. EXAMPLE 2

Five mRNA transcripts were measured, GAPDH, cyclophilin, LIF, SOCS3, and VGF. A matrix plot of the five genes can be seen in **Fig. 2**. A PCA was performed, and the scores and loadings plots are presented in **Fig. 4**. The scores plot (**Fig. 4A**) shows a clear separation in the MCAO left hemisphere samples in the first component (*x*-axis), with the latter time-points the most extreme. In the center of the first component, the MCAO right hemisphere samples are clustered, and to the right both hemispheres from the sham-operated animals are clustered together. This treatment pattern explains 57% of the variability in the dataset. The second component does not demonstrate any clear separation between treatment groups. The first principal component in the loadings plots (**Fig. 4B**) directs us to the genes that are upregulated in the treatment pattern described above. LIF and SOCS3 are clearly upregulated by the MCAO treatment and are highly correlated, as they are very close to each other in the loadings plot. This effect can also be clearly seen in the matrix plot (**Fig. 2**). VGF is also affected by MCAO treatment but to a lesser degree. GAPDH and cyclophilin, on the other hand, show some downregulation in the MCAO animals, as their loadings in the first component are in the opposite direction. The loadings in the second component are all positive, indicating that all the genes have a common trend; as this is not related to treatment, we can associate it with the reference gene effect or nonspecific variability. As we would expect, cyclophilin and GAPDH have large loadings in this component. The second component explains 33% of the variability in the dataset. In this example, we have used five genes to demonstrate this methodology; however, increasing the number of genes will improve the power of the method. In addition, in studies that involve large numbers of genes, summarizing univariate analyses becomes difficult, and a multivariate approach can alleviate this.

4. Notes

1. There are a number of new kits available for RNA isolation, many of which are based on centrifugation through a column. In our own experience, the RNEasy kits (Qiagen) can be used to isolate total or polyA-enriched RNA of sufficient quality for real-time RT-PCR. We have found the RNEasy kit preferable for RNA isolation from tissue culture preparations. A useful adjunct to the RNEasy kits is RNAlater reagent, which can be used to store small (<5 mm) pieces of tissue prior to RNA isolation, avoiding the use of liquid nitrogen.
2. One of the additional uses of an internal oligonucleotide probe is that it acts as an internal control for specificity. When using the SYBR green method, it is advis-

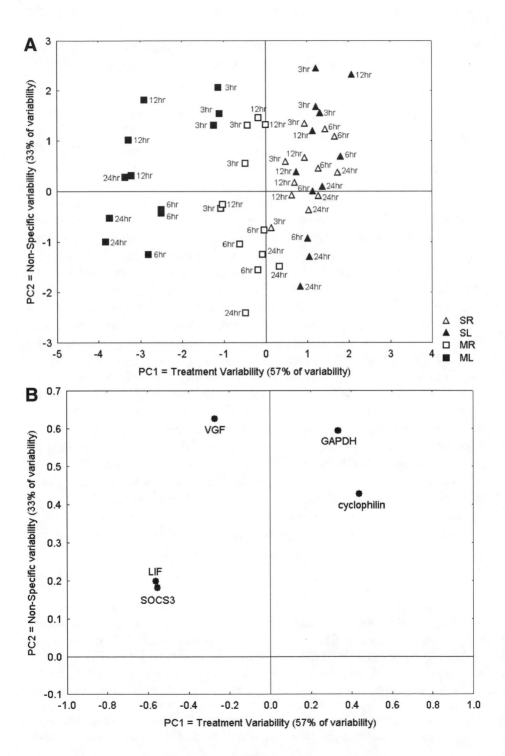

able to confirm specificity by checking the size and identity of the PCR product prior to the main experiment. It may also be useful to check primer specificity by running a BLAST search at the design stage.

3. PolyA enrichment of RNA is integral to some of the column-based RNA isolation methods discussed above. Enrichment techniques that are carried out following isolation of total RNA (e.g., using TRIzol) are also available. We have avoided this, as additional manipulation of RNA tends to be at the expense of RNA integrity and results in an increase in threshold cycle. DNAse treatment is also incorporated into many of the newer RNA isolation methods. Again, additional manipulations of RNA post isolation are best avoided, and may result in an increased threshold cycle.

4. RT substrate quantity will depend on availability of material, and the abundance of the transcripts being detected. In our experience, the cDNA yield in the reverse transcription method described here starts to plateau when more than 1 µg of total RNA is used. Use of less than 10 ng of total RNA may require extension of the number of PCR cycles employed, and we have observed that as threshold cycle increases, its reproducibility decreases.

5. There are three possible methods of priming the reverse transcription.

    a. *Gene specific primers.* These will give the most efficient reverse transcription of the mRNA to be tested, but one is then restricted regarding using the cDNA to measure multiple transcripts. If these have been decided in advance, a mixture of gene-specific reverse primers could be used.

Fig. 4. *(facing page)* **(A)** The scores plot from a principle component analysis (PCA) analysis of five genes in a stroke model experiment. The figure shows a clear separation in the MCAO left hemisphere samples in the first component (x-axis), with the latter time-points the most extreme. In the center of the first component, the MCAO right hemisphere samples are clustered, and to the right both hemispheres from the sham-operated animals are clustered together. The second component does not demonstrate any clear separation between treatment groups. SR, sham, right (contralateral) hemisphere; SL, sham, left (ipsilateral) hemisphere; MR, MCAO, right hemisphere; ML, MCAO, left hemisphere. **(B)** The loadings plot from the same PCA analysis of five genes in a stroke model experiment. The first principal component (x-axis) directs us to the genes that are upregulated in the treatment pattern described above. Leukemia-inhibitory factor (LIF) and suppressor of cytokine signaling-3 (SOCS-3) are clearly upregulated by the MCAO treatment and are highly correlated, as they are very close to each other in the loadings plot. GAPDH and cyclophilin, on the other hand, show some downregulation in the MCAO animals, as their loadings in the first component are in the opposite direction. The loadings in the second component (y-axis) are all positive, indicating that all the genes have a common trend and as this is not related to treatment, we can associate it with the reference gene effect or nonspecific variability.

b. *Oligo(dT) primer.* This will only reverse-transcribe mRNA, but with differing efficiencies, depending on the length and structure of 3' UTR of different transcripts.

c. *Random oligomers, most commonly 6–9 bases in length.* These will reverse-transcribe all RNA within a sample including ribosomal RNA. The contribution of mRNA to the resulting cDNA pool will thus be much smaller, so threshold cycle will increase.

References

1. Read, S. J., Parsons, A. A., Harrison, D. C., et al. (2001) Stroke genomics: approaches to identify, validate and understand ischemic stroke gene expression. *J. Cereb. Blood Flow Metab.* **21,** 755–778.
2. Heid, C. A., Stevens, J., Livak, K. J., and Williams, P. M. (1996) Real time quantitative PCR. *Genome Res.* **6,** 986–994.
3. Gibson, U. E. M., Heid, C. A., and Williams, P. M. (1996) A novel method for real time quantitative RT-PCR. *Genome Res.* **6,** 995–1001.
4. Harrison, D. C., Davis, R. P., Bond, B. C., et al. (2001) Caspase mRNA expression in a rat model of focal cerebral ischemia. *Mol. Brain Res.* **89,** 133–146.
5. Bates, S., Read, S. J., Harrison, D. C., et al. (2001) Characterisation of gene expression changes following permanent MCAO in the rat using subtractive hybridisation. *Mol. Brain Res.* **93,** 70–80.
6. Zea-Longa, E., Weinstein, P. R., Carlson, S., and Cummins, R. (1989) Reversible middle cerebral artery occlusion without craniectomy in rats. *Stroke* **20,** 84–91.
7. Harrison, D. C., Medhurst A. D., Bond, B. C., Campbell, C. A., Davis, R. P., and Philpott, K. L. (2000) The use of quantitative RT-PCR to measure mRNA expression in a rat model of focal ischemia—caspase-3 as a case study. *Mol. Brain Res.* **75,** 143–149.
8. Festing, M. F. W., Overend, P., Das, R. G., Borja, M. C., and Berdoy, M. (2002) *The Design of Animal Experiments.* Royal Society of Medicine Press, London, pp. 68–70.
9. Bond, B. C., Virley, D. J., Cairns, N. J., et al. (2002) The quantification of gene expression in an animal model of brain ischaemia using Taqman™ real-time RT-PCR. *Mol. Brain Res.* **106,** 101–116.
10. Pattyn, F., Speleman, F., De Paepe, A., and Vandesompele, J. (2002) RTPrimerDB: the real-time PCR primer and probe database. *Nucleic Acids Res.* **31,** 122–123.
11. Spanakis, E. (1993) Problems related to the interpretation of autoradiographic data on gene expression using common constitutive transcripts as controls. *Nucleic Acids Res.* **21,** 3809–3819.
12. Spanakis, E. and Brouty-Boyé, D. (1994) Evaluation of quantitative variation in gene expression. *Nucleic Acids Res.* **22,** 799–806.
13. Kaiser, L. (1989) Adjusting for baseline: change or percentage change? *Stat. Med.* **8,** 1183–1190.

14. Cochran, W. G. and Cox, G. M. (1957) *Experimental Designs*. Wiley, New York.
15. Chatfield, C. and Collins, A. J. (1980) *An Introduction to Multivariate Analysis*. Chapman and Hall, London.
16. Hole, S. J. W., Howe, P. W. A., Stanley, P. D., and Hadfield, S. T. (2000) Pattern recognition analysis of endogenous cell metabolites for high-throughput mode of action identification: removing the post-screening dilemma associated with whole-organism high throughput screening. *J. Biomol. Screen.* **5,** 335–342.

13

Effective Analysis of Genomic Data

Paul R. Nelson, Andrew B. Goulter, and Richard J. Davis

Summary

High-throughput biotechnology has enabled genome-wide investigation of gene expression and has the potential to identify genes that have a role to play in focal cerebral ischemia, as well as many other interventions. The advent of this technology has also led to the generation of large amounts of expensive and complex expression data. One of the major problems with the generation of so much data is locating and extracting the relevant information to aid target identification and interpretation effectively and reliably. Statistical involvement is vital. Not only does it help to ensure effective extraction of information from the data, it also increases the likelihood that the data collected will embody the information about the differential expression of interest in the first place. The goal of this chapter is to recommend an effective process for investigating gene expression data. There are five stages in this process that we believe lead to reliable results when routinely applied to an expression dataset, once it has been appropriately generated and collected: (1) biological problem definition and design selection; (2) data examination, "preprocessing," and reexamination; (3) data analysis step I: screening for differentially expressed genes; (4) data analysis step II: verifying differential expression; and (5) biological verification, interpretation, and communication.

KEY WORDS

Differential expression; experimental design; data examination; visualization; preprocessing; baseline features; data analysis; multivariate; principal components analysis; hierarchical cluster analysis; partial least squares discriminant analysis; regression coefficients; variable influence on projection; univariate; analysis of variance; covariate; fold difference.

1. Introduction

The format of **Subheading 3.**, Methods, follows the recommended stages and grew out of our experience in applying statistical methods at each stage of investigation to gene expression data in a pharmaceutical drug discovery setting.

From: *Methods in Molecular Medicine, Vol. 104: Stroke Genomics: Methods and Reviews*
Edited by: S. J. Read and D. Virley © Humana Press Inc., Totowa, NJ

Stage 1 begins with the need to define the goals of the research clearly and then to tailor the experiment to collect data that will help achieve those goals. Stage 2 suggests ways to examine and recognize features in the data collected and offers methods to help reduce or remove unwanted artifacts and variations from the data prior to the analysis. Stage 3 presents tools that can be used to separate important genes that are worthy of further study from the many trivial genes that are not differentially expressed. Stage 4 focuses on building models, knowledge, and understanding of the roles those genes surviving the initial screening stage play in focal cerebral ischemia. Stage five involves verification, selection of therapeutically relevant genes, and interpretation and communication of the results.

This chapter is not intended to be a comprehensive tutorial for the theory and application of the tools discussed. It serves only as an introduction to the methods involved at each stage of investigation of the data. To ensure that the techniques outlined are appropriately applied, the reader is encouraged to refer to the publications listed in the reference section for detailed technical descriptions, applications, and understanding of the methods.

2. Materials

For each stage of the investigations discussed in **Subheading 3.**, Methods, two examples are used to help illustrate and explain the approaches and methods presented. The two datasets, supplied by Pharmagene, are from gene-chip arrays and TaqMan™ real-time reverse-transcriptase polymerase chain reaction (RT-PCR), to measure selected mRNA expression in nondiseased and diseased human tissues. These diseases included heart failure and cystic fibrosis. However, the approach and methods presented here are equally applicable to other technologies, such as 2D-gel electrophoresis and liquid chromatography–mass spectometry or biofluid nuclear magnetic resonance (NMR) protein profiling.

The key multivariate output—principal component analysis (PCA) and partial least squares discriminant analysis (PLS-DA)—presented in this chapter was generated using the SIMCA-P package available from Umetrics UK (www.umetrics.com). The remaining multivariate and univariate statistical output, such as analysis of variance and covariance, was generated using the Microsoft® Excel™ add-in Cellula developed by Prism Training & Consultancy (www.prismtc.co.uk).

3. Methods

3.1. Stage 1: Define the Biological Problem and Select a Design to Match

Spending time defining the research problem helps ensure that appropriate protocols and a good experimental design are developed to gather relevant and precise data. If the correct methods of analysis to match the design are adhered to, then a good design will increase the power and precision to detect differential expression, thus turning an experiment with no significant conclusions (in which it is impossible to distinguish any observed differences in expression from those that might arise by chance) into one that shows expressional differences clearly.

Designing a study in such a way as to ensure that the information you need (to effectively examine, preprocess, analyze, and answer the biological questions you pose) is contained within the data you collect cannot be overemphasized. No amount of data manipulation after the experiment will compensate for bad design. We cannot recommend the exact type of experimental design you should choose. This will depend on the biological question(s) you pose, the precision or variability of your technology, and the constraints imposed by the management and resources of the study.

However, the experiments we are mostly concerned with are comparative, and it is simply a matter of signal vs noise. The signal (i.e., change in expression owing to treatment or condition) should be stronger than the noise in order to detect it. If it isn't, then you can attack either side of the ratio. The signal can be increased or the noise can be decreased by adhering to the simple fundamental principles established by R.A. Fisher in 1925 and 1926 *(1,2)*, namely:

1. Replication—increasing the precision and power of your design to detect changes of a specified size, enabling calculation of the uncertainty in the estimates of the differences in expression to be made, and thus the ability to measure, understand, and manage the experimental noise
2. Randomization—avoids systematic biases that can arise from subjective allocation of treatments or conditions to arrays or plates and helps provide valid estimates of treatment effects and experimental noise
3. Blocking—grouping the samples, plates, and arrays into "blocks" (e.g., by operator, run, batch, day) to capture large potential sources of noise between blocks that can be measured and set aside from the changes of interest.

A useful additional means of improving the noise and precision is to ensure the absence of systematic error by including, when feasible, additional information on a measurement variable (e.g., by using a housekeeping gene) thought to influence or covary with the test gene. Inclusion of a covariate, such as a housekeeping gene, may well help to explain some of the variability in the measured response unrelated to the biology under investigation, such as the degradation of the tissue sample, and hence will decrease the noise and in turn increase precision. It is therefore assumed that the treatments under investigation do not affect the covariate; a measure of this can be given by a covariance efficiency factor in an analysis of covariance (*see* **Subheading 4.**, Discussion).

Finally, the treatments or conditions in a gene expression study may be "structured" in some way. They may, for example, include one or more control groups. This enables the experimenter to make relative or comparative statements about the effects of treatments or conditions on gene expression. Perhaps the most important type of treatment structure is given by a factorial structure, which permits simultaneous study of different treatment factors each at different levels *(3)*. Thus, in a study with three different drugs, including a control, these might each be tried in combination with each of two disease conditions. The six combinations then constitute the treatment groups.

	Control (D1)	Drug 1 (D2)	Drug 2 (D3)
Non-diseased Tissue (T1)	T1, D1	T1, D2	T1, D3
Diseased Tissue (T2)	T2, D1	T2, D2	T2, D3

Factorial experimentation has various advantages over experimentation on one treatment factor at a time: greater precision yields more information, especially on possible interactions (i.e., the magnitude of effect of one factor, say drug, on expression depends on another factor, such as phenotype or hemisphere) and provides a wider range of validity for the conclusions that are to be drawn.

The TaqMan™ study, one of the examples used to illustrate the methods in this chapter, comprised 36 human tissue samples obtained to study differential expression of genes in heart failure (12 samples from each of three disease groups). Group 1 was a control group, with no specific clinical or pathological history of heart disease, so the specific interest was the detection of changes in expression in group 2 (tissues from donors with idiopathic dilated cardiomyopathy) and group 3 (tissues from donors with ischemic cardiomyopathy). In total, mRNA levels for 78 test genes were investigated, together with 78 replicates of the housekeeping gene, GAPDH, to accompany each test gene.

In the second example, a microarray experiment was conducted using human donor information for six nonsmoking control (Con) and six cystic fibrosis (CyF) extracted RNA tertiary bronchial samples. The extracted RNA from each donor was run on duplicate chips to allow assessment of chip-to-chip variation.

3.2. Stage 2: Visualize the Data and Preprocess, if Needed

The primary purpose of this stage is to look for patterns and features in the data that warrant further investigation. It is also important to look for obvious errors, anomalies, or unwanted sources of variation, which are likely to affect the analysis adversely by masking the intentional variation of interest. These sources of noise will either need to be removed prior to analysis (i.e., by preprocessing), using the type of methods presented later, or accounted for during the analysis stages (i.e., by modeling).

Since high-throughput biotechnology can produce an overabundance of data from a great many genes, a major problem is visualizing, locating, and extracting the relevant information from so much data and interpreting it easily and reliably. Unsupervised pattern recognition techniques, such as PCA and hierarchical cluster analysis (HCA), are ideally suited to this task. They reduce the complexity of high-dimensional data to just a few relevant axes. Information can then be displayed simply for easy, but comprehensive interpretation. HCA gives a broader view of the data, whereas PCA can be used to investigate further samples and clusters that are highlighted in HCA. Our recommendation therefore is to use both of the methods.

HCA *(4)* calculates the interpoint distances (i.e., similarity) between all the samples, or all the genes, to form clusters of samples or genes with similar patterns of expression. It then represents that information in the form of a hierarchical tree, called a dendrogram, as illustrated in the samples and a selection of genes from the microarray dataset in **Fig. 1**.

Clustering both genes and samples simultaneously produces a heat map (also known as an intensity, matrix, or two-way joining plot). This can be used to identify clusters of samples with similar values of expression across the genes (displayed as "areas" of similar color) to uncover genes with similar or discriminatory patterns of intensity. **Figure 2** is a segment of a heat map corresponding to the two separate dendrograms shown in **Fig. 1**.

PCA *(5)* is a favorite tool for data compression and overview. By analyzing all the genes simultaneously, PCA utilizes the correlations that exist between gene expression profiles to compress the high-dimensional coordinate system efficiently to one comprising only a few relevant uncorrelated variables, called *principal component scores*.

These new variables are simply a weighted linear combination of the genes, and they describe the maximum amount of variation in the data with the minimum number of axes in which to plot the samples. Usually only a few are needed to summarize a large data table and convey the clustering of the samples. The weights or "loadings" that combine the original variables in the data table, to form the new variables or scores, describe the pattern of the genes and their similarities and differences.

The scores and their corresponding loadings can then be used in simple displays to visualize complex high-dimensional data in just a few relevant axes for biological understanding and interpretation. They can help to identify patterns, relationships, and groupings for classification. **Figure 3** demonstrates how PCA represents the variation between groups and samples for the TaqMan data and between groups, samples, and duplicate chips for the microarray

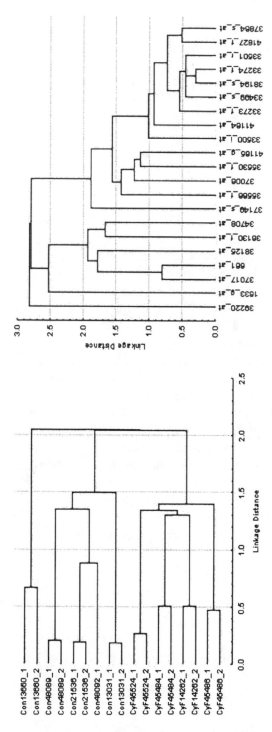

Fig. 1. (A) Dendrogram for the microarray data identifies the similarity in expression of the duplicate chips, which are joined at small linkage distances, and then of the two large clusters of control (Con) and cystic fibrosis (CyF) samples. Note that one control sample, 13660, appears to be different from the others. Later we discuss preprocessing steps to adjust microarray data for differences that may not have arisen owing to intentional biological comparisons. (B) The dendrogram for the genes comprises two large clusters, clustering together genes of a similar function and regulation owing to the disease.

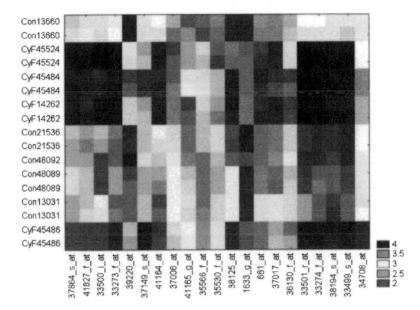

Fig. 2. Heat map corresponding to the two-way clustering of the samples and genes in **Fig. 1**, highlighting simultaneously similarities and differences between the two groups of samples and the genes.

example. Using the loadings, the contribution of each gene to the separation between samples and groupings, owing to their changes in expression, can also be observed (**Fig. 4**).

Generally a great deal of correlated or redundant information can be generated by gene expression experiments, and thus PCA can reduce the dimensionality of the datasets to a significantly reduced number without losing much information. For instance, a PCA performed on the reference gene data from the TaqMan study confirmed what was expected, that the many replicates of the reference gene GAPDH generated similar expression (**Fig. 5A**). Therefore a weighted average expression for the housekeeping gene, derived from the PCA loadings (**Fig. 5B**), would be representative of all the replicates and could be used to adjust the data either prior to or during an analysis (see later).

PCA and HCA provide insightful graph overviews of the data in order to pick out both desirable and undesirable patterns, features, and trends in the data. However, discovery often involves working with different views of the same data. Profile charts and radial plots offer alternative views and insights into the complexity of gene expression data. For example, by displaying the average expression of the two groups of microarray samples (control vs cystic fibrosis) as profiles or "spectra" in **Fig. 6A**, genes or clusters of genes are

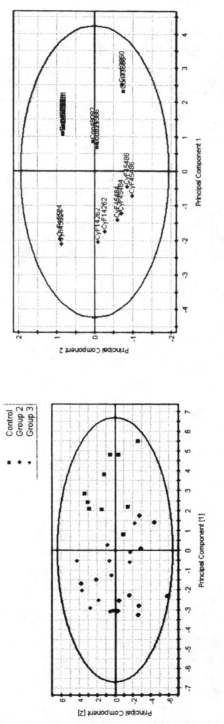
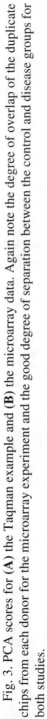

Fig. 3. PCA scores for (**A**) the Taqman example and (**B**) the microarray data. Again note the degree of overlap of the duplicate chips from each donor for the microarray experiment and the good degree of separation between the control and disease groups for both studies.

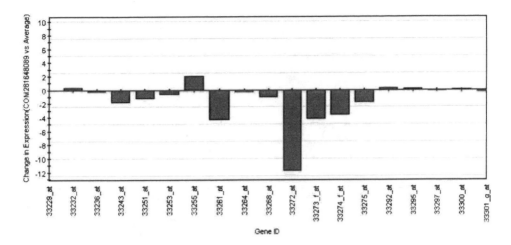

Fig. 4. Contribution plot from a block of genes from the microarray example show-ing there is a reduction in expression for genes 33272_at, 33273_f_at, and 33274_f_at when comparing a control sample with the average of the samples.

identified with differences in expression between the two groups. Patching and zooming in on a particular region of the profile chart (**Fig. 6B**) facilitates closer inspection of individuals or groups of genes worthy of further and closer investigation.

The data from TaqMan or gene-chip arrays have similar characteristics and problems as do those of imaging and chemometric data, e.g., spectroscopy *(6)*, particularly biofluid NMR. The datasets are very large and often show large systematic variations that are not related to the biology under investigation. There is also a great deal of random high-frequency noise, mostly contributed by low-abundance genes, in the case of gene expression. Preprocessing of the data is often necessary; otherwise these unwanted sources of variation, which can be large relative to the changes of interest, will dominate the analysis unless removed.

However, keep in mind that preprocessing changes the data and will influ-ence the results either positively or negatively. Examination of the preprocessed data is essential to see whether the result of preprocessing is optimal and what was expected. The scatter plot in **Fig. 7A** shows evidence of a linear relation-ship between a test gene's expression and that of the housekeeping gene, GAPDH. If the housekeeping gene is not linked to the biology under investiga-tion, which can be tested, then there is evidence of a baseline feature unrelated to the biology, which may dominate the analysis unless it is removed or accounted for in the analysis (i.e., using the housekeeping gene as a covariate in an analysis of covariance).

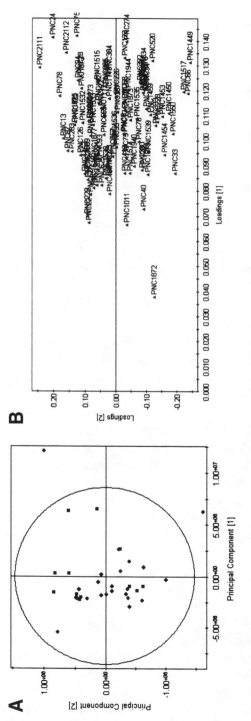

Fig. 5. PCA scores plot of the housekeeper replicates to the test genes (**A**) and loadings plot (**B**) resulting from the analysis performed on Vic data from the TaqMan study.

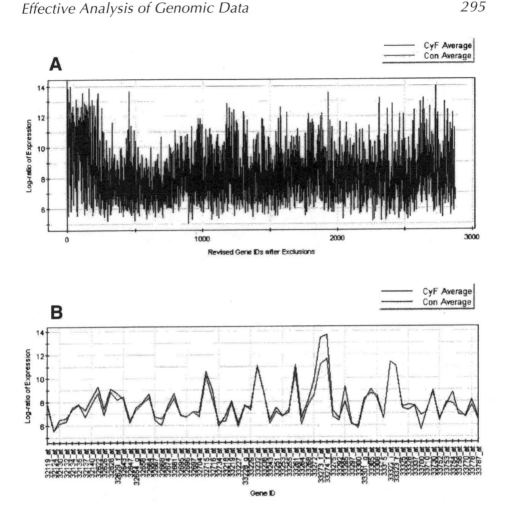

Fig. 6. (**A**) Profile chart providing a contrast in gene expression profiles between the two groups of microarray samples. (**B**) Changes in expression profiles for genes 33272_f_at, 33273_f_at, and 33274_f_at are emphasized using patch and zoom facilities. Con, control; CyF, cystic fibrosis.

Figure 7B demonstrates that taking the ratio of each test gene result to the weighted average GAPDH expression i.e., representing fold changes from the housekeeping gene, resulted in the removal of the relationship with GAPDH and hence corrected for the baseline features in this example. Understanding the biology and the physics underlying unwanted sources of variation aids selection of the preprocessing techniques.

There are two categories of preprocessing tools. There are those that operate on the samples, such as normalization to remove systematic variation between

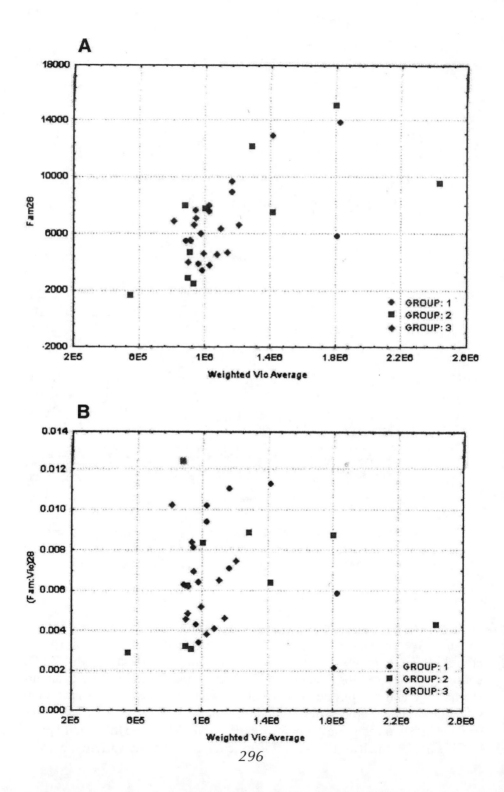

the samples or conditions, e.g., dye bias *(7)*, or to handle low-abundance genes *(8)*. Sample weighting can also be applied when information is available about the relative reliability of some samples over others, e.g., from replicated experiments *(9)*. Another use of these tools is to satisfy assumptions on the methods of analysis, e.g., homogeneity of variance in regression analysis *(10)*.

Then there are preprocessing tools operating on the genes such as transforming to log ratios of expression, which, apart from aiding interpretation, i.e., by representing fold changes in expression, usually also helps to stabilize the variance associated with increasing gene expression. Gene weighting or scaling to correct for baseline features, as in **Fig. 7**, or to give some genes more influence on the analysis than others. In the case of replicated experiments, this could take the form of weighting genes according to their non-reproducibility, although more often unreliable genes, e.g., low-abundance genes, are given a weight of zero and excluded from the analysis. Additionally, genes that ultimately exhibit the greatest change in expression between groups do not always have the greatest overall expression and therefore variation as a consequence. Commonly used unit variance scaling puts all genes on an equal footing prior to analysis but can increase the noise of lower abundance genes to match that of their high-abundance counterparts and thus mask their discriminatory influence. Pareto scaling—giving each gene a variance equal to its standard deviation—has become more popular in recent years, as it offers an intermediate alternative scale between the extremes of no scaling and unit variance scaling. Finally, mean centering of the data prior to analysis emphasises the contrast in gene expression between groups and is accomplished by subtracting the mean expression from each gene from all its entries. In **Fig. 8B** the log-ratio expression data for two samples from the microarray experiment have been mean-centered. As a result, the relative differences in intensity of expression for each of the genes surviving screening are easier to discern.

Selecting the preprocessing tools to best suit your data commonly involves iteration between the examination, preprocessing, and analysis steps. However, try to ensure that you adopt a preprocessing strategy because it removes the undesirable features from your data that you know about, while retaining known desirable features. An understanding of the biology and/or physics underlying these unwanted sources of variation aids selection of the preprocessing techniques.

Fig. 7. *(facing page)* (**A**) number of test gene expressions appeared to be linearly related to the housekeeper gene GAPDH; hence adjustment of the GAPDH expression from test gene expression was required. (**B**) This was achieved by taking the ratio of each test gene with a weighted average expression corresponding to a PCA carried out on replicates of GAPDH.

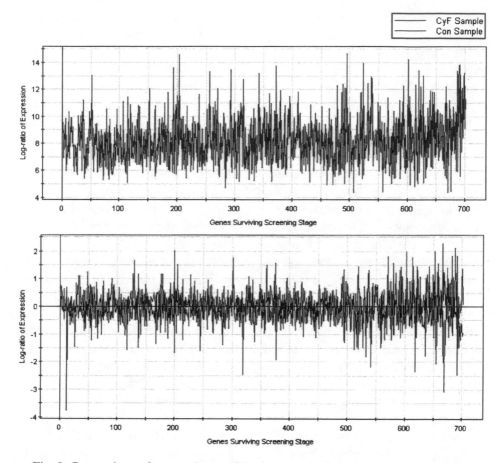

Fig. 8. Comparison of expression profiles for a cystic fibrosis microarray sample vs a control sample before (**A**) and after (**B**) mean centering. Con, control; CyF, cystic fibrosis.

The two multivariate pattern recognition methods presented in this section, HCA and PCA, are called "unsupervised" methods, i.e., the class identifications are not used in deriving these graphs. They provide valuable overviews of gene expression datasets: elucidating and confirming natural clusters without regard to assumed class membership. In the next stage of investigation, supervised methods, and in particular PLS-DA, are discussed, whereby known class membership is used to focus on the changes in gene expression between groups.

3.3. Stage 3: Separate the Vital Few Genes From the Trivial Many

In Stage 2 we showed how multivariate methods are ideally suited to the task of handling and visualizing vast amounts of genomic data. In this stage we describe how multivariate classification techniques can be used to observe groupings and separations of the samples or conditions, as well as to rapidly identify the genes governing these groupings and separations. This allows screening of important genes to carry them forward for further study, distinguishing them from the many trivial genes that are not differentially expressed above the noise of the experiment.

There are just two key steps involved in locating the key genes at this stage. First, statistics resulting from multivariate classification techniques, in particular PLS-DA, substantially facilitate the screening of test genes for further analysis. These statistics can be displayed graphically to aid biological insight, understanding, and inference, as well as selection.

Although the goal of PCA is to maximize the amount of variation explained, PLS-DA combines PCA with regression to maximize the separation between clusters of samples (**Figs. 9** and **10**) and identifies genes influential in this separation. The regression coefficients from PLS-DA help to differentiate between genes that have either been up- or downregulated as a result of a condition, disease, or treatment (**Fig. 11**).

Finally, from a PLS-DA, the variable influence on projection (VIP) provides the most condensed summary statistic of each gene's overall influence on the discrimination between all the different groups of samples. It ignores whether a gene is up- or downregulated according to disease to focus on the magnitude of the change in expression, accumulated over all PLS dimensions. (For each test, the VIP value is a weighted sum of squares of the PLS test gene weights, taking into account the amount of discrimination between the categories of samples explained by each dimension. This weighted sum of squares is divided by the total percent SS explained by the PLS model and multiplied by the number of the terms in the model.) The higher the VIP value, the stronger the overall discriminatory power of the test gene and the change in expression.

The second step in separating the genes most likely to have changed in expression involves using the statistics derived from PLS-DA and associated with the genes, such as the loadings, regression coefficients, and VIP statistics, to arrange genes according to size and similarity in pattern, function, and regulation of gene expression.

It is useful that the squared sum of all VIP values is equal to the number of test genes in the model. This suggests that one approach toward selecting a natural threshold, to distinguish important from unimportant test genes, would be to choose those genes with VIPs greater than 1 as the most relevant for explaining the discrimination between the tissues. The VIP statistic is therefore an obvious candidate to for ranking and screening genes according to their importance. **Figure 12** displays the VIP plots from the PLS-DA conducted on **Fig. 12A** the TaqMan and (**Fig. 12B)** the microarray data.

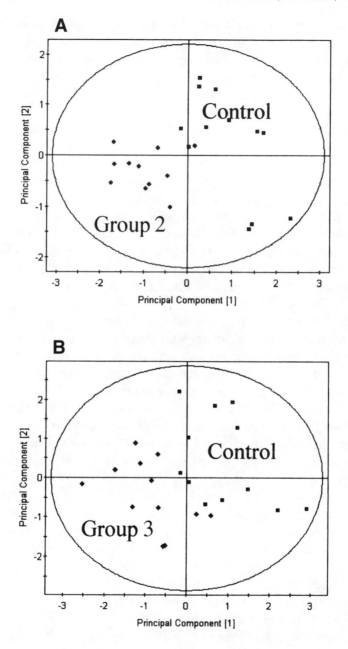

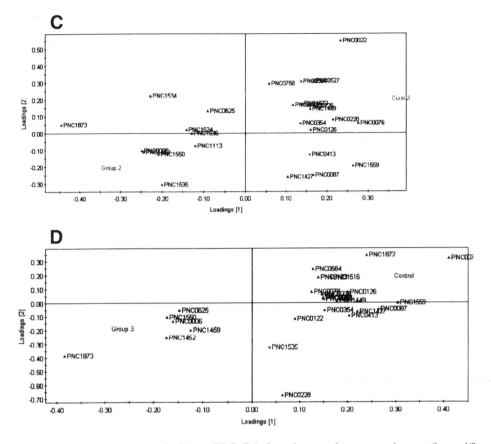

Fig. 9. *(facing page and above)* PLS-DA focusing on the comparisons of specific interest in a TaqMan example: disease groups 2 and 3, relative to the control group 1, provided good discrimination, as can be seen in the score plots (**A** and **B**, respectively). The PLS-DA loading plots (**C** and **D**) helped to "graphically" identify the test genes governing the discrimination between groups 2 and 3 compared with the control group.

In addition to the natural VIP threshold, there are alternative approaches you can adopt to use the VIP values to locate "cutoffs," which enable you to separate the important test genes from the unimportant ones with a measurable degree of confidence. In addition to classic histogram-based thresholding, a simple but effective visual tool for identifying which test genes are really likely to be discriminatory is to use a probability or quantile plots of the VIP values (**Fig. 13**).

To interpret these plots, values that are likely to come from unimportant genes will form a reference distribution of genes and plot as a straight line close to 0, whereas points falling away from the line to the right are likely to be

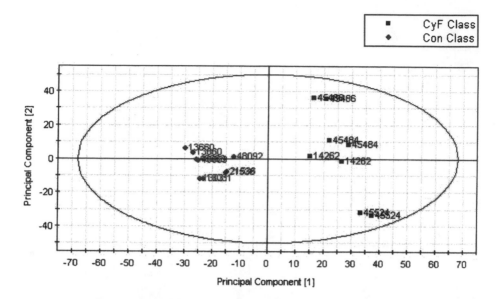

Fig. 10. PLS-DA was also used to improve the discrimination between the diseased vs control samples for the microarray data and make it easier to identify those test genes that brought about the discrimination. Con, control, CyF, cystic fibrosis.

from genes with large VIP values and therefore high discriminatory power. It is possible to estimate the optimum threshold(s) or breakpoint(s) objectively using breakpoint regression *(10)*.

In this stage we have focused on one supervised pattern recognition technique, namely, PLS-DA. Other methods such as K-nearest neighbor and linear discriminant analysis work well provided that all samples are homogeneous and that the number of genes is much smaller than the number of samples *(11)*. These constraints are clearly violated in the case of gene expression data at the stage of screening genes to identify those worthy of further carrying forward to the next stage of analysis. These methods, along with univariate analysis of variance approaches, whereby a gene is often treated as a factor in the linear model, are more readily embraced in stage 4.

It is important to note that these different techniques and others, in particular soft independent modeling of class analogies (SIMCA; clearly explained in Beebe et al. *[12]*) are complementary, not competitive. It is obviously a useful verification if the same result is obtained by the different techniques, but it is more informative if different techniques give different perspectives on the same problem, thus enhancing our biological understanding.

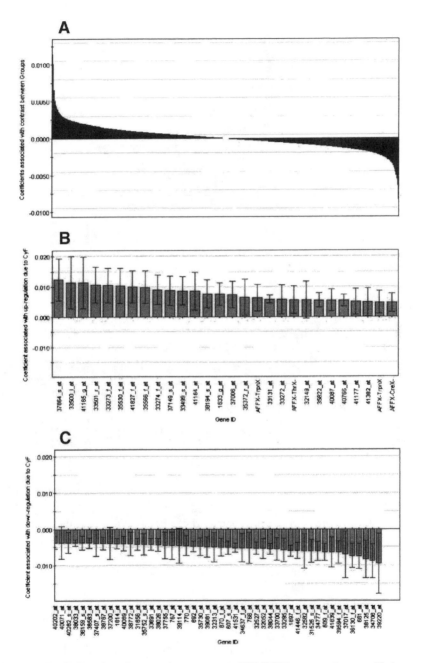

Fig. 11. (**A–C**) Patch and zoom-in on a sorted PLS-DA regression coefficients plot identifies which test genes have been up or downregulated for each category of sample. CyF, cystic fibrosis.

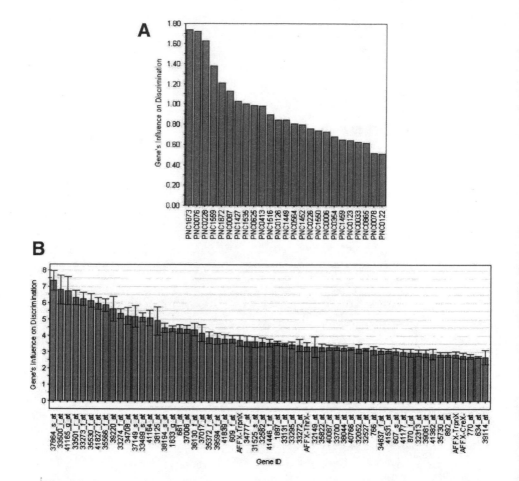

Fig. 12. Test genes carrying the greatest amount of class discriminating information and thus experiencing the greatest change in expression for (A) the TaqMan data control vs group 3 and (B) the microarray data. For the TaqMan data, genes 1873, 0076, and 0228 exhibit the strongest discriminatory influence, whereas genes 38764_s_at, 33500_i_at, 33501_r_at, and 41165_g_at show evidence of the largest changes in expression from the microarray dataset.

3.4. Stages 4 and 5: Focus, Infer, and Verify

More detailed follow-up "univariate" statistical routines, such as analysis of variance or covariance, can be used on the genes surviving stages 2 and 3 to verify their importance and to provide a risk assessment (i.e., a p-value) and/or a level of confidence associated with the estimated changes in expression.

It is important at this stage to choose the right model and analysis to suit the structure of the design you selected for the study at stage 1. The combinatorial

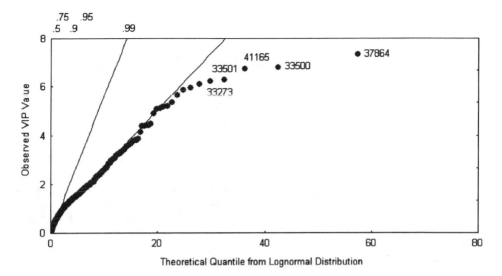

Fig. 13. In the case of the microarray data, the majority of test genes exhibiting no differential expression can be used to define a straight line of "nonexpression." There are a number of test genes with VIP values between 2 and 4 lying on a further line with a shallower slope, and there would be yet another line for those test genes with values of 4 and above. VIP, variable influence on projection.

structure of factors, both intentional "treatment" factors (such as disease, phenotype, drug, and so on) and nonintentional "block" factors (such as donor, plate, replicate, and so on) has long been analyzed using mixed models and analysis of variance (ANOVA)-type analyses. Designs and subsequent analyses can accommodate approaches that range from the simple comparison of two groups using a T statistic to a multifactorial setup resulting in an F statistic.

By matching the correct model and analysis to the design of your study, you will analyze the data more effectively, so as to extract the information contained within the data more efficiently, and draw more sound conclusions concerning your pathophysiological hypotheses.

For the TaqMan study, the objective was to identify differential expression of genes associated with the failing heart by comparing diseased with nondiseased groups. Of specific interest was the detection of changes in regulation owing to the two disease groups, 2 and 3, relative to control group 1. Separate ANOVAs were performed on data from each of the selected test genes from stage 3 to derive and display their group differences as fold differences, p-values, and 95% confidence intervals for comparisons of groups 2 and 3 back to group 1, using Dunnett's multiple comparison procedure. Together with the VIP and coefficient values resulting from the PLS-DA, the results are displayed in **Table 1A** for group 2 vs 1 and **Table 1B** for group 3 vs 1.

Table 1A and B
The Key Genes From the TaqMan Study[a]

A

Gene	VIP	Coeff	p-value	Geometric Mean Group1	Group2	Fold Difference	95% Confidence Limits Lower	Upper
PNC1873	2.127	-0.147	0.028	0.036046	0.243112	6.74	1.25	36.41
PNC0006	1.169	-0.109	0.028	0.000002	0.000005	2.89	1.13	7.39
PNC1452	1.149	-0.109	0.017	0.000856	0.001863	2.84	1.23	6.55
PNC1534	1.212	-0.035	0.176	0.000497	0.001319	2.65	0.62	11.31
PNC1550	0.990	-0.099	0.011	0.000766	0.001872	2.44	1.26	4.76
PNC1535	1.054	-0.133	0.210	0.000020	0.000046	2.35	0.59	9.34
PNC1524	0.676	-0.045	0.022	0.007131	0.013053	1.83	1.10	3.05
PNC1536	0.607	-0.045	0.065	0.000148	0.000256	1.73	0.96	3.11
PNC1113	0.567	-0.057	0.018	0.028900	0.048216	1.67	1.10	2.53
PNC0625	0.521	-0.005	0.527	0.000104	0.000153	1.47	0.43	5.03
PNC0758	0.605	0.081	0.538	0.000843	0.000658	0.78	0.34	1.77
PNC1427	0.751	-0.017	0.365	0.000540	0.000346	0.64	0.24	1.74
PNC0123	0.609	0.075	0.009	0.039745	0.024056	0.61	0.42	0.87
PNC0354	0.637	0.059	0.304	0.000195	0.000109	0.56	0.18	1.76
PNC0564	0.829	0.111	0.207	0.000145	0.000080	0.56	0.22	1.42
PNC1872	0.706	0.085	0.000	10.520623	5.781718	0.55	0.44	0.69
PNC0226	0.757	0.087	0.039	0.002630	0.001360	0.52	0.28	0.96
PNC0413	0.821	0.030	0.161	0.000007	0.000004	0.51	0.19	1.34
PNC1449	0.775	0.086	0.082	0.001064	0.000537	0.50	0.23	1.10
PNC0126	0.757	0.060	0.064	0.000201	0.000101	0.50	0.24	1.05
PNC0087	0.977	0.008	0.182	0.000336	0.000165	0.49	0.17	1.44
PNC0527	0.943	0.123	0.248	0.000004	0.000002	0.49	0.14	1.72
PNC0228	1.013	0.092	0.357	0.000028	0.000011	0.40	0.05	3.04
PNC0022	1.444	0.195	0.051	0.000004	0.000001	0.37	0.13	1.00
PNC1559	1.352	0.053	0.101	0.001108	0.000357	0.32	0.08	1.27
PNC0076	1.306	0.110	0.008	0.000518	0.000158	0.30	0.13	0.71

B

Gene	VIP	Coeff	p-value	Geometric Mean Group1	Group3	Fold Difference	95% Confidence Limits Lower	Upper
PNC1873	1.740	-0.145	0.032	0.036046	0.169260	4.696	1.152	19.133
PNC1452	0.795	-0.133	0.046	0.000656	0.001324	2.018	1.013	4.022
PNC1550	0.737	-0.050	0.026	0.000766	0.001536	2.005	1.096	3.666
PNC0006	0.726	-0.048	0.094	0.000002	0.000003	1.910	0.886	4.115
PNC0625	0.984	0.043	0.236	0.000104	0.000189	1.814	0.859	4.995
PNC1459	0.845	-0.137	0.012	0.001482	0.002455	1.656	1.128	2.430
PNC1535	1.004	-0.008	0.846	0.000020	0.000017	0.865	0.187	3.998
PNC0228	1.630	-0.170	0.812	0.000028	0.000022	0.773	0.084	7.114
PNC0122	0.510	-0.004	0.303	0.001483	0.001033	0.697	0.342	1.419
PNC0078	0.516	0.079	0.023	0.000683	0.000413	0.605	0.395	0.926
PNC0564	0.803	0.180	0.205	0.000145	0.000086	0.596	0.262	1.356
PNC0123	0.643	0.126	0.003	0.039745	0.022727	0.572	0.401	0.815
PNC0226	0.756	0.127	0.054	0.002630	0.001453	0.553	0.302	1.012
PNC0033	0.625	0.076	0.040	0.001905	0.001043	0.547	0.309	0.970
PNC0865	0.614	0.051	0.037	0.000877	0.000479	0.546	0.310	0.961
PNC0354	0.680	0.024	0.079	0.000195	0.000105	0.541	0.271	1.081
PNC1516	0.897	0.185	0.189	0.000003	0.000001	0.515	0.186	1.423
PNC1449	0.840	0.112	0.054	0.001064	0.000520	0.489	0.236	1.015
PNC0126	0.841	0.116	0.009	0.000201	0.000090	0.447	0.248	0.803
PNC0413	0.978	-0.015	0.106	0.000007	0.000003	0.441	0.161	1.208
PNC1427	1.028	-0.006	0.060	0.000540	0.000223	0.413	0.163	1.043
PNC1872	1.207	0.254	0.000	10.520623	4.011519	0.381	0.314	0.463
PNC0087	1.130	0.061	0.043	0.000336	0.000116	0.346	0.125	0.962
PNC1559	1.379	0.013	0.030	0.001108	0.000326	0.294	0.098	0.879
PNC0076	1.723	0.282	0.005	0.000518	0.000099	0.191	0.063	0.579

The statistics accompanying each gene help to determine the practical as well as statistical significance of the contrasts between the groups for each gene and to highlight or confirm which genes to target for future investment. Of particular interest here is the upregulated gene 1873, which has provided a good contrast between both groups 2 and 3 vs control group 1.

In the second microarray study, CyF-extracted RNA tertiary bronchial samples were compared with six nonsmoking Con, and the statistics associated with the contrast between the two groups for each gene are displayed in **Table 2**.

A great deal has already been written on the univariate analysis techniques briefly touched on in this section. In fact, these techniques are typically used in place of the multivariate methods proposed in stage 3, to provide multi-univariate tests either for the change in expression separately for each and every test gene, or—by including gene as a factor in the analysis—for their interaction (i.e., dependency) with a change in treatment or disease condition, as well as any other factors accounted for in the study design. Statistical inferences are then made about excursions of the test statistics, probability values, or interval estimates above a specified threshold using either distributional assumptions or not (i.e., a nonparametric approach).

4. Discussion

The approach outlined in this chapter is first to employ multivariate methods to provide a better overview of the patterns in the data and to help filter or screen out the unimportant genes, and only then to use common univariate statistical methods on those genes surviving the initial screen. This alleviates the multiplicity problem (for review, *see* **ref. *13***) testing an extremely large number of contrasts while having to maintain strong control over the experiment-wise type I error by applying a possibly untenable correction to the large number of multiple comparisons.

Although it is true that some statistical approaches are better than others in a particular context, it is nearly always the case that many different approaches

Table 1 *(continued from page 306)* [a]Genes were targeted after stages 1–3 as being worthy of further study and are listed together with their influence on the contrast between groups 2 and 3 vs group 1, type of regulation, p-value (i.e., risk assessment), fold difference (i.e., ratio of expression for group 1 or 2 relative to control group 1), and interval providing a degree of confidence in the estimated change in expression. Genes with results shaded gray were upregulated (i.e., fold difference > 1), whereas unshaded genes were downregulated. A number of genes generated differences that were statistically significantly greater than the noise at a 5% level of significance (% risk of a false-positive result). In this respect the columns of p-values provide a likelihood of falsely classifying a gene as important and worthy of further study, taking into consideration the variation of its own results.

Table 2
Genes Carried Forward to Final Stage of Analysis for the Microarray Study[a]

Gene	Fold Difference	p-value	95% Confidence Limits Lower	Upper
37864_s_at	6.610	4.76E-06	3.700	11.810
33500_i_at	5.739	0.0004	2.531	13.016
41165_g_at	5.645	4.52E-05	2.942	10.832
33501_r_at	5.056	7.66E-06	3.009	8.494
33273_f_at	4.948	3.66E-06	3.061	7.999
35530_f_at	4.821	5.12E-05	2.648	8.777
41827_f_at	4.619	4.33E-06	2.897	7.363
35566_f_at	4.516	0.0002	2.324	8.775
33274_f_at	3.952	2.60E-05	2.418	6.458
37149_s_at	3.780	0.0032	1.684	8.488
33499_s_at	3.727	5.80E-05	2.244	6.190
41164_at	3.688	1.38E-05	2.374	5.728
38194_s_at	3.173	0.0001	1.964	5.124
1633_g_at	3.121	5.75E-07	2.327	4.185
37006_at	3.095	0.0012	1.691	5.664
39220_at	0.234	0.0005	0.117	0.469
34708_at	0.262	0.0050	0.110	0.625
38125_at	0.284	0.0405	0.086	0.940
661_at	0.321	0.0003	0.190	0.543
35130_f_at	0.325	0.0050	0.157	0.674
37017_at	0.344	0.0159	0.149	0.794

[a]Genes are presented together with the resulting fold difference, test of significance, and 95% confidence limits for estimated fold difference

apply equally well. This is particularly true of gene expression methods and the analyses of their results at the present stage of development, and one can perhaps draw parallels with the dawn of statistical methods applied to the analysis of functional neuroimages. As with the statistical assessment of functional neuroimages, until a unifying statistical approach to the treatment of gene expression data is made available, one may best adopt a pragmatic and pluralistic attitude toward explaining gene expression data. Therefore, statistics resulting from both the multivariate and univariate data analysis steps are used to evaluate the importance of each gene. (By considering data from all genes simultaneously, statistics resulting from a multivariate data analysis, such as the regression coefficients and VIP values from PLS-DA, are likely to be more reliable measurements of a gene's regulation or expression, rather than the statistics resulting from a single gene, which are strongly influenced by the variation of their own results.)

An additional benefit of performing initial multivariate data analyses is that you can also make use of the results generated from this step to aid the subsequent univariate analyses. For example, by using a PCA on the 78 replicates of the housekeeping gene from the TaqMan study, a more stable "weighted average" estimate of the reference gene was established, and potential bias from just a single (and possibly unreliable) realization of the reference gene was removed.

For the TaqMan study, adjustment (normalization) according to the housekeeping gene had been made prior to the analysis. However, adjustment for the reference gene can also be made during the analysis by treating it as a covariate. An analysis of covariance (ANCOVA) compares the adjusted means for each group against the variability of individual samples relative to any linear relationship that may exist between expression levels of the test genes with the housekeeping gene. If there is a linear relationship, then an ANCOVA makes for a far more appropriate test to assess whether the groups are different, as the comparisons will be adjusted for any relationship between the test and housekeeping gene or any other factor confounded with it (for review, *see* **ref. *14***).

Inclusion of the housekeeping gene as a covariate may well help to explain some of the variability in the measured response and hence will decrease the residual variation and in turn increase precision. It is, however, assumed that the treatments under investigation do not affect the covariate; a measure of this, and therefore of the effectiveness of the housekeeping, is given by the covariance efficiency factor (CEF).

For example, a principal component analysis (PCA) performed on the GAPDH data of the TaqMan study confirmed that replicates of the reference gene generated similar expression. The first principal component therefore provided a weighted average expression of the reference gene that could be used in any further analysis as representative of GAPDH.

Figure 14 provides ANCOVA output from the Excel add-on Cellula for gene 1873, identified as an important gene by PLS-DA. The scatter plot with regression lines describing the relationship between the identified test gene and weighted average expression for groups 1 and 2 is provided. The accompanying statistical summary provides no graphical or statistical evidence that the slopes for the two groups are different (i.e., they are parallel and the differences between the two groups do not appear to depend on GAPDH expression). Also, there does not appear to be a strong linear relationship between gene 1873 and GAPDH expression, as shown by the small F ratio for the covariate (GAPDH) term in the ANCOVA table and the resulting p-value (>0.1).

The group term in the ANCOVA table informs us that the two groups are significantly different, adjusted for the covariate (weighted average of GAPDH from PCA) having an effect. The CEF for group measures the degree of independence between the treatments under study and the covariate. Values close to zero indicate that the covariate has indeed been affected by treatments, and

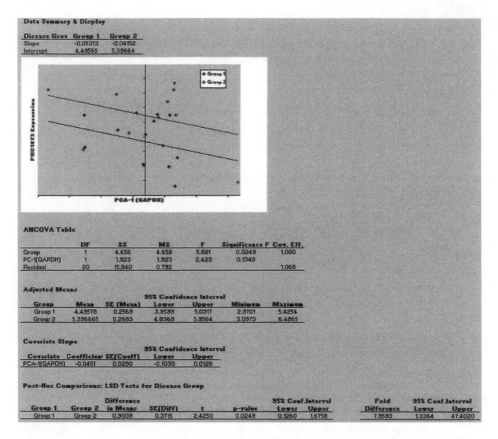

Fig. 14. ANCOVA output for gene 1873 adjusted for the covariate (housekeeping gene GAPDH).

hence use of the housekeeping as a covariate is questionable. When equal to unity (the maximum value), as is the case here, the CEF indicates total independence between treatments and the covariate.

A listing is also given of the means and the difference between means for the two groups adjusted for the housekeeping gene and the "Covariate Slope" together with standard errors and 95% confidence intervals on the both logarithmic and backtransformed to the original scale to reflect fold differences.

The field of genomics has come a long way in a relatively short period. Developments in statistical approaches have made a significant contribution, providing methods of design, preprocessing, visualization, and analysis. The underlying technology and associated methodology are still in the development stage in terms of the interrogation and interpretation of genomic data.

References

1. Fisher, R. A. (1925) *Statistical Methods for Research Workers*. Oliver & Boyd, Edinburgh.
2. Fisher, R. A. (1926) The arrangement of field experiments. *J. Minis. Agric.* **33,** 503–513.
3. Yates, F. (1937) *The Design and Analysis of Factorial Experiments. Technical Communication No. 35*. Imperial Bureau of Soil Science, Harpenden, Hertfordshire, UK.
4. Eisen, M. B., Spellman, P. T., Brown, P. O., and Botstein, D. (1998) Cluster analysis and display of genome-wide expression patterns. *Proc. Natl. Acad. Sci. USA* **95,** 14863–14868.
5. Jackson, J. E. (1980) Principal components and factor analysis: part I – principal components. *J. Qual. Technol.* **12,** 201–213.
6. Wold, S., Albano, C., Dunn, W. J., et al. (1984) Multivariate data analysis in chemistry, in: *Chemometrics: Mathematics and Statistics in Chemistry* (Kowalski, B. R., ed.), D. Reidel, Dordrecht.
7. Smyth, G. K. and Speed, T. (2003) Normalization of cDNA microarray data. *Methods* **31,** 265–273.
8. Lin, Y., Nadler, S. T., Attie, A. D., and Yandell, B. S. (2001) Mining for low-abundance transcripts in microarray data. Department of Statistics Technical Report #1031, University of Wisconsin, Madison, WI.
9. Dudoit, S., Yang, Y. H., Callow, M. J., and Speed, T. P. (2002) Statistical methods for identifying differentially expressed genes in replicated cDNA microarray experiments. *Stat. Sin.* **12,** 111–140.
10. Draper, N. and Smith, H. (1981) *Applied Regression Analysis*, 2nd ed. Wiley, New York.
11. Albano, C., Dunn, W. J. III, Edlund, U., et al. (1978) Four levels of pattern recognition. *Anal. Chim. Acta* **103,** 429–443.
12. Beebe, K. R., Pell, R. J., and Seasholtz, M. B. (1998) *Chemometrics: A Practical Guide*. Wiley, New York.
13. Hsu, J. C. Multiple Comparisons. Chapman and Hall, London.
14. Wetherill, G. B. *Intermediate Statistical Methods* (1981) Chapman and Hall, London, UK.

14

Bioinformatic Approaches to Assigning Protein Function From Novel Sequence Data

David Michalovich and Richard Fagan

Summary

The current pace of functional genomic initiatives and genome sequencing projects has provided researchers with a bewildering array of sequence and biological data to analyze. The disease system-driven approach to identifying key genes frequently identifies nucleotide and protein sequences for which the gene and protein function are not known in sufficient detail to allow informed follow-up. Using a range of bioinformatic tools and sequence-based clues, most of unassigned sequences can now be annotated. This chapter takes as an example an unannotated expressed sequence tag, describing how to identify its related gene, and how to annotate the encoded protein using sequence, profile, and structure-based annotation methodologies.

Key Words

Bioinformatics; genome; gene; protein; threading; transcript; kinase.

1. Introduction

Large-scale cDNA and genome sequencing projects have redefined how we identify novel genes and their encoded proteins. Combining the exponential growth of biological sequence data with high-throughput experimental data from genetic, RNA expression, and proteomic screens provides a vast array of data to analyze and mine. Computational analysis of these high-throughput data sources is therefore clearly required and has established bioinformatics as an essential component of industrial and academic genomic research projects.

A feature of large-scale experimental screens is that they often identify biologically interesting sequence targets for which the associated protein and its biochemical function is unknown. This chapter describes how bioinformatics

From: *Methods in Molecular Medicine, Vol. 104: Stroke Genomics: Methods and Reviews*
Edited by: S. J. Read and D. Virley © Humana Press Inc., Totowa, NJ

tools can be applied to the identification, characterization, and annotation of high-throughput sequence information. The chapter is intended to provide an overview of the applications, tools, and methodologies required to identify and characterize novel sequences at both the DNA and protein level. For this purpose we annotate an expressed sequence tag (EST) through to its associated gene and encoded protein sequence. We then describe how that protein can be annotated using sequence, profile, and structure-based bioinformatic annotation tools.

One common route to characterize a previously unannotated sequence is to use computational and statistical approaches to allow annotation transfer between defined sequences or structures and the novel query sequence. To achieve this end, two essential components are required, namely, a well-maintained and annotated source database and computational methods to search and assign similarities from that database.

1.1. Data Resources

The primary source databases for nucleotide sequences are held at three major collaborative centers: The European Molecular Biology Laboratory (EMBL) data library located at the European Bioinformatics Institute (EBI), GenBank®, the National Institutes of Health (NIH) database located at the National Center for Biotechnology Information (NCBI), and the DNA Data Bank of Japan (DDBJ). The centers make up the International Nucleotide Sequence Database Collaboration, which feeds the protein databases such as SwissProt/ TrEMBL *(1)* and the Protein Information Resource (PIR) *(2)*. The information is shared freely among centers, nominally negating the need to query each database individually. The database centers also provide retrieval services; these allow users to query the databases via text searches using user interfaces such as NCBI's Entrez system or the EBI's Sequence Retrieval System (SRS).

Sitting logically on top of the primary sequence databases, a set of secondary annotation databases and resources exist with precalculated analysis of genome assemblies, gene structure, and protein annotation. Applicable to this chapter are the genome annotation projects such as UCSC's Golden Path *(3)* and the EBI's ENSEMBL genome annotation resource *(4)*. Both resources annotate human genome assemblies and those of other species such as mouse and rat with known and predicted genes.

For protein annotation a variety of multiple alignment, protein sequence patterns, and regular expression databases exist. They can be used to aid function annotation through the grouping of proteins into evolutionary families. PFAM *(5)* and SMART *(6)* are both multiple alignment-based databases, whereas PRINTS *(7)* uses sets of smaller motif alignments or "fingerprints" to annotate novel proteins. These secondary annotation systems have been combined into INTERPRO *(8)*, which provides a valuable integrated overview of a protein's features.

1.2. Bioinformatic Search Tools and Websites

The most commonly used sequence query tool is the Basic Local Alignment Search Tool (BLAST) program. This is based on a heuristic sequence comparison algorithm used to search sequence databases for optimal local alignments to a query *(9,10)*. The BLAST suite of programs supports a number of different query options: DNA sequence query to DNA database (BLASTN), protein sequence to protein database (BLASTP), protein to six-frame translated DNA database (TBLASTN), and translated DNA sequence against a translated DNA sequence database (TBLASTX). BLAST can be downloaded for local use or can be run at a number of public web sites such as the NCBI or DDBJ.

A statistical scoring system exists within BLAST. This is known as the E-value and is based on the probability of finding an exact match of the query in the database by chance. As a rough yardstick, an E-value of 10–3 is a useful cutoff for describing significant matches between soluble proteins.

Functional annotation can be reliably transferred between proteins that share 30% or more sequence identity, confidence increasing with identity. For matches below 30% identity, more sophisticated tools are required. These use conserved features of a protein family to produce a statistical profile of the sequence and/or use structural features of the family to link annotation.

A hybrid of BLAST is Position-Specific Iterated Blast (PSI-BLAST) *(10)*. This combines the speed of the BLAST alignment algorithm with the advantages of searching with a sequence profile. On the first iteration of the PSI-BLAST program, a normal BLASTP-type search is carried out. Sequences, which are found within a predetermined significance threshold, are used to build a profile—an empirical description of the amino acid residues found in homologs at each point of the query sequence. On subsequent iterations, the profile is used to search the database, and the profile is refined based on the new matches identified. This method has the advantage that more distant sequence relationships can be found.

A major consideration when using PSI-BLAST is the masking of the query sequence prior to a PSI-BLAST search. Owing to the iterative process of PSI-BLAST, regions of low amino acid sequence complexity, biased composition, and repetitive subsequences can result in false matches being brought into the PSI-BLAST profile. This then corrupts future iterations of the search, and false sequence relationships are formed. Notable regions of low complexity or redundancy are coiled-coil domains and transmembrane spanning regions.

A variation of the PSI-BLAST algorithm has been employed in the NCBI's Conserved Domain Database (CDD); this is called reverse-position-specific

BLAST (RPS-BLAST) *(11)*. In this tool PFAM and SMART multiple sequence alignments are use to build a PSI-BLAST profile known as a Position-Specific Score Matrix (PSSM). A query sequence can be searched against the precalculated PSSMs, which are stored in the database *(11)*.

Profile-based search tools need accurate alignments (whether these are generated automatically or by hand) to be successful. Since the 3D structure of a protein is more strongly conserved than the primary sequence, using the 3D structure information in building multiple alignments leads to the building of better profiles and hence provides the best discriminators of homology *(12,13)*. This type of approach is embodied in algorithms such as 3D-PSSM and FUGUE, which can extend the identification of homologous relationships beyond the scope of a PSI-BLAST-based search *(14,15)* using structurally defined alignments and search methods.

Finally, structure can also be used as a template on which protein sequences can be matched and validated. This fitting or "threading" of a protein sequence onto structure often uses empirically defined energy potentials derived from alignment of sequences and structures. The first algorithm to employ this technique was THREADER *(16)*. THREADER can try to predict a fold of a protein even when there is little or no sequence identity, using a large library of protein folds as its database. However, there is no assessment of significance of the match, requiring careful expert interpretation of the data. Practically, the program is slow and compute intensive. A revision of the original THREADER algorithm is called GenThreader. GenThreader has changed significantly from THREADER and incorporates aspects of profile-based searching in conjunction with structural alignments *(17)*. When it was applied to annotation of the *Mycoplasma genitalium* genome, the authors were able to assign a 3D structure to at least one domain of 46% of the open reading frames (ORFs) *(18)*.

2. Materials

Most of the tools described here are available on the Internet for free public use. For more sophisticated analysis, a unix-based computer system tends to be the workstation of choice. This can greatly enhance the depth of bioinformatic analysis that can be achieved, as many of the tools have been developed for implementation on this system. A good starting guide to both bioinformatic tools and setting up a bioinformatic unix workstation can be found in *Developing Bioinformatics Computer Skills (19)* .

A list of websites for bioinformatics databases and tools mentioned and used in the chapter is given in **Table 1**.

3. Methods

The methods described below detail the process of mapping an unannotated EST sequence to the genome, assigning the relevant protein sequence to the

Table 1
URLs for Databases and Bioinformatic Resources

Resource	URL
Biological sequence and structure repositories	
EMBL	http://www.ebi.ac.uk/Databases/index.html
EMBL SRS	http://srs6.ebi.ac.uk/
GenBank®	http://www.ncbi.nlm.nih.gov/GenBank/index.html
NCBI Entrez	http://www.ncbi.nlm.nih.gov/Entrez/
DDBJ	http://www.ddbj.nig.ac.jp/
SwissProt	http://www.ebi.ac.uk/swissprot/
	http://www.expasy.ch/sprot/sprot-top.html
TrEMBL	http://www.ebi.ac.uk/trembl/index.html
PIR-PSD	http://pir.georgetown.edu/
PDB	http://www.rcsb.org
Genome annotation projects	
UCSC Genome Browser	http://genome.ucsc.edu/cgi-bin/hgGateway
UCSC BLAT interface	http://genome.ucsc.edu/cgi-bin/hgBlat
ENSEMBL	http://www.ensembl.org/Homo_sapiens/
Protein annotation resources	
PFAM	http://www.sanger.ac.uk/Software/Pfam/
SMART	http://smart.embl-heidelberg.de/
PRINTS	http://www.bioinf.man.ac.uk/dbbrowser/PRINTS/
PROSITE	http://us.expasy.org/prosite/
PRODOM	http://prodes.toulouse.inra.fr/prodom/current/html/
home.php	
Search tools	
BLAST and PSI-BLAST	http://www.ncbi.nlm.nih.gov/BLAST/
NCBI CDD	http://www.ncbi.nlm.nih.gov/Structure/cdd/cdd.shtml
3D-PSSM	http://www.sbg.bio.ic.ac.uk/~3dpssm/
FUGUE	http://www-cryst.bioc.cam.ac.uk/~fugue/
Commercial resources	
Biopendium™	http://www.inpharmatica.co.uk

EST (*see* **Subheading 3.1.**), and then annotating that protein using sequence, profile, and structure-based annotation tools (*see* **Subheading 3.2.**).

3.1. Mapping an EST Onto the Human Genome

Human EST AW771894 originates from a kidney cDNA and has not been associated with a protein-coding region. The GenBank entry for this EST can be viewed via the Entrez retrieval system at NCBI (**Fig. 1**). The GenBank entry provides further information as to the origins of the sequence, cloning infor-

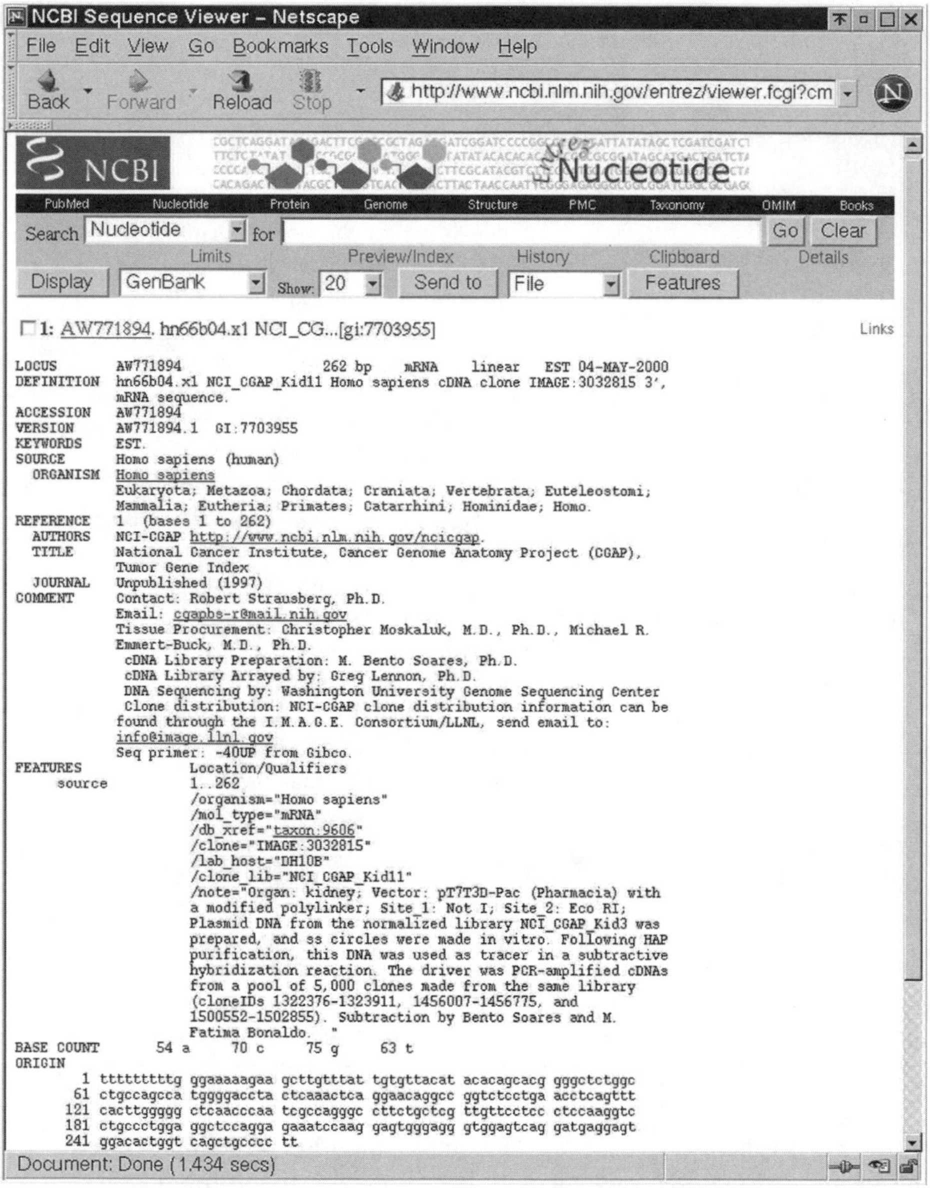

Fig. 1. Genbank Entry for EST AW771894 viewed in the NCBI Entrez retrieval system.

mation, and whether the EST is derived from the 5' or 3' end of the cDNA clone (*see* **Note 1**). In this case the EST is a 3'end read and will therefore be in reverse

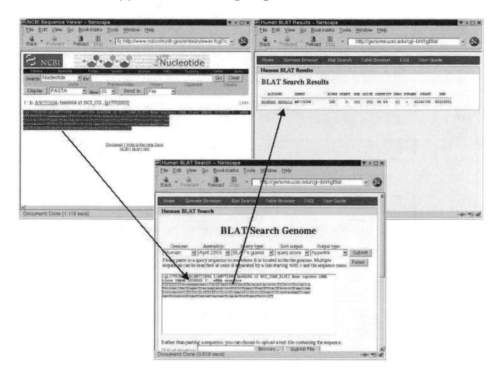

Fig. 2. Querying the human genome with an EST sequence using BLAT.

complement orientation with respect to the original cDNA clone.

There are two excellent resources that can be used to map the EST on to the human genome, ENSEMBL Genome Browser and UCSC Genome Browser. In this example the UCSC browser will be used.

1. The browser can be queried directly with a DNA or protein sequence using a BLAT search *(20)* (*see* **Note 2**).
2. From the NCBI Entrez interface, the FASTA format is selected from the display options.
3. The FASTA-formatted sequence of AW771894 is copied into the UCSC BLAT interface (**Fig. 2**), selecting the latest human genome assembly to search against.
4. Submitting the query returns the search results page, providing further links to the browser itself, and a details page, which provides alignment information. The extent of the match, percentage identity, chromosomal region, and DNA strand are also given on the results page.
5. The details page (**Fig. 3**) provides the opportunity to check the quality of the match by eye, showing the alignment of the query sequence to the genome plus any mismatches (*see* **Notes 3** and **4**).

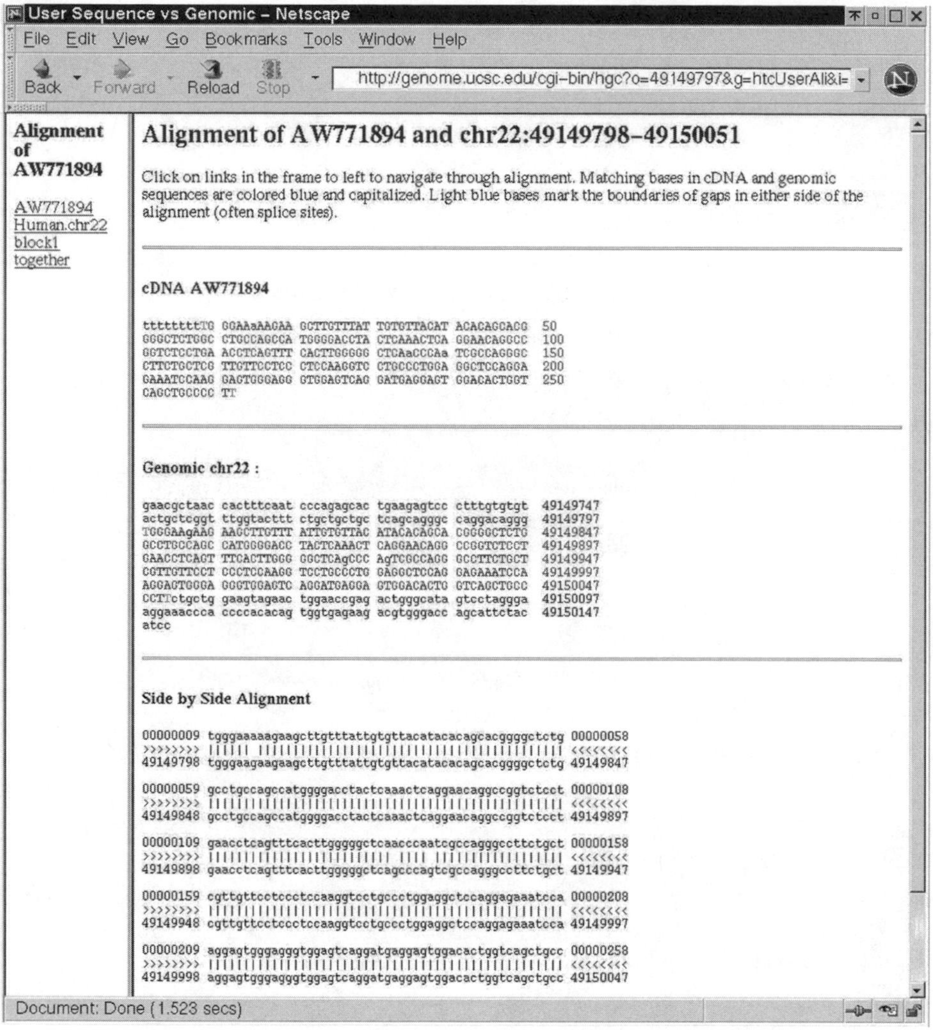

User Sequence vs Genomic – Netscape

File   Edit   View   Go   Bookmarks   Tools   Window   Help

Back   Forward   Reload   Stop   http://genome.ucsc.edu/cgi–bin/hgc?o=49149797&g=htcUserAli&i=

Alignment
of
AW771894

AW771894
Human.chr22
block1
together

Alignment of AW771894 and chr22:49149798–49150051

Click on links in the frame to left to navigate through alignment. Matching bases in cDNA and genomic
sequences are colored blue and capitalized. Light blue bases mark the boundaries of gaps in either side of the
alignment (often splice sites).

cDNA AW771894

```
tttttttTG GGAAaAAGAA GCTTCTTTAT TGTCTTACAT ACACAGCACG  50
GGGCTCTGGC CTGCCAGCCA TGGGGACCTA CTCAAACTCA GGAACAGGCC  100
GGTCTCCTGA ACGTCAGTTT CACTTGGGGG CTCAaCCCAa TCGCCAGGGC  150
CTTCTGCTCG TTGTTCCTCC CTCCAAGGTC CTGCCCTGGA GGCTCCAGGA  200
GAAATCCAAG GAGTGGGAGG GTGGAGTCAG GATGAGGAGT GGACACTGGT  250
CAGCTGCCCC TT
```

Genomic chr22 :

```
gaacgctaac cactttcaat cccagagcac tgaagagtcc ctttgtgtgt  49149747
actgctcggt ttggtacttt ctgctgctgc tcagcagggc caggacaggg  49149797
TGGGAAgAAG AAGCTTGTTT ATTGTGTTAC ATACACAGCA CGGGGCTCTG  49149847
GCCTGCCAGC CATGGGGACC TACTCAAACT CAGGAACAGG CCGGTCTCCT  49149897
CAACCTCAGT TTCACTTGGG GGCTCAgCCC AgTCGCCAGG GCCTTCTGCT  49149947
CGTTGTTCCT CCCTCCAAGG TCCTGCCCTG GAGGCTCCAG GAGAAATCCA  49149997
AGGAGTGGGA GGGTGGAGTC AGGATGAGGA GTGGACACTG GTCAGCTGCC  49150047
CCTTctgctg gaagtagaac tggaaccgag actgggcata gtcctaggga  49150097
aggaaaccca ccccacacag tggtgagaag acgtgggacc agcattctac  49150147
atcc
```

Side by Side Alignment

```
00000009 tgggaaaaagaagcttgtttattgtgttacatacacagcacggggctctg 00000058
>>>>>>>> |||||| |||||||||||||||||||||||||||||||||||||| <<<<<<<<
49149798 tgggaagaagaagcttgtttattgtgttacatacacagcacggggctctg 49149847

00000059 gcctgccagccatggggacctactcaaactcaggaacaggccggtctcct 00000108
>>>>>>>> |||||||||||||||||||||||||||||||||||||||||||||| <<<<<<<<
49149848 gcctgccagccatggggacctactcaaactcaggaacaggccggtctcct 49149897

00000109 gaacctcagtttcacttggggggctcaacccaatcgccagggccttctgct 00000158
>>>>>>>> ||||||||||||||||||||| ||| ||||||||||||||||||| <<<<<<<<
49149898 gaacctcagtttcacttgggggctcagcccagtcgccagggccttctgct 49149947

00000159 cgttgttcctccctccaaggtcctgccctggaggctccaggagaaatcca 00000208
>>>>>>>> |||||||||||||||||||||||||||||||||||||||||||||| <<<<<<<<
49149948 cgttgttcctccctccaaggtcctgccctggaggctccaggagaaatcca 49149997

00000209 aggagtgggagggtggagtcaggatgaggagtggacactggtcagctgcc 00000258
>>>>>>>> |||||||||||||||||||||||||||||||||||||||||||||| <<<<<<<<
49149998 aggagtgggagggtggagtcaggatgaggagtggacactggtcagctgcc 49150047
```

Document: Done (1.523 secs)

Fig. 3. Alignment of EST query sequence AW771894 to the human genome via the
UCSC BLAT interface.

6. In this case the EST is 262 bp in length; the match excludes the first 9 bp, as this
    is part of the polyA tail (*see* **Note 4**). The match is 98.9% identical over 262 bp.
    Allowing for possible sequence errors in the EST, this is the level of identity
    required for a one-to-one match. Equally, the EST is only matching one region of
    the genome. Therefore confidence that this EST is correctly mapped is high.
7. Moving back to the results page and selecting the browser option (**Fig. 4**) allows
    the query sequence to be viewed visually in the context of other genomic fea-
    tures. The initial view is limited to just the region of the genome that matches
    the query (**Fig. 4A**).

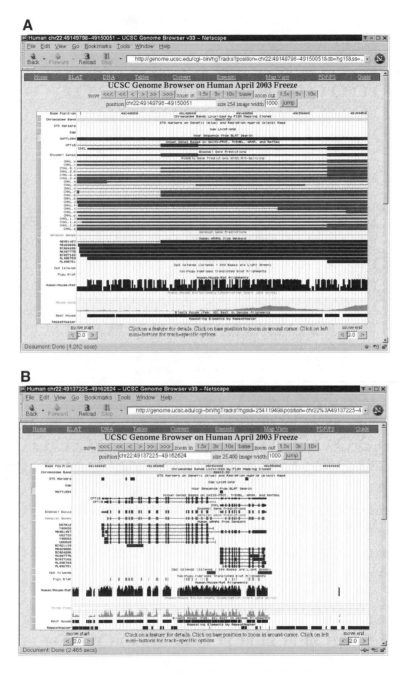

Fig. 4. (**A**) Genome browser view of EST AW771894. (**B**) The same match zoomed out 100X to show full gene content.

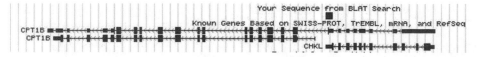

Fig. 5. Zoomed-in view of UCSC genome browser showing context of EST query with respect to CPT1B and CHKL. Note that both CPT1B and CHKL are on the reverse strand, as denoted by chevrons pointing right to left. The coding exons are displayed taller than 5' and 3'noncoding exons.

8. The UCSC browser provides the option to expand the view of the matching region; in **Fig. 4B** the region is expanded 100× from 254 to 25.4 kb. Expanding the view allows the context of the match with respect to other gene features to be assessed. The different tracks display a range of genomic features such as known genes, EST matches, results from gene prediction algorithms, regions of identity to the mouse genome, repetitive elements, and so on. These can be turned on and off as required.
9. In this example, the genomic context is complex; two genes are located on the minus strand, Choline ethanolamine kinase (CHKL) and Carnitine palmitoyltransferase 1B isoform a (CPT1B).
10. Unusually the 5' noncoding exons of CPT1B overlap most of the coding exons of CHKL, although the exact splicing pattern is different (**Fig. 5**). The EST AW771894 overlaps the terminal 3' exon of CHKL.
11. Since the EST AW771894 contains a polyA signal and part of a polyA tail, it represents the 3'end read of the CHKL gene as opposed to being a splicing variant of CPT1B. Therefore, using contextual information from both the EST and genome browser, EST AW771894 can be assigned to the gene CHKL.
12. The protein sequence for CHKL can be obtained by clicking on the gene model for CHKL in the browser and then following the SwissProt/TrEMBL link from the sequence report page.

3.2. Annotating Protein Using Sequence, Profile, and Structure-Based Tools

A number of methods exist in the public domain that can be used to put biochemical annotation onto a protein sequence of interest. The methods used here exemplify a subset of those that are available and give a good range in magnitude of capabilities.

Obtain the protein sequence of interest, in this case by following the link to the SwissProt/TrEMBL database, which takes one to the record for CHKL, otherwise known as KICE_Human or choline/ethanolamine kinase (**Fig. 6**). The SwissProt record for this sequence already gives us a clue as to the biochemical function of this protein. It catalyzes the ATP-dependent phosphorylation of choline, the first committed step in the CDP-choline pathway for the

General information	
Entry name	**KICE_HUMAN**
Accession number	Q9Y259, O13388
Created	Rel. 39, 30-MAY-2000
Sequence update	Rel. 39, 30-MAY-2000
Annotation update	Rel. 42, 15-SEP-2003

Description and origin of the Protein	
Description	Choline/ethanolamine kinase [Includes: Choline kinase (EC 2.7.1.32) (CK); Ethanolamine kinase (EC 2.7.1.82) (EK)].
Gene name(s)	CHKL OR CHETK.
Organism source	Homo sapiens (Human).
Taxonomy	Eukaryota; Metazoa; Chordata; Craniata; Vertebrata; Euteleostomi; Mammalia; Eutheria; Primates; Catarrhini; Hominidae; Homo.
NCBI TaxID	9606

References		
	[1]	Yamazaki,N., **Human gene for choline/ethanolamine kinase.Submitted JUL-1999 to the EMBL GenBank DDBJ databases**
		Position SEQUENCE FROM N.A.
	[2]	Smink,L.J., Huckle,E.J., **Submitted JUL-1999 to the EMBL GenBank DDBJ databases**
		Position SEQUENCE FROM N.A.
	[3]	Adams,M.D., **Submitted JUN-1996 to the EMBL GenBank DDBJ databases**
		Position SEQUENCE FROM N.A.

Comments	
CATALYTIC ACTIVITY	ATP + CHOLINE = ADP + O-PHOSPHOCHOLINE.
CATALYTIC ACTIVITY	ATP + ETHANOLAMINE = ADP + O- PHOSPHOETHANOLAMINE.
SIMILARITY	BELONGS TO THE CHOLINE/ETHANOLAMINE KINASE FAMILY.

Fig. 6. SwissProt record for Q9Y259, choline kinase.

biosynthesis of phosphatidylcholine. Although the name suggests that these proteins function as phosphotransfer enzymes, they bear no sequence similarity to the family of eukaryotic protein serine/threonine and tyrosine kinases. This is not unexpected, since the former phosphorylate small molecule substrates whereas the latter phosphorylate protein substrates. The structural determinants of the phosphorylation reaction catalyzed by CHKL remained elusive until recently, when the first crystal structure of a choline kinase was solved *(21)*, revealing a structure remarkably similar to those of eukaryotic protein kinases and aminoglycoside phosphotransferases (**Fig. 7**). Although this particular protein's biochemical function is well known, we will show how the use of profile and structure-based bioinformatics could have identified the distant structural relationship of the choline/etanolamine kinases with the eukaryotic protein kinases, enabling identification of key functional residues and giving the most comprehensive annotation possible.

1. Obtain the sequence in a FASTA format from the sequence record, and go to the EBI.
2. At the EBI under the services section, go to InterPro *(8)*.
3. At InterPro go to the sequence search, where you can paste the sequence and initiate an InterPro scan.

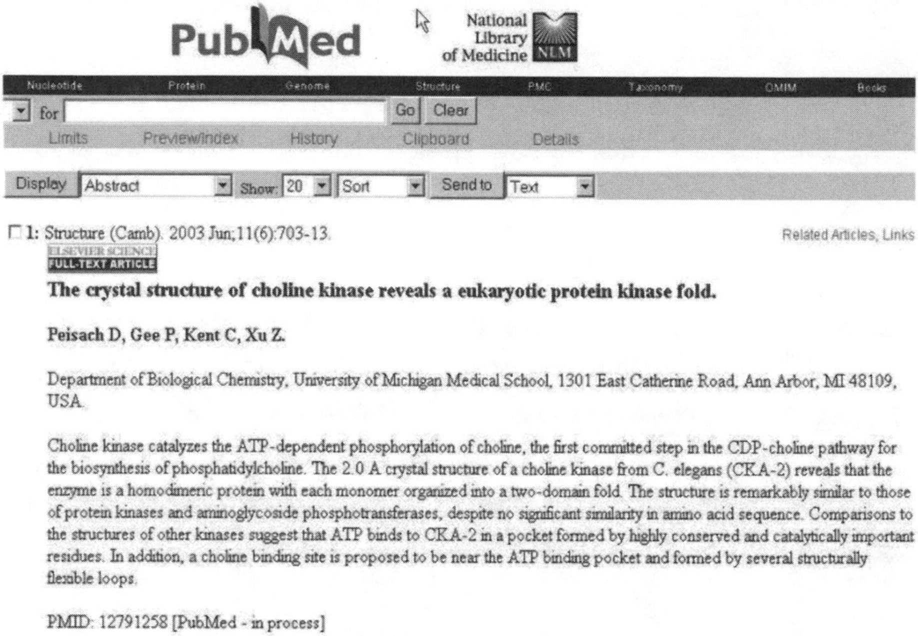

Pub**M**ed   National ▨
                      Library
                      of Medicine **NLM**

| Nucleotide | Protein | Genome | Structure | PMC | Taxonomy | OMIM | Books |

▾ for [_____] Go Clear
   Limits   Preview/Index   History   Clipboard   Details

Display [Abstract ▾] Show: [20 ▾] [Sort ▾] Send to [Text ▾]

☐ 1: Structure (Camb). 2003 Jun;11(6):703-13.   Related Articles, Links
ELSEVIER SCIENCE
FULL-TEXT ARTICLE

The crystal structure of choline kinase reveals a eukaryotic protein kinase fold.

Peisach D, Gee P, Kent C, Xu Z.

Department of Biological Chemistry, University of Michigan Medical School, 1301 East Catherine Road, Ann Arbor, MI 48109, USA.

Choline kinase catalyzes the ATP-dependent phosphorylation of choline, the first committed step in the CDP-choline pathway for the biosynthesis of phosphatidylcholine. The 2.0 Å crystal structure of a choline kinase from C. elegans (CKA-2) reveals that the enzyme is a homodimeric protein with each monomer organized into a two-domain fold. The structure is remarkably similar to those of protein kinases and aminoglycoside phosphotransferases, despite no significant similarity in amino acid sequence. Comparisons to the structures of other kinases suggest that ATP binds to CKA-2 in a pocket formed by highly conserved and catalytically important residues. In addition, a choline binding site is proposed to be near the ATP binding pocket and formed by several structurally flexible loops.

PMID: 12791258 [PubMed - in process]

Fig. 7. PubMed abstract for paper reporting the crystal structure of choline kinase.

This is an excellent resource that uses several methods to annotate a sequence. InterPro Scan uses the Prosite *(22)*, Prints *(7)*, PFAM *(5)*, ProDom *(23)*, and SMART *(6)* databases. Diagnostically, these resources have different areas of optimum application owing to the different underlying analysis methods. In terms of family coverage, the protein signature databases are similar in size but differ in content. Although all the methods share a common interest in protein sequence classification, some focus on divergent domains (e.g., PFAM), some focus on functional sites (e.g., PROSITE), and others focus on families, specializing in hierarchical definitions from superfamily down to subfamily levels in order to pinpoint specific functions (e.g., PRINTS). **Figure 8** shows the InterPro Scan results for Q9Y259, Choline Kinase. The results show a match to the PFAM entry for choline kinase, PF01633. This identifies the Q9Y259 sequence as having a choline kinase domain. This is the only identified domain in this sequence. Another piece of information that is shown is that Q9Y259 belongs to the IPR002573 family of proteins. Hyperlink to this page and there is information on the choline kinase family of proteins. Note that the InterPro scan does not identify the relationship to the family of eukaryotic protein kinases; if the relationship was identified, it would show up on the results page.

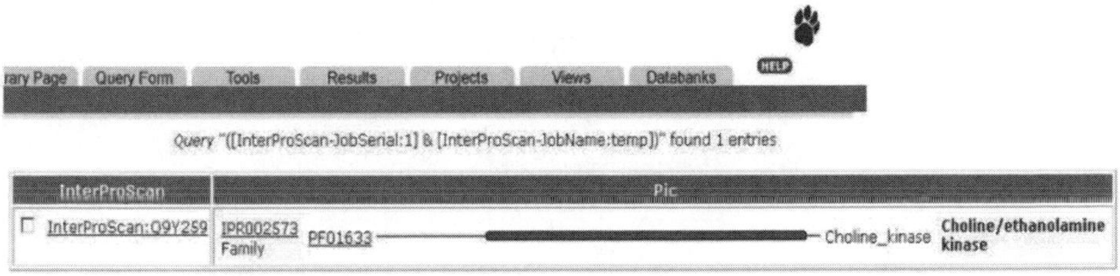

Fig. 8. InterPro results for Q9Y259, choline kinase.

1. Next go to NCBI Entrez and follow the links from BLAST to the PSI-BLAST site.
2. Paste the choline kinase sequence in a FASTA format into the search box.
3. Underneath the search box there are several options. Choose a database to search your sequence against (the preset database is the NCBI nonredundant protein database [nr]).

This option gives a choice of five databases, nr, swissprot, pdb, pat, and month to search against. The nr database contains all nonredundant GenBank CDS translations+RefSeq Proteins+PDB+SwissProt+PIR+PRF. The SwissProt database contains the last major release of the SwissProt protein sequence database. The pat database contains the proteins from the Patent division of GenPept. The pdb database contains the sequences derived from the 3D structure from the Brookhaven Protein Data Bank. The month database contains all new or revised GenBank CDS translation+PDB+SwissProt+PIR+PRF released in the last 30 d. The type of information you are looking for largely determines the database that you choose to search against (*see* **Note 5**). In this case we want to know whether our sequence of interest is similar to any other well-characterised sequences. As that is the case we will search against SwissProt. The SwissProt Protein database is a highly curated protein sequence database that provides a high level of annotation (such as the description of protein function, domain structure, posttranslational modifications, and variants) and a minimal level of redundancy, unlike nr, which gives maximal coverage but suffers from much redundancy and little curation *(1)*.

1. Another option available is "do CD search"; leave this on. This compares protein sequences with the Conserved Domain Database (CDD) *(11)*. The CDD is a database containing a collection of functional and/or structural domains derived from two sites, SMART and PFAM. This is similar to the InterPro scan but can differ subtly, as there are personal contributions from people at the NCBI.
2. Leave all other options at the preset position and initiate the search.
3. **Figure 9** shows the results of the CDD search that is carried out automatically. Again, CDD identifies a choline kinase domain but fails to identify the relationship to eukaryotic protein kinases.

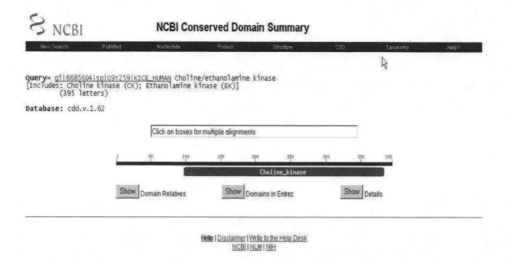

Fig. 9. CDD results for Q9Y259, choline kinase.

4. One can hyperlink from the domain GIF to the detailed results of the CDD search.

Here you can see the detailed alignment of your sequence to the identified domain and the assigned E-value. This is often very useful, as sometimes two different but overlapping domains may be identified in the search. To assess which is the more likely, one can look at the alignments and, more importantly, the E-values. The domain with an E-value of greater significance is more likely to be the correct annotation (*see* **Note 6**).

1. To view the results from the first iteration of PSI-BLAST, hit the format button. **Figure 10** shows these results. In this case the first iteration has identified choline kinase homologs in alternative organisms such as mouse, yeast, and fly. There are no sequences here that fall outside the choline kinase family.
2. When using PSI-BLAST, keep iterating the search until PSI-BLAST identifies similarity to a family of proteins that are biochemically characterized. In this example, we know the result we are looking for, eukaryotic protein kinases, so we can keep iterating the search to see if PSI-BLAST can identify this relationship.
3. If one has a sequence of unknown function you may stop searching when PSI-BLAST identifies a family of well-characterised function, with significant E-values.
4. In the case of choline kinase, it was not until iteration 18 that eukaryotic protein kinases started to appear in the search with any confidence (**Fig. 11**, *see* **Note 7**).

It may be that PSI-BLAST is unable to identify similarity to a family of biochemically characterized proteins. Attention must be paid to domain delineation when one is using PSI-BLAST. If a protein has more than one domain,

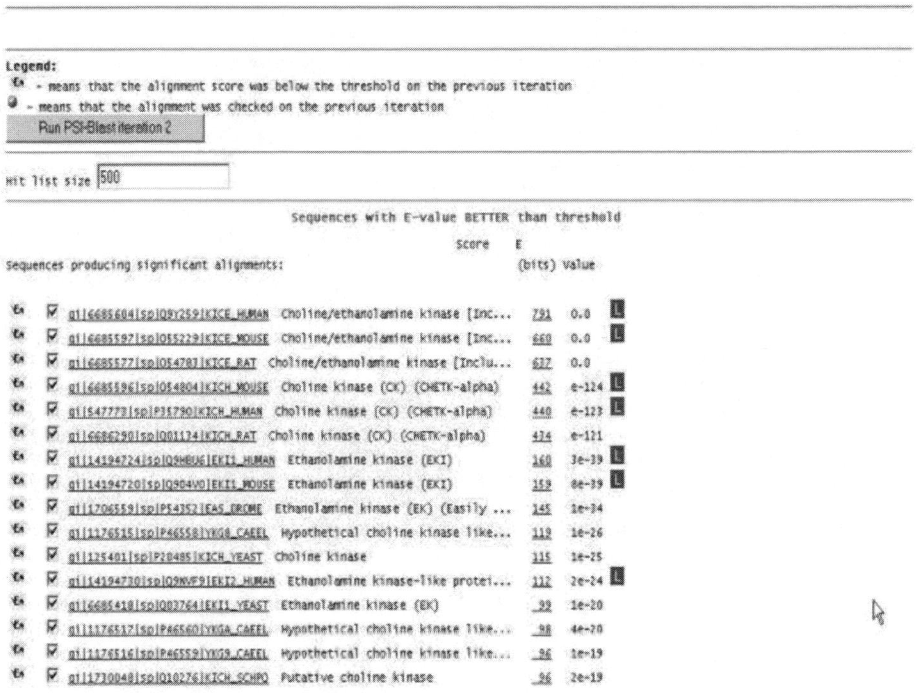

Legend:
ℰ₊ - means that the alignment score was below the threshold on the previous iteration
◑ - means that the alignment was checked on the previous iteration

[Run PSI-Blast iteration 2]

Hit list size 500

Sequences with E-value BETTER than threshold

                                                    Score   E
Sequences producing significant alignments: (bits) Value

ℰ₊ ☑ gi|6685604|sp|Q9Y259|KICE_HUMAN Choline/ethanolamine kinase [Inc... 791 0.0 Ⓛ
ℰ₊ ☑ gi|6685597|sp|O55229|KICE_MOUSE Choline/ethanolamine kinase [Inc... 660 0.0 Ⓛ
ℰ₊ ☑ gi|6685577|sp|O54787|KICE_RAT Choline/ethanolamine kinase [Inclu... 637 0.0
ℰ₊ ☑ gi|6685596|sp|O54804|KICH_MOUSE Choline kinase (CK) (CHETK-alpha)   442 e-124 Ⓛ
ℰ₊ ☑ gi|1547773|sp|P35790|KICH_HUMAN Choline kinase (CK) (CHETK-alpha)   440 e-123 Ⓛ
ℰ₊ ☑ gi|6686290|sp|Q01134|KICH_RAT Choline kinase (CK) (CHETK-alpha)   474 e-121
ℰ₊ ☑ gi|14194724|sp|Q9HBU6|EKI1_HUMAN Ethanolamine kinase (EKI)   160 3e-39 Ⓛ
ℰ₊ ☑ gi|14194720|sp|Q904V0|EKI1_MOUSE Ethanolamine kinase (EKI)   153 8e-39 Ⓛ
ℰ₊ ☑ gi|1206559|sp|P54352|EAS_DROME Ethanolamine kinase (EK) (Easily ... 145 1e-34
ℰ₊ ☑ gi|1176515|sp|P46558|YKG8_CAEEL Hypothetical choline kinase like... 119 1e-26
ℰ₊ ☑ gi|125401|sp|P20485|KICH_YEAST Choline kinase   115 1e-25
ℰ₊ ☑ gi|14194730|sp|Q9NVF9|EKI2_HUMAN Ethanolamine kinase-like protei... 112 2e-24 Ⓛ
ℰ₊ ☑ gi|6685418|sp|Q03764|EKI1_YEAST Ethanolamine kinase (EK)   99 1e-20
ℰ₊ ☑ gi|1176517|sp|P46560|YKGA_CAEEL Hypothetical choline kinase like... 98 4e-20
ℰ₊ ☑ gi|1176516|sp|P46559|YKG9_CAEEL Hypothetical choline kinase like... 96 1e-19
ℰ₊ ☑ gi|1710048|sp|Q10276|KICH_SCHPO Putative choline kinase   96 2e-19

Fig. 10. NCBI PSI-BLAST results for iteration 1 using Q9Y259, choline kinase, as the query sequence.

PSI-BLAST will identify sequences that have homology either to one of the domains or to both. For example, a protein of interest may have a kinase domain and an SH2 domain. PSI-BLAST will identify homologs of this particular protein as well as unrelated proteins that happen to have an SH2 domain or happen to have a kinase domain.

To increase our confidence that the relationship to eukaryotic protein kinases identified by PSI-BLAST is reliable, as 18 iterations were required to identify this relationship, we move to structural based methods for protein annotation. These methods use structure-information to predict the fold of a sequence, even in the absence of any significant sequence similarity. 3D-PSSM is a method for protein fold recognition using 1D and 3D sequence profiles coupled to secondary structure and solvation potential information *(14)*.

1. To access the 3D-PSSM site, go to the 3D-PSSM server at Imperial College, and follow the links to recognize a fold.
2. Here you can paste your sequence of interest and initiate the search.

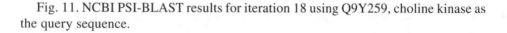

```
ᴱᴬ  ☑  gi|3025193|sp|O43924|UBIB_AZOCH   Probable ubiquinone biosynthesis...    73   8e-13
ᴱᴬ  ☑  gi|12230858|sp|Q9K7G2|YX99_BACHD  Hypothetical protein BH3399           72   3e-12
ᴱᴬ  ☑  gi|17366408|sp|P83101|KG3H_DROME  Putative glycogen synthase kina...    71   4e-12
ᴱᴬ  ☑  gi|13626380|sp|Q92630|DYR2_HUMAN  Dual-specificity tyrosine-phosp...    56   1e-07  L
ᴱᴬ  ☑  gi|13124808|sp|P18431|SGG_DROME   Protein kinase shaggy (Protein z...   55   3e-07
ᴱᴬ  ☑  gi|18203050|sp|Q9J523|V212_FOWPV  Probable serine/threonine prote...    52   3e-06
ᴱᴬ  ☑  gi|17366413|sp|P83102|DYR3_DROME  Putative dual-specificity tyros...    50   1e-05
ᴱᴬ  ☑  gi|12643714|sp|Q10452|SKP1_SCHPO  Protein kinase skp1                   47   5e-05
ᴱᴬ  ☑  gi|13626369|sp|O43781|DYR3_HUMAN  Dual-specificity tyrosine-phosp...    47   5e-05
ᴱᴬ  ☑  gi|20455502|sp|P49841|KG3B_HUMAN  Glycogen synthase kinase-3 beta...    46   1e-04  L
ᴱᴬ  ☑  gi|11133187|sp|Q9WV60|KG3B_MOUSE  Glycogen synthase kinase-3 beta...    46   1e-04  L
ᴱᴬ  ☑  gi|125374|sp|P18266|KG3B_RAT    Glycogen synthase kinase-3 beta (GS...  46   1e-04
ᴱᴬ  ☑  gi|125373|sp|P18265|KG3A_RAT    Glycogen synthase kinase-3 alpha (G... 45   2e-04
ᴱᴬ  ☑  gi|12644292|sp|P49840|KG3A_HUMAN  Glycogen synthase kinase-3 alph...    45   3e-04  L
ᴱᴬ  ☑  gi|13626405|sp|Q9NR20|DYR4_HUMAN  Dual-specificity tyrosine-phosp...    44   5e-04  L
ᴱᴬ  ☑  gi|15214310|sp|P74745|SPKC_SYNY3  Probable serine/threonine-prote...    43   8e-04
ᴱᴬ  ☑  gi|11133532|sp|Q40518|MSK1_TOBAC  shaggy-related protein kinase N...    43   0.001
ᴱᴬ  ☑  gi|140201|sp|P14680|YAK1_YEAST   Protein kinase YAK1                    42   0.003
ᴱᴬ  ☑  gi|585649|sp|P08018|PBS2_YEAST   POLYMYXIN B RESISTANCE PROTEIN KI...   41   0.003
ᴱᴬ  ☑  gi|585472|sp|P38615|MDS1_YEAST   Serine/threonine-protein kinase M...   41   0.004
ᴱᴬ  ☑  gi|9296963|sp|Q9Y463|DYRB_HUMAN  Dual-specificity tyrosine-phosph...    41   0.005  L
ᴱᴬ  ☑  gi|29611832|sp|Q9S2C0|PKSC_STRCO  Serine/threonine protein kinase...    41   0.005
```

[Run PSI-Blast iteration 19]

Fig. 11. NCBI PSI-BLAST results for iteration 18 using Q9Y259, choline kinase as the query sequence.

Figure 12 shows the top results from the 3D-PSSM search on human choline kinase, Q9Y259. The top hit and highest confidence is to the solved choline kinase structure. The similarity to the aminoglycoside phosphotransferases fold is predicted with a similar high confidence. 3D-PSSM then identifies the similarity to the eukaryotic protein kinase family fold, with a lesser confidence but still significant. As can be seen, the percentage sequence identity is very low, but by using structural information, one can, with confidence, detect the relationship of the family of choline kinases to the family of eukaryotic kinases.

Threading is another technology that uses structural information to predict the fold/function of a protein sequence *(17)*. At Inpharmatica we have used this technology to produce the Biopendium database. The Biopendium is a precalculated database that has comprehensively analyzed every publicly available sequence using proprietary PSI-BLAST and Threading algorithms *(24)*. **Figure 13** depicts the output of the Biopendium for choline kinase. As can be seen, the Biopendium Genome Threader results for choline kinase predict the relationship to the eukaryotic protein kinases such as cyclin-dependent kinase 2 with 100% confidence, despite a very low sequence identity of 8%. The relationship is predicted to a large number of the eukaryotic protein kinases, not just one or two.

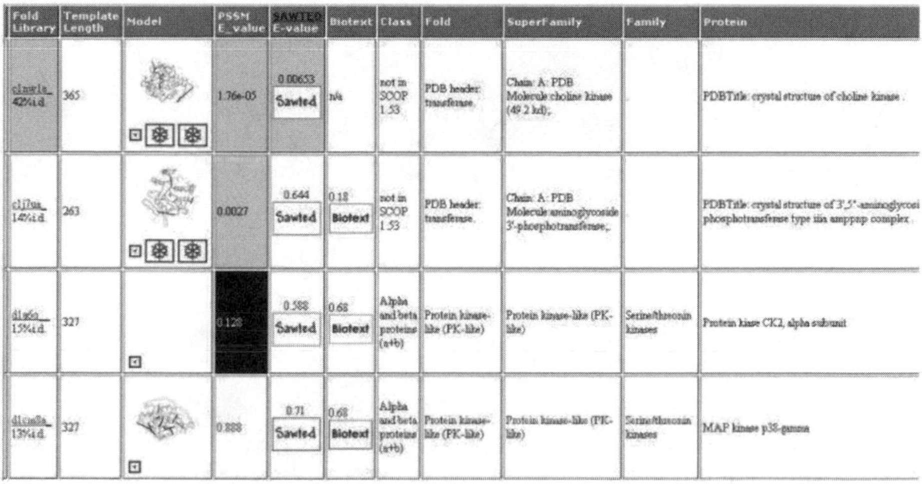

Fold Library	Template Length	Model	PSSM E_value	SAWTED E-value	Biotext	Class	Fold	Superfamily	Family	Protein
c1awla 42%i.d.	365		1.76e-05	0.00653 Sawted	n/a	not in SCOP 1.53	PDB header: transferase	Chain: A: PDB Molecule choline kinase (49.2 kd);		PDBTitle: crystal structure of choline kinase .
c1j7ua 14%i.d.	263		0.0027	0.644 Sawted	0.18 Biotext	not in SCOP 1.53	PDB header: transferase	Chain: A: PDB Molecule aminoglycoside 3'-phosphotransferase;		PDBTitle: crystal structure of 3',5'-aminoglycosi phosphotransferase type iiia anppapp complex
d1x6o 15%i.d.	327		0.128	0.588 Sawted	0.68 Biotext	Alpha and beta proteins (a+b)	Protein kinase-like (PK-like)	Protein kinase-like (PK-like)	Serine/threonin kinases	Protein kinase CK2, alpha subunit
d1csna 13%i.d.	327		0.888	0.71 Sawted	0.68 Biotext	Alpha and beta proteins (a+b)	Protein kinase-like (PK-like)	Protein kinase-like (PK-like)	Serine/threonin kinases	MAP kinase p38-gamma

Fig. 12. 3D-PSSM results for Q9Y259, choline kinase.

The methods shown allow one to move from sequence-based annotation methods all the way to structure-based methods. Depending on the protein being analysed, sequence-based annotation methods may be sufficient. What we have seen with choline kinase is that, although the biochemical function of this protein is easily ascertainable, it is only really the structure-based methods that identify its very distant relationship to the eukaryotic protein kinase family.

4. Notes

1. The EST sequence itself also gives clues as to its origin. In this case the EST starts with a run of Ts. Most cDNA libraries used for EST sequencing are Oligo dT primed, and the 3'end sequence read is in reverse complement orientation to its corresponding cDNA; therefore a run of Ts at the beginning of the EST represents part of the polyA tail of the cDNA. A further sequence clue is found at nucleotides 26–31; the sequence 5'-TTTATT-3', when read as the reverse complement, is 5'-AATAAA-3', the polyA signal required for polyadenylation. Therefore by sequence alone the 3'end assignment can be confirmed.

2. BLAT is an algorithim designed to search genomic sequences rapidly for matches at greater than 95% identity over 40 bases or more at nucleotide level and greater than 80% sequence identity over 20 amino acids or more at the protein level *(20)*. Its main application is therefore to find exact or near to exact matches in large sequence databases rapidly. It is not the algorithm of choice to identify distant homologs in the genome; a TBLASTN-based search would be more applicable for that purpose.

3. Again, further information can be gathered as to gene assignment of the query sequence based on how it matches to the genome. A 3'end EST, since it is the

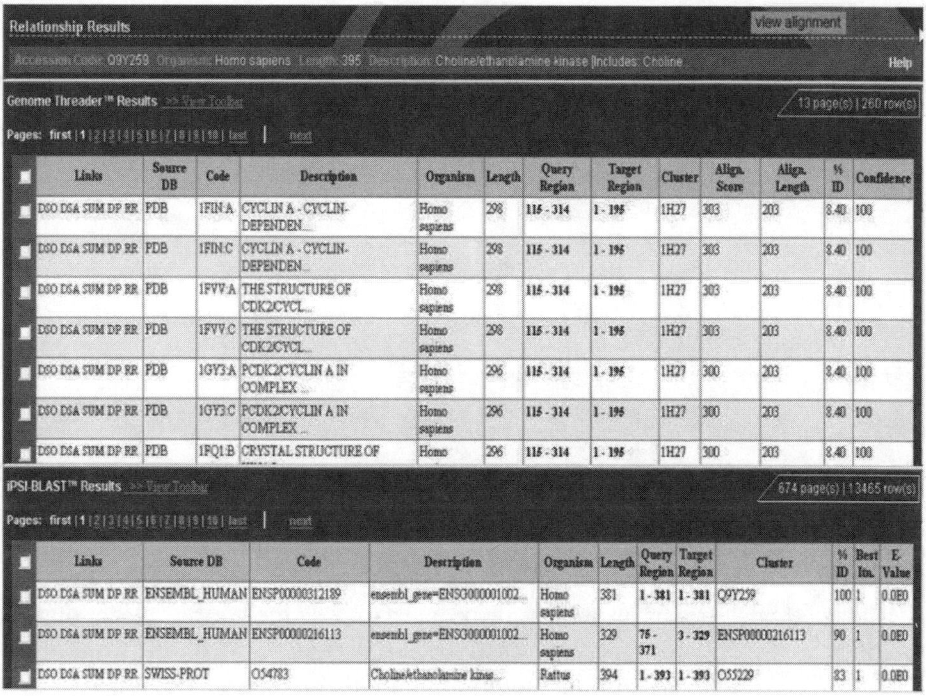

Relationship Results | View alignment

Accession Code: Q9Y259 Organism: Homo sapiens Length: 395 Description: Choline/ethanolamine kinase [Includes: Choline ... | Help

Genome Threader™ Results >> View Toolbar | 13 page(s) | 260 row(s)

Pages: first | 1 | 2 | 3 | 4 | 5 | 6 | 7 | 8 | 9 | 10 | last | next

Links	Source DB	Code	Description	Organism	Length	Query Region	Target Region	Cluster	Align. Score	Align. Length	% ID	Confidence
DSO DSA SUM DP RR	PDB	1FIN:A	CYCLIN A - CYCLIN-DEPENDEN...	Homo sapiens	298	115 - 314	1 - 195	1H27	303	203	8.40	100
DSO DSA SUM DP RR	PDB	1FIN:C	CYCLIN A - CYCLIN-DEPENDEN...	Homo sapiens	298	115 - 314	1 - 195	1H27	303	203	8.40	100
DSO DSA SUM DP RR	PDB	1FVV:A	THE STRUCTURE OF CDK2/CYCL...	Homo sapiens	298	115 - 314	1 - 195	1H27	303	203	8.40	100
DSO DSA SUM DP RR	PDB	1FVV:C	THE STRUCTURE OF CDK2/CYCL...	Homo sapiens	298	115 - 314	1 - 195	1H27	303	203	8.40	100
DSO DSA SUM DP RR	PDB	1GY3:A	PCDK2/CYCLIN A IN COMPLEX ...	Homo sapiens	296	115 - 314	1 - 195	1H27	300	203	8.40	100
DSO DSA SUM DP RR	PDB	1GY3:C	PCDK2/CYCLIN A IN COMPLEX ...	Homo sapiens	296	115 - 314	1 - 195	1H27	300	203	8.40	100
DSO DSA SUM DP RR	PDB	1FQ1:B	CRYSTAL STRUCTURE OF	Homo	296	115 - 314	1 - 195	1H27	300	203	8.40	100

iPSI-BLAST™ Results >> View Toolbar | 674 page(s) | 13465 row(s)

Pages: first | 1 | 2 | 3 | 4 | 5 | 6 | 7 | 8 | 9 | 10 | last | next

Links	Source DB	Code	Description	Organism	Length	Query Region	Target Region	Cluster	% ID	Best Itn.	E Value
DSO DSA SUM DP RR	ENSEMBL_HUMAN	ENSP00000312189	ensembl_gene=ENSG00001002...	Homo sapiens	381	1 - 381	1 - 381	Q9Y259	100	1	0.0E0
DSO DSA SUM DP RR	ENSEMBL_HUMAN	ENSP00000216113	ensembl_gene=ENSG00001002...	Homo sapiens	329	75 - 371	3 - 329	ENSP00000216113	90	1	0.0E0
DSO DSA SUM DP RR	SWISS-PROT	O54783	Choline/ethanolamine kinas...	Rattus	394	1 - 393	1 - 393	O55229	83	1	0.0E0

Fig. 13. Inpharmatica Biopendium results for Q9Y259, choline kinase.

reverse complement of its cDNA, will match the reverse strand of a gene located on the plus strand of the genomic DNA. A match on the plus strand with a 3' EST will represent a gene on the minus strand. The equivalent is true for the 5' read; as it represents the plus strand of a cDNA, it will be on the same strand of the transcript when matched to genomic DNA.

4. Since the polyA tail is added post-transcriptionally, this should not be present in the genomic DNA and will be excluded from the match. Its presence in the match indicates that the cDNA clone from which the EST was sequenced is likely to have originated from genomic contamination of the source cDNA library.
5. SwissProt is a very highly curated database of nonredundant protein sequences. This is in comparison with GenBank, which has very little curation and is highly redundant. When one is trying to annotate a protein, it is important to use a database that has as much coverage of "sequence space" as possible, for instance if one searched against the PDB only, one will inevitably miss certain relationships, as the PDB database is relatively limited in size. SwissProt gives good coverage of sequence space coupled with curation.
6. Sometimes more than one domain is predicted by CDD. That is fine if they are nonoverlapping. If they do overlap, go to the relevant site that describes the domain and read up on it. For example, CDD predicts the choline kinase domain

with very high confidence. If you look at detailed results, it also predicts a DUF227 domain. The e-value of this prediction is very poor, so we disregard it. There is another prediction for a CotS domain. Upon reading, a CotS domain is a choline kinase domain with a different name.

7. To run 18 iterations of PSI-BLAST is unrealistic. It is recommended not to do more than five iterations for fear of false positives. When one is running a PSI-BLAST search, the ultimate end of the search is when no more new sequences are identified at the next iteration; this is called convergence.

Acknowledgments

The authors would like to thank colleagues at Inpharmatica for comments and discussions on this chapter.

References

1. Boeckmann B., Bairoch A., Apweiler R., et al. (2003) The Swiss-Prot protein knowledgebase and its supplement TrEMBL in 2003. *Nucleic Acids Res.* **31,** 365–370.
2. Barker, W. C., Garavelli, J. S., Huang, H., et al. (2001) Protein Information Resource: a community resource for expert annotation of protein data. *Nucleic Acids Res.* **29,** 29–32.
3. Karolchik, D., Baertsch, R., Diekhans, M., et al. (2003) The UCSC Genome Browser Database. *Nucleic Acids Res.* **31,** 51–54.
4. Hubbard, T., Barker, D., Birney, E., et al. (2002) The Ensembl genome database project. *Nucleic Acids Res.* **30,** 38–41.
5. Bateman A., Birney E., Cerruti L., et al. (2002) The Pfam Protein Families Database. *Nucleic Acids Res.* **30,** 276–280.
6. Letunic I., Goodstadt L., Dickens N. J., et al. (2002) Recent improvements to the SMART domain-based sequence annotation resource. *Nucleic Acids Res.* **30,** 242–244.
7. Attwood T. K., Bradley P., Flower D. R., et al. (2003) PRINTS and its automatic supplement, prePRINTS. *Nucleic Acids Res.* **31,** 400–402.
8. Mulder N. J., Apweiler R., Attwood T. K., et al. (2003) The InterPro Database, 2003 brings increased coverage and new features. *Nucleic Acids Res.* **31,** 315–318.
9. Altschul, S. F., Gish, W., Miller, W., Myers, E. W., and Lipman D. J (1990) Basic local alignment search tool. *J. Mol. Biol.* **215,** 403–410.
10. Altschul, S. F., Madden, T. L., Schaffer, A. A., et al. (1997) Gapped BLAST and PSI-BLAST: a new generation of protein database search programs. *Nucleic Acids Res.* **25,** 3389–3402.
11. Marchler-Bauer, A., Anderson, J. B., DeWeese-Scott, C., et al. (2003) CDD: a curated Entrez database of conserved domain alignments. *Nucleic Acids Res.* 31, 383–387
12. Murzin, A. G. (1993) Sweet-tasting protein monellin is related to the cystatin family of thiol proteinase inhibitors. *J. Mol. Biol.* **230,** 689–694.

13. Russell, R. B., Saqi, M. A., Sayle, R. A., Bates, P. A., and Sternberg, M. J. (1997) Recognition of analogous and homologous protein folds: analysis of sequence and structure conservation. *J. Mol. Biol.* **269,** 423–439.
14. Kelley, L. A., MacCallum, R. M., and Sternberg, M. J. (2000) Enhanced genome annotation using structural profiles in the program 3D-PSSM. *J. Mol. Biol.* **299,** 499–520.
15. Shi, J., Blundell, T. L., and Mizuguchi, K. (2001) FUGUE: sequence-structure homology recognition using environment-specific substitution tables and structure-dependent gap penalties. *J. Mol. Biol.* **310,** 243–257.
16. Jones, D. T., Taylor, W. R., and Thornton, J. M. (1992) A new approach to protein fold recognition. *Nature* **358,** 86–89
17. Jones, D. T. (1999) GenTHREADER: an efficient and reliable protein fold recognition method for genomic sequences. *J. Mol. Biol.* **287,** 797–815.
18. Jones, D. T., Tress, M., Bryson, K., and Hadley, C. (1999) Successful recognition of protein folds using threading methods biased by sequence similarity and predicted secondary structure. *Proteins* **37,** 104–111.
19. Gibas, C. and Jambeck, P. (2001). *Developing Bioinformatic Computer Skills.* O'Reilly & Associates, Sebastopol, CA.
20. Kent, W. J. (2002) BLAT-The BLAST-Like Alignment Tool. *Genome Res.* **12,** 656–664.
21. Peisach, D., Gee, P., Kent, C., and Xu, Z. (2003) The crystal structure of choline kinase reveals a eukaryotic protein kinase fold. *Structure (Camb.)* **6,** 703–713.
22. Falquet L., Pagni M., Bucher P., et al. (2002) The PROSITE database, its status in 2002. *Nucleic Acids Res.* **30,** 235–238.
23. Corpet F., Servant F., Gouzy J., and Kahn D. (2000) ProDom and ProDom-CG: tools for protein domain analysis and whole genome comparisons. *Nucleic Acids Res.* **28,** 267–269.
24. Michalovich, D., Overington, J., and Fagan, R. (2002) Protein sequence analysis in silico: application of structure-based bioinformatics to genomics initiatives. *Curr. Opin. Pharmacol.* **2,** 574–580.

15

Pragmatic Target Discovery From Novel Gene to Functionally Defined Drug Target

The Interleukin-1 Story

Stuart McRae Allan

Summary

The pro-inflammatory cytokine interleukin-1 (IL-1) has been proposed as a mediator of the acute neurodegenerative changes that occur following stroke. This is based largely on experimental studies in rodents, in particular the marked reduction in ischemic cell death seen when IL-1 receptor antagonist (IL-1ra) is administered. Mechanisms of IL-1 action remain largely unknown, but they may involve complex effects on many cells including microglia, astrocytes, neurons, and endothelial cells. In light of this, IL-1ra is currently under consideration as a potential treatment for stroke and other neurodegenerative conditions.

KEY WORDS

Interleukin-1; cytokines; inflammation; neurodegeneration; cell death; cerebral ischemia; proinflammatory; neuroprotection.

1. Introduction

It has become increasingly apparent that in many neurodegenerative conditions, including stroke, there is an involvement of inflammatory processes in disease progression. Whether this inflammation is primary or secondary to the pathology is not clear and is the subject of intense research. Cytokines are key inflammatory mediators in the periphery and have also been shown to exert a number of actions in the central nervous system (CNS), including an active role in diverse forms of neurodegeneration. The precise role of cytokines in neurodegeneration depends on a number of factors, including the relative expression of pro-inflammatory and anti-inflammatory molecules, the former

From: *Methods in Molecular Medicine, Vol. 104: Stroke Genomics: Methods and Reviews*
Edited by: S. J. Read and D. Virley © Humana Press Inc., Totowa, NJ

largely being considered detrimental and the latter deemed beneficial, although this is not always the case, and the precise role of a particular cytokine can depend on a number of factors. Most evidence to date supports a role for the pro-inflammatory cytokine interleukin-1 (IL-1) in neurodegeneration, and this chapter focuses on this in more detail, with particular reference to stroke.

2. Interleukin-1 Family

IL-1, often referred to as a "prototypical" pro-inflammatory cytokine, was the first cytokine described to have effects on the brain, in which it induced fever. Subsequently, many actions of IL-1 (both endogenous and exogenous) in the CNS have been identified *(1)*.

The IL-1 family (**Fig. 1**) actually encompasses three closely related proteins, IL-1α, IL-1β, and IL-1ra, which are the products of separate genes that share limited sequence homology. IL-1α and IL-1β are agonists that largely produce the same biological effects through binding with an 80-kDa cell surface receptor, IL-1RI (*see* **Subheading 2.1.**), although some differential actions have been reported *(2)*. IL-1ra is a highly selective, competitive receptor antagonist that is thought to block all actions of IL-1 *(3)* and is produced as either a secreted (sIL-1ra) or intracellular form (icIL-1ra) *(4)*. All IL-1 family members are formed as precursors but differ in that pro-IL-1α and pro-IL-1ra are biologically active, whereas pro-IL-1β is not and requires cleavage by caspase 1, the first identified member of the cysteine aspartate group of proteases associated with apoptosis, to produce the mature active protein *(5)*.

Several putative new members of the IL-1 family have been described *(6)*, based largely on sequence homology, and this has led to adoption of a new nomenclature *(7)*. IL-18 (IL-1F4) was initially identified as interferon-γ-inducing factor, and although it shares homology with IL-1, it is cleaved by caspase 1 and binds to a receptor and accessory protein within the same family, it does not seem to share many biological actions, including a role in stroke *(8)*. Very little is known about the other novel ligands, and no clear function has yet been demonstrated, although IL-1F9 has been shown to activate nuclear factor-κB (NF-κB) in a cell line transfected with IL-1RrP2 (IL-1R6) *(9)*, which is a member of the extended IL-1 receptor family *(10)*, and IL-1F7 is cleaved by caspase 1 and binds to the IL-18 receptor *(11,12)*.

Lack of leader sequences means the mechanisms by which IL-1α and -β are released from cells is not clear. IL-1α is thought to remain largely intracellular, whereas release of IL-1β from the cell may be linked directly to cleavage by caspase 1, which is induced by the activation of purinergic $P2X_7$ receptors by extracellular ATP *(13)*.

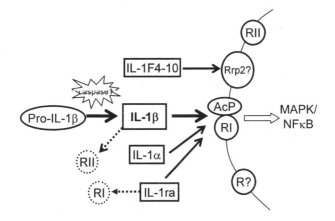

Fig. 1. The interleukin-1 family. AcP, accessory protein; NFκB, nuclear factor κB; MAPK, mitogen-activated protein kinase; Rrp2, receptor-related protein 2.

2.1. Interleukin-1 Receptors

There are two specific receptors for IL-1, the 80-kDa type I (IL-1RI) and the 65-kDa type II (IL-1RII) *(14)*. All signaling, however, is believed to be mediated through IL-1RI *(15)*, which requires association in the membrane with the IL-1 accessory protein (IL-1AcP) in order to initiate signal transduction *(16)*. IL-1RII lacks an intracellular domain and therefore functions as a decoy receptor that binds IL-1 but does not trigger downstream signaling pathways *(17)*. IL-1RI and IL-1RII can be shed from the plasma membrane and exist in soluble form, thereby acting to reduce the effectiveness of extracellular IL-1 by limiting its binding to the membrane bound functional IL-1RI.

IL-1RI and IL-1RII, as well as several mammalian homologs, are part of the IL-1R/toll-like receptor (TLR) superfamily *(18)*. Many of these receptors have no known ligand, so it may be that additional pathways exist for IL-1 signaling in the brain.

2.2. Interleukin-1 Signaling

Following IL-1 binding and formation of the IL-1RI/IL-1RAcP complex, two cytoplasmic proteins, MyD88 and Tollip, are recruited, which leads to subsequent addition of the IL-1 receptor kinases IRAK and IRAK4 and then tumor necrosis factor receptor-associated factor 6 (TRAF6) *(18)*. TIFA and Pellino I have recently been identified as important adaptor proteins in the link between IRAK and TRAF6 *(19,20)*. This then leads to the activation of the NF-κB signaling pathway through a series of phosphorylation and kinase steps involving transforming growth factor-β-activated kinase 1 (TAK-1) and TAK-1 binding proteins (TAB1 and TAB2) *(21)*. TRAF6 also activates mitogen-activated protein kinase (MAPK) signalling pathways via a different mechanism that does not involve phosphorylation *(22)*. Both pathways ultimately result in the activation of nuclear transcription factors that subsequently induce the expression of many genes.

3. Expression Studies

For any factor to be implicated in a disease process, it must be present in sufficient amounts and at an appropriate time in the temporal evolution of the disease. For many conditions it is not possible to determine this easily in the clinical setting, and appropriate preclinical paradigms must be established to investigate it further. Using various models of stroke, marked changes in expression (gene and/or protein) of IL-1 are observed in response to ischemic insults. The major findings from these studies are described below, as well as the particular approach used, since each can have specific advantages/disadvantages, depending on the type of information required. It is important to be aware that IL-1 is expressed at very low, virtually undetectable levels in normal human and rodent brain *(23)*. Therefore a relatively small increase could be important, especially since very few IL-1 receptors need to be occupied to produce a biological response *(24)*. Expression of IL-1 has been shown in most cells in the brain, including neurons, astrocytes, oligodendrocytes, and endothelial cells, although activated microglia appear to be the major source in pathological states such as stroke. Another source of IL-1 is invading macrophages, particularly when there is blood–brain barrier breakdown.

Finally, for any factor to be deemed causal to the neurodegeneration that occurs after stroke, it needs to be present before significant pathology occurs. It is important therefore to consider the temporal profile of changes in IL-1 expression in relation to the progression of neuronal death, which only a limited number of studies have actually done. It should also be noted that by far the majority of studies investigating IL-1 expression and actions in stroke have focused on IL-1β rather than IL-1α. Therefore, for most of this chapter, when IL-1 is referred to it usually means the β form, unless the opposite is stated.

3.1. Gene Expression

A number of studies report increases in IL-1 mRNA in stroke, using a multitude of approaches and detection techniques. Rather than provide a detailed and exhaustive list, some of the key developments and main findings will be presented.

The first report of changes in expression of IL-1 in stroke was in the early 1990s; using Northern blot analysis, increases of IL-1β mRNA as early as 15 min after reperfusion in a model of transient forebrain ischaemia were observed, with peak levels being seen at 30 min and 4 h *(25)*. One disadvantage of Northern blot analysis is that one is unable to determine the cell type expressing the mRNA. This was overcome in subsequent studies that used *in situ* hybridization (ISH) after transient forebrain ischemia *(26)* and permanent middle cerebral artery occlusion (pMCAO) *(27)*. Both reported early increases in IL-1β mRNA expression in a number of brain regions that subsequently show degeneration. The mRNA appeared to be mainly localized to activated microglial cells, although some was observed early on in the meninges and also around the vascular walls. Subsequent studies concur with the finding that microglia and macrophages represent the major source of IL-1 after ischemic *(28)*, and other forms of injury *(29,30)*, which is in agreement with the idea that microglia act as sensors of CNS pathology *(31)*.

A limiting factor in these early, and subsequent, studies using Northern blot analysis and ISH to look at IL-1 mRNA expression is the nonquantitative nature of the data and an inability to detect low levels of mRNA. This has been overcome by the use of real-time polymerase chain reaction (PCR) *(32)* which has confirmed that early increases (3–6 h) in IL-1β mRNA, as well as that of other inflammatory genes, are seen in the ischemic hemisphere after transient MCAO (tMCAO) *(33)*.

3.2. Protein Expression

Since raised mRNA levels of IL-1 are not necessarily translated to increased protein expression *(34,35)*, it is important to consider changes in the latter after ischemia. Immunoreactive IL-1 has been determined in several reports, using immunohistochemistry or enzyme-linked immunosorbent assay (ELISA), in tissue sections or brain homogenates respectively.

Using ELISA, Saito and colleagues observed a peak increase in immunoreactive IL-1β 6 h after global ischemia in the gerbil *(36)*, whereas increases in IL-1β as early as 30 min after pMCAO in the rat were reported by Hillhouse and coworkers *(37)*. Exactly how this early rise in IL-1 contributes to cell death is not clear, since similar changes at this time point were observed in the undamaged contralateral cortex, differences between the two hemispheres not emerging until 4 h *(37)*. Such an early change was not observed after pMCAO in the spontaneously hypertensive rat, although a clear increase was seen at 4 h post MCAO, with a delayed peak at 3 d *(38)*. Interestingly, this latter study is one of the few to assess changes in IL-1α levels after stroke, with a peak of expression similar to that of IL-1β at 3 d, although no earlier increase was found.

Like Northern blot and PCR analysis of mRNA, ELISA detection techniques do not allow one to determine the cell type that expresses the protein. This can, however, be achieved using immunohistochemistry (IHC) techniques to label brain sections with an appropriate antibody.

In a model of pMCAO in the rat early (1 h) expression of IL-1β protein is observed in meningeal macrophages, followed (at 6 h) by positive staining of microglia in the region of the emerging infarct *(39)*. By the time of full resolution of the ischemic damage (48 h), there is widespread immunoreactive IL-1β expression in microglia and macrophages throughout the brain, including the contralateral hemisphere and other undamaged regions. A very rapid (15-min) induction of IL-1β-immunoreactive cells is reported after tMCAO in the mouse, with a peak at 1–2 h postocclusion *(40)*. The early response is mainly in endothelial cells, with later expression in microglia, both ipsilateral and contralateral. Although contralateral and ipsilateral expression of IL-1 in undamaged brain regions is widely reported, its precise role remains unclear. IL-1 itself does not induce cell death *(41,42)*, so it may be that IL-1 is detrimental only when exposed to compromised cells, such as in the ischemic territory.

One drawback of immunodetection studies to date is the inability of the antibodies to distinguish between the pro and mature forms of the IL-1β protein. It is possible though to determine bioactive IL-1β levels using a bioassay. Although this has not been reported after ischemia in the adult, transient changes in bioactive IL-1β (as well as IL-6) have been seen after hypoxia–ischemia in the neonate *(43)*.

Expression of other members of the IL-1 family can also modified by different insults. Increases in IL-1RI expression are seen after global ischemia *(28)*. In this study, using ISH and IHC, IL-1β is primarily expressed in glial cells, whereas IL-1RI is mainly found on neurons, although some expression is observed on the vascular endothelium, suggesting that although the IL-1 is produced in glial cells, its primary site of action may be neurons.

In summary, significant levels of IL-1 appear to be present during the evolution of ischemic brain injury, in a number of different experimental paradigms, which supports a possible role in development of the damage.

4. Functional Role

The presence of a factor in a disease is important but does not necessarily reflect an active involvement in the ongoing pathology. For this to be the case, functional data must be obtained, and such evidence is presented below for IL-1. This provides both direct and indirect evidence of a role for IL-1 in stroke and broadly speaking can be broken down into the following categories: (1)

studies looking at the effects of exogenous IL-1; (2) studies looking at the effects of blocking endogenous IL-1 action (e.g., by using IL-1ra); and (3) studies using transgenic animals that have modifications in the IL-1 system.

4.1. Exogenous Interleukin-1 Administration

Direct administration of IL-1 itself into the brain parenchyma does not appear to cause neuronal death *(41,42)*. In contrast, when given at the time of an ischemic insult, it can dramatically worsen the resultant brain damage *(44–46)*. Determination of the sites and mechanisms of action of IL-1 in exacerbating injury have proved difficult, not least because in vivo actions do not always translate to in vitro effects. Indeed, IL-1 has been shown to be neuroprotective in cell culture systems *(47,48)* and to promote repair and recovery *(49)*. Further evidence for a protective role of IL-1 under certain conditions comes from studies on preconditioning, in which a subthreshold insult protects against subsequent ischemia. In such situations, increased expression of IL-1 is seen, and this has been proposed to play an active role in offering protection against the ensuing challenge *(50,51)*. Why IL-1 is capable of a dual role is not clear, but it may be because of its diverse range of effects, through both direct and indirect actions, on neurons, glia, and the brain vasculature (**Fig. 2**) *(1,52)*. Activation of glia by IL-1 will trigger the release of many factors that have the potential to be either beneficial or detrimental, and it is more than likely that the contribution of IL-1 to ischemic cell death depends on the balance between these factors. IL-1 can directly affect neurons, having been shown to produce changes in neurotransmitter release and channel activity *(53)*, as well as having a number of effects on the brain vasculature, including increased blood–brain barrier permeability, recruitment of neutrophils *(54–56)*, and induction of cyclo-oxygenase and IκB (a marker of NF-κB activation) *(57)*. Such effects have the potential to alter ischemic brain damage significantly; since IL-1 receptors have unequivocally been shown on endothelial and other barrier-related cells *(58,59)*, the brain vasculature probably represents a primary site of IL-1 action.

4.2. Blockers of Interleukin-1 Action

Injection of an exogenous agent into the brain does not necessarily mimic the actions of the endogenous molecule. The IL-1 system is perhaps unique therefore in that there is a valuable experimental tool readily available to investigate actions of the endogenous protein, namely, IL-1ra, which, as far as is known, has no other effect other than antagonism of IL-1 actions *(60)*. Administration of IL-1ra to experimental animals will thus indicate the functional role of IL-1.

This was first demonstrated in 1992 when intracerebroventricular (icv) IL-1ra was shown to significantly protect against ischemic brain damage resulting

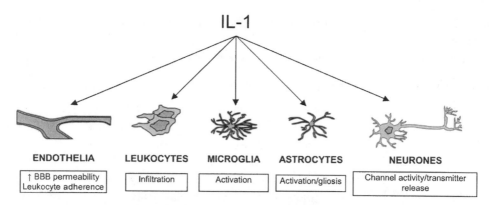

Fig. 2. Multiple cell types affected by interleukin-1.

from pMCAO in the rat *(61)*. Subsequently this has been confirmed after central *(44,62–64)* or peripheral *(65,66)* administration of IL-1ra in the rat and mouse. Furthermore, recent data indicate that delayed administration (up to 3 h) of IL-1ra can offer significant protection against cell death after tMCAO in the rat (Ross et al., unpublished observations), which is important when one considers the therapeutic implications (*see* **Subheading 5.**). Also, protective actions are sustained since IL-1ra-treated animals still show reduced damage up to 1 wk after occlusion *(65,66a,67)*, as well as improvements in neurological score *(65)*.

In addition to IL-1ra, several other methods of blocking endogenous IL-1 action have been shown to offer significant neuroprotection against ischemic brain injury, including knockout *(68)* or inhibition of caspase 1 *(69,70)*, administration of neutralizing antibody *(44)*, or overexpression of IL-1ra *(71,72)*.

4.3. Knockout Animals

Increasing numbers of studies are making use of animals that have a specific gene knocked out, in order to demonstrate a functional role. The IL-1 system is no different, and genetically modified mice for most of the IL-1 family are available *(73)*. Deletion of either the IL-1α or IL-1β gene in mice has no effect on ischemic injury, whereas deletion of both confers about 80% neuroprotection against the tMCAO damage *(64)*. Interestingly, this degree of neuroprotection is somewhat greater than that offered by IL-1ra (usually around 50%; *see* **Subheading 4.2.**), which suggests that IL-1ra may not fully block all actions of IL-1. Mice deficient in IL-1RI or IL-1RAcP show no change in ischemic injury, whereas those deficient in IL-1ra show an increased amount of damage, compared with the wild type *(74,75)*. The latter finding suggests that endogenous IL-1ra is neuroprotective, which is in agreement with an earlier study using a neutralizing antibody against IL-1ra in the rat, in which increased damage was seen after pMCAO *(62)*. In IL-1RI knockout mice icv administration of IL-1β exacerbates ischemic injury to a similar degree as in the wild type *(74)*. This indicates that IL-1 might enhance or contribute to brain injury via a novel receptor, independent of IL-1RI. As reported earlier, mice with deletion of caspase 1 also show reduced (~45%) cell death after tMCAO *(69)*.

5. Translating to the Clinic and Future Studies

The available preclinical evidence, as presented above, strongly implicates IL-1 as a mediator of ischemic brain damage. However, many factors over the years have been proposed to play a role in stroke, with appropriate interventions being neuroprotective in preclinical studies. Even so, all these strategies failed to show benefit in clinical trials, and, with the exception of tissue plasminogen activator, there is still no treatment for stroke. The reasons for this are many and have led to the recommendation that certain criteria be met during preclinical development *(76)*. Whether IL-1ra will meet all these criteria and prove successful is not clear, but IL-1ra certainly satisfies many of them already and as a result is currently in a phase II trial for acute stroke *(77)*.

There are, however, some limitations relating to the use of IL-1ra as a potential future therapy, not least the fact it is a large protein with a short half-life and therefore may not penetrate the brain quickly enough to offer major benefit. In this respect several other approaches might be taken to block the effects of IL-1. These might include inhibition of its expression and/or release or selective targeting of the intracellular signaling cascades triggered by IL-1. Another approach could be to promote the actions of anti-inflammatory cytokines such as IL-10, which inhibits IL-1 release *(78)* and has been shown to be neuroprotective *(79)*.

In summary, strong evidence has been obtained over the years, from early expression studies to more recent experiments making use of genetically modified animals, that implicates IL-1 as a mediator of cell death in stroke and other neurodegenerative conditions. However, it remains to be seen whether targeting the IL-1 system in these diseases will prove to be an effective therapeutic strategy.

References

1. Rothwell, N. J. and Luheshi, G. N. (2000) Interleukin 1 in the brain: biology, pathology and therapeutic target. *Trends Neurosci.* **23,** 618–625.
2. Anforth, H. R., Bluthe, R. M., Bristow, A., et al. (1998) Biological activity and brain actions of recombinant rat interleukin-1α and interleukin-1β. *Eur. Cytokine Netw.* **9,** 279–288.
3. Hannum, C. H., Wilcox, C. J., Arend, W. P., et al. (1990) Interleukin-1 receptor antagonist activity of a human interleukin-1 inhibitor. *Nature* **343,** 336–340.

 4. Arend, W. P. (2002) The balance between IL-1 and IL-1Ra in disease. *Cytokine Growth Factor Rev.* **13,** 323–340.
 5. Thornberry, N. A., Bull, H. G., Calaycay, J. R., et al. (1992) A novel heterodimeric cysteine protease is required for interleukin-1β processing in monocytes. *Nature* **356,** 768–774.
 6. Dunn, E., Sims, J. E., Nicklin, M. J., and O'Neill, L. A. (2001) Annotating genes with potential roles in the immune system: six new members of the IL-1 family. *Trends Immunol.* **22,** 533–536.
 7. Sims, J. E., Nicklin, M. J., Bazan, J. F., et al. (2001) A new nomenclature for IL-1-family genes. *Trends Immunol.* **22,** 536–537.
 8. Wheeler, R. D., Boutin, H., Touzani, O., Luheshi, G. N., Takeda, K., and Rothwell, N. J. (2003) No role for interleukin-18 in acute murine stroke-induced brain injury. *J. Cereb. Blood Flow Metab.* **23,** 531–535.
 9. Debets, R., Timans, J. C., Homey, B., et al. (2001) Two novel IL-1 family members, IL-1delta and IL-1epsilon, function as an antagonist and agonist of NF-κB activation through the orphan IL- 1 receptor-related protein 2. *J. Immunol.* **167,** 1440–1446.
10. Sims, J. E. (2002) IL-1 and IL-18 receptors, and their extended family. *Curr. Opin. Immunol.* **14,** 117–122.
11. Kumar, S., Hanning, C. R., Brigham-Burke, M. R., et al. (2002) Interleukin-1F7B (IL-1H4/IL-1F7) is processed by caspase-1 and mature IL-1F7B binds to the IL-18 receptor but does not induce IFN-gamma production. *Cytokine* **18,** 61–71.
12. Pan, G., Risser, P., Mao, W., et al. (2001) IL-1H, an interleukin 1-related protein that binds IL-18 receptor/IL-1Rrp. *Cytokine* **13,** 1–7.
13. Ferrari, D., Chiozzi, P., Falzoni, S., Hanau, S., and Di Virgilio, F. (1997) Purinergic modulation of interleukin-1 β release from microglial cells stimulated with bacterial endotoxin. *J. Exp. Med.* **185,** 579–582.
14. Sims, J. E., Giri, J. G., and Dower, S. K. (1994) The two interleukin-1 receptors play different roles in IL-1 actions. *Clin. Immunol. Immunopathol.* **72,** 9–14.
15. Sims, J. E., Gayle, M. A., Slack, J. L., et al. (1993) Interleukin 1 signalling occurs exclusively via the type 1 receptor. *Proc. Natl. Acad. Sci. USA* **90,** 6155–6159.
16. Greenfeder, S. A., Nunes, P., Kwee, L., Labow, M., Chizzonite, R. A., and Ju, G. (1995) Molecular cloning and characterization of a second subunit of the interleukin 1 receptor complex. *J. Biol. Chem.* **270,** 13757–13765.
17. Colotta, F., Re, F., Muzio, M., et al. (1993) Interleukin-1 type II receptor: a decoy target for IL-1 that is regulated by IL-4. *Science* **261,** 472–475.
18. Dunne, A. and O'Neill, L. A. (2003) The interleukin-1 receptor/Toll-like receptor superfamily: signal transduction during inflammation and host defense. *Sci. STKE* **2003,** re3.
19. Takatsuna, H., Kato, H., Gohda, J., et al. (2003) Identification of TIFA as an adapter protein that links tumor necrosis factor receptor-associated factor 6 (TRAF6) to interleukin-1 (IL-1) receptor-associated kinase-1 (IRAK-1) in IL-1 receptor signaling. *J. Biol. Chem.* **278,** 12144–12150.

20. Jiang, Z., Johnson, H. J., Nie, H., Qin, J., Bird, T. A., & Li, X. (2003) Pellino 1 is required for interleukin-1 (IL-1)-mediated signaling through its interaction with the IL-1 receptor-associated kinase 4 (IRAK4)-IRAK-tumor necrosis factor receptor-associated factor 6 (TRAF6) complex. *J. Biol. Chem.* **278**, 10952–10956.

21. Ninomiya-Tsuji, J., Kishimoto, K., Hiyama, A., Inoue, J., Cao, Z., and Matsumoto, K. (1999) The kinase TAK1 can activate the NIK-I κB as well as the MAP kinase cascade in the IL-1 signalling pathway. *Nature* **398**, 252–256.

22. Li, X., Commane, M., Jiang, Z., and Stark, G. R. (2001) IL-1-induced NFκ B and c-Jun N-terminal kinase (JNK) activation diverge at IL-1 receptor-associated kinase (IRAK). *Proc. Natl. Acad. Sci. USA* **98**, 4461–4465.

23. Vitkovic, L., Bockaert, J., and Jacque, C. (2000) "Inflammatory" cytokines: neuromodulators in normal brain? *J. Neurochem.* **74**, 457–471.

24. Dinarello, C. A. (1996) Biologic basis for interleukin-1 in disease. *Blood* **87**, 2095–2147.

25. Minami, M., Kuraishi, Y., Yabuuchi, K., Yamazaki, A., and Satoh, M. (1992) Induction of interleukin-1b mRNA in rat brain after transient forebrain ischemia. *J. Neurochem.* **58**, 390–392.

26. Yabuuchi, K., Minami, M., Katsumata, S., Yamazaki, A., & Satoh, M. (1994) An in situ hybridization study on interleukin-1 β mRNA induced by transient forebrain iscemia in the rat brain. *Mol. Brain Res.* **26**, 135–142.

27. Buttini, M., Sauter, A., and Boddeke, H. W. G. M. (1994) Induction of interleukin-1b mRNA after focal cerebral ischaemia in the rat. *Mol. Brain Res.* **23**, 126–134.

28. Sairanen, T. R., Lindsberg, P. J., Brenner, M., and Siren, A.-L. (1997) Global forebrain ischemia results in differential cellular expression of interleukin-1b (IL-1b) and its receptor at mRNA and protein level. *J. Cerebr. Blood Flow Metab.* **17**, 1107–1120.

29. Pearson, V. L., Rothwell, N. J., and Toulmond, S. (1999) Excitotoxic brain damage in the rat induces interleukin-1β protein in microglia and astrocytes: correlation with the progression of cell death. *Glia* **25**, 311–323.

30. Eriksson, C., Van Dam, A. M., Lucassen, P. J., Bol, J. G., Winblad, B., & Schultzberg, M. (1999) Immunohistochemical localization of interleukin-1β, interleukin-1 receptor antagonist and interleukin-1β converting enzyme/caspase-1 in the rat brain after peripheral administration of kainic acid. *Neuroscience* **93**, 915–930.

31. Kreutzberg, G. W. (1996) Microglia: a sensor for pathological events in the CNS. *Trends Neurosci.* **19**, 312–318.

32. Medhurst, A. D., Harrison, D. C., Read, S. J., Campbell, C. A., Robbins, M. J., and Pangalos, M. N. (2000) The use of TaqMan RT-PCR assays for semiquantitative analysis of gene expression in CNS tissues and disease models. *J. Neurosci. Methods* **98**, 9–20.

33. Berti, R., Williams, A. J., Moffett, J. R., et al. (2002) Quantitative real-time RT-PCR analysis of inflammatory gene expression associated with ischemia-reperfusion brain injury. *J. Cereb. Blood Flow Metab.* **22**, 1068–1079.

34. Schindler, R., Clark, B. D., and Dinarello, C. A. (1990) Dissociation between interleukin-1 β mRNA and protein synthesis in human peripheral blood mononuclear cells. *J. Biol. Chem.* **265,** 10232–10237.

35. Roux-Lombard, P. (1998) The interleukin-1 family. *Eur. Cytokine Netw.* **9,** 565–576.

36. Saito, K., Suyama, K., Nishida, K., Sei, Y., and Basile, A. S. (1996) Early increases in TNF-a, IL-6 and IL-1b levels following transient cerebral ischemia in gerbil brain. *Neurosci. Lett.* **206,** 149–152.

37. Hillhouse, E. W., Kida, S., and Iannotti, F. (1998) Middle cerebral artery occlusion in the rat causes a biphasic production of immunoreactive interleukin-1β in the cerebral cortex. *Neurosci. Lett.* **249,** 177–179.

38. Legos, J. J., Whitmore, R. G., Erhardt, J. A., Parsons, A. A., Tuma, R. F., and Barone, F. C. (2000) Quantitative changes in interleukin proteins following focal stroke in the rat. *Neurosci. Lett.* **282,** 189–192.

39. Davies, C. A., Loddick, S. A., Toulmond, S., Stroemer, R. P., Hunt, J., and Rothwell, N. J. (1999) The progression and topographic distribution of interleukin-1β expression after permanent middle cerebral artery occlusion in the rat. *J. Cereb. Blood Flow Metab.* **19,** 87–98.

40. Zhang, Z., Chopp, M., Goussev, A., and Powers, C. (1998) Cerebral vessels express interleukin 1β after focal cerebral ischemia. *Brain Res.* **784,** 210–217.

41. Lawrence, C. B., Allan, S. M., and Rothwell, N. J. (1998) Interleukin-1β and the interleukin-1 receptor antagonist act in the striatum to modify excitotoxic brain damage in the rat. Eur. *J. Neurosci.* **10,** 1188–1195.

42. Andersson, P.-B., Perry, V. H., and Gordon, S. (1992) Intracerebral injection of proinflammatory cytokines or leukocyte chemotaxins induces minimal myelomonocytic cell recruitment to the parenchyma of the central nervous system. *J. Exp. Med.* **176,** 255–259.

43. Hagberg, H., Gilland, E., Bona, E., et al. (1996) Enhanced expression of interleukin (IL)-1 and IL-6 messenger RNA and bioactive protein after hypoxia-ischemia in neonatal rats. *Pediatr. Res.* **40,** 603–609.

44. Yamasaki, Y., Matsuura, N., Shozuhara, H., Onodera, H., Itoyama, Y., and Kogure, K. (1995) Interleukin-1 as a pathogenetic mediator of ischemic brain damage in rats. *Stroke* **26,** 676–681.

45. Loddick, S. A. and Rothwell, N. J. (1996) Neuroprotective effects of human recombinant interleukin-1 receptor antagonist in focal cerebral ischaemia in the rat. *J. Cereb. Blood Flow Metab.* **16,** 932–940.

46. Stroemer, R. P. and Rothwell, N. J. (1998) Exacerbation of ischemic brain damage by localized striatal injection of interleukin-1β in the rat. *J. Cereb. Blood Flow Metab.* **18,** 833–839.

47. Strijbos, P. J. & Rothwell, N. J. (1995) Interleukin-1β attenuates excitatory amino acid-induced neurodegeneration in vitro: involvement of nerve growth factor. *J. Neurosci.* **15,** 3468–3474.

48. Akaneya, Y., Takahashi, M., and Hatanaka, H. (1995) Interleukin-1b enhances survival and interleukin-6 protects against MPP+ neurotoxicity in cultures of fetal rat dopaminergic neurons. *Exp. Neurol.* **136,** 44–52.

49. Mason, J. L., Suzuki, K., Chaplin, D. D., and Matsushima, G. K. (2001) Interleukin-1β promotes repair of the CNS. *J. Neurosci.* **21,** 7046–7052.

50. Jander, S., Schroeter, M., Peters, O., Witte, O. W., and Stoll, G. (2001) Cortical spreading depression induces proinflammatory cytokine gene expression in the rat brain. *J. Cereb. Blood Flow Metab.* **21**, 218–225.
51. Ohtsuki, T., Ruetzler, C. A., Tasaki, K., and Hallenbeck, J. M. (1996) Interleukin-1 mediates induction of tolerance to global ischemia in gerbil hippocampal CA1 neurons. *J. Cereb. Blood Flow Metab.* **16**, 1137–1142.
52. Allan, S. M. and Rothwell, N. J. (2001) Cytokines and acute neurodegeneration. *Nat. Rev. Neurosci.* **2**, 734–744.
53. O'Connor, J. J. & Coogan, A. N. (1999). Actions of the pro-inflammatory cytokine IL-1β on central synaptic transmission. *Exp. Physiol.* **84**, 601–614.
54. Quagliarello, V. J., Wispelwey, B., Long Jr, W. J., & Scheld, W. M. (1991) Recombinant human interleukin-1 induces meningitis and blood-brain barrier injury in the rat. Characterization and comparison with tumor necrosis factor. *J. Clin. Invest.* **87**, 1360–1366.
55. Anthony, D. C., Bolton, S. J., Fearn, S., and Perry, V. H. (1997) Age-related effects of interleukin-1 β on polymorphonuclear neutrophil-dependent increases in blood-brain barrier permeability in rats. *Brain* **120**, 435–444.
56. Blamire, A. M., Anthony, D. C., Rajagopalan, B., Sibson, N. R., Perry, V. H., and Styles, P. (2000) Interleukin-1β-induced changes in blood-brain barrier permeability, apparent diffusion coefficient, and cerebral blood volume in the rat brain: a magnetic resonance study. *J. Neurosci.* **20**, 8153–8159.
57. Proescholdt, M. G., Chakravarty, S., Foster, J. A., Foti, S. B., Briley, E. M., and Herkenham, M. (2002) Intracerebroventricular but not intravenous interleukin-1β induces widespread vascular-mediated leukocyte infiltration and immune signal mRNA expression followed by brain-wide glial activation. *Neuroscience 112,* 731–749.
58. Ericsson, A., Liu, C., Hart, R. P., and Sawchenko, P. E. (1995) Type 1 interleukin-1 receptor in the rat brain: distribution, regulation, and relationship to sites of IL-1-induced cellular activation. *J. Comp. Neurol.* **361**, 681–698.
59. Van Dam, A. M., De Vries, H. E., Kuiper, J., et al. (1996) Interleukin-1 receptors on rat brain endothelial cells: a role in neuroimmune interaction? *FASEB J.* **10**, 351–356.
60. Dinarello, C. A. and Thompson, R. C. (1991) Blocking IL-1: interleukin-1 receptor antagonist in vivo and in vitro. *Immunol. Today* **12**, 404–410.
61. Relton, J. K. and Rothwell, N. J. (1992) Interleukin-1 receptor antagonist inhibits ischaemic and excitotoxic neuronal damage in the rat. *Brain Res. Bull.* **29**, 243–246.
62. Loddick, S. A., Wong, M. L., Bongiorno, P. B., Gold, P. W., Licinio, J., and Rothwell, N. J. (1997) Endogenous interleukin-1 receptor antagonist is neuroprotective. *Biochem. Biophys. Res. Commun.* **234**, 211–215.
63. Stroemer, R. P. and Rothwell, N. J. (1997) Cortical protection by localized striatal injection of IL-1ra following cerebral ischemia in the rat. *J. Cereb. Blood Flow Metab.* **17**, 597–604.
64. Boutin, H., LeFeuvre, R. A., Horai, R., Asano, M., Iwakura, Y., and Rothwell, N. J. (2001) Role of IL-1alpha and IL-1β in ischemic brain damage. *J. Neurosci.* **21**, 5528–5534.
65. Garcia, J. H., Liu, K.-F., and Relton, J. K. (1995) Interleukin-1 receptor antagonist decreases the number of necrotic neurons in rats with middle cerebral artery occlusion. *Am. J. Pathol.* **147**, 1477–1486.

66. Relton, J. K., Martin, D., Thompson, R. C., and Russell, D. A. (1996) Peripheral administration of interleukin-1 receptor antagonist inhibits brain damage after focal cerebral ischemia in the rat. *Exp. Neurol.* **138,** 206–213.

66a. Mulcahy, N. J., Ross, J., Rothwell, N. J. and Loddick, S. A. (2003) Delayed administration of interleukin-1 receptor antagonist protects against transient cerebral ischaemic in the rat. *Br. J. Pharmacol.* **140,** 471–476.

67. Loddick, S. A. and Rothwell, N. J. (1996) Neuroprotective effects of human recombinant interleukin-1 receptor antagonist in focal cerebral ischaemia in the rat. *J. Cereb. Blood Flow Metab.* **16,** 932–940.

68. Schielke, G. P., Yang, G. Y., Shivers, B. D., and Betz, A. L. (1998) Reduced ischemic brain injury in interleukin-1 β converting enzyme-deficient mice. *J. Cereb. Blood Flow Metab.* **18,** 180–185.

69. Hara, H., Friedlander, R. M., Gagliardini, V., et al. (1997) Inhibition of interleukin 1β converting enzyme family proteases reduces ischemic and excitotoxic neuronal damage. *Proc. Natl. Acad. Sci. USA* **94,** 2007–2012.

70. Loddick, S. A., MacKenzie, A., and Rothwell, N. J. (1996) An ICE inhibitor, z-VAD-DCB attenuates ischaemic brain damage in the rat. *Neuroreport* **7,** 1465–1468.

71. Betz, A. L., Yang, G.-Y., and Davidson, B. L. (1995) Attenuation of stroke size in rats using an adenoviral vector to induce over expression of interleukin-1 receptor antagonist in brain. *J. Cereb. Blood Flow Metab.* **15,** 547–551.

72. Yang, G.-Y., Zhao, Y.-J., Davidson, B. L., and Betz, A. L. (1997) Overexpression of interleukin-1 receptor antagonist in the mouse brain reduces ischemic brain injury. *Brain Res.* **751,** 181–188.

73. Fantuzzi, G. (2001) Lessons from interleukin-deficient mice: the interleukin-1 system. *Acta Physiol. Scand.* **173,** 5–9.

74. Touzani, O., Boutin, H., LeFeuvre, R., et al. (2002) Interleukin-1 influences ischemic brain damage in the mouse independently of the interleukin-1 type I receptor. *J. Neurosci.* **22,** 38–43.

75. Boutin, H. and Rothwell, N. J. (2002) Cerebral ischaemic processes and cytokines: can transgenic mice help?, in *Pharmacology of Cerebral Ischaemia* (Krieglstein, J. and Klumpp, S., eds.), Medpharm, Stuttgart, pp. 183–190.

76. Stroke Therapy Academic Industry Roundtable (1999) Recommendations for standards regarding preclinical neuroprotective and restorative drug development. *Stroke* **30,** 2752–2758.

77. Rothwell, N. (2003) Interleukin-1 and neuronal injury: mechanisms, modification, and therapeutic potential. *Brain Behav. Immun.* **17,** 152–157.

78. Ledeboer, A., Brevꞁ, J. J. P., Poole, S., Tilders, F. J. H., and van Dam, A. M. (2000) Interleukin-10, interleukin-4, and transforming growth factor-β differentially regulate lipopolysaccharide-induced production of pro-inflammatory cytokines and nitric oxide in co-cultures of rat astroglial and microglial cells. *Glia* **30,** 134–142. 79. Spera, P. A., Ellison, J. A., Feuerstein, G. Z., and Barone, F. C. (1998) IL-10 reduces rat brain injury following focal stroke. *Neurosci. Lett.* **251,** 189–192.

Index